Principles of Mathematical Petrophysics

International Association for Mathematical Geosciences

STUDIES IN MATHEMATICAL GEOSCIENCES

1. **William B. Size, Editor**
   Use and Abuse of Statistical Methods in the Earth Sciences

2. Lawrence J. Drew
   Oil and Gas Forecasting: Reflections of a Petroleum Geologist

3. **Ricardo A. Olea, Editor**
   Geostatistical Glossary and Multilingual Dictionary

4. **Regina L. Hunter and C. John Mann, Editors**
   Techniques for Determining Probabilities of Geologic Events and Processes

5. **John C. Davis and Ute Christina Herzfeld, Editors**
   Computers in Geology—25 Years of Progress

6. **George Christakos**
   Modern Spatiotemporal Geostatistics

7. **Vera Pawlowsky-Glahn and Ricardo Olea**
   Geostatistical Analysis of Compositional Data

8. **P.J. Lee**
   Statistical Methods for Estimating Petroleum Resources

9. **John H. Doveton**
   Principles of Mathematical Petrophysics

PRINCIPLES OF MATHEMATICAL PETROPHYSICS

John H. Doveton
Kansas Geological Survey

Oxford University Press is a department of the University of

Oxford. It furthers the University's objective of excellence in research, scholarship, and education by publishing worldwide.

Oxford   New York
Auckland   Cape Town   Dar es Salaam   Hong Kong   Karachi
Kuala Lumpur   Madrid   Melbourne   Mexico City   Nairobi
New Delhi   Shanghai   Taipei   Toronto

With offices in
Argentina   Austria   Brazil   Chile   Czech Republic   France   Greece
Guatemala   Hungary   Italy   Japan   Poland   Portugal   Singapore
South Korea   Switzerland   Thailand   Turkey   Ukraine   Vietnam

Oxford is a registered trademark of Oxford University Press
in the UK and certain other countries.

Published in the United States of America by
Oxford University Press
198 Madison Avenue, New York, NY 10016

© Oxford University Press 2014

All rights reserved. No part of this publication may be reproduced, stored in
a retrieval system, or transmitted, in any form or by any means, without the prior
permission in writing of Oxford University Press, or as expressly permitted by law,
by license, or under terms agreed with the appropriate reproduction rights organization.
Inquiries concerning reproduction outside the scope of the above should be sent to the
Rights Department, Oxford University Press, at the address above.

You must not circulate this work in any other form
and you must impose this same condition on any acquirer.

Library of Congress Cataloging-in-Publication Data

Doveton, John H., 1944–
Principles of mathematical petrophysics / John H. Doveton.
    pages cm
Includes bibliographical references and index.
ISBN 978–0–19–997804–5 (alk. paper)
1. Petrology—Mathematical models.   2. Mineralogy—Mathematical
models.   3. Rocks.   I. Title.
QE431.6.P5D68 2014
552.00151—dc23
    2014001402

9 8 7 6 5 4 3 2 1
Printed in China
on acid-free paper

Dedicated with thanks and remembrance of good times to the Mathematical Geology Section and the Energy Research Section of the Kansas Geological Survey, without whose work and inspiration, this book would not have been possible.

CONTENTS

Foreword to the Series   xi
Preface   xiii
Acknowlegements   xv

1. Fluid Saturation Evaluation   1
    The Archie equations   1
    The Humble equation and its variants   4
    Sensitivity analysis of Archie equation parameters   8
    Non-Archie sandstones   10
    Shaly sandstone analysis   13
    Double-layer shaly sandstone models   15
    Dual-water shaly sandstone models   20
    The Archie equation in carbonate rocks   23
    Porosity partitioning in dual-porosity systems   26
    The porosity exponent in a triple-porosity system   31
    Dielectric logging measurement of the porosity exponent   31
    Petrographic evaluations of the porosity exponent in carbonates   32
    The saturation exponent, n 33
    Wettability effects on the saturation exponent   36
    Archie redux   38
2. Porosity Volumetrics and Pore Typing   44
    Porosity of spherical packs   44
    Clastic "effective" porosity   44
    Neutron-density shale volumetrics   45
    Gamma-ray estimations of shale volume   47
    Correction of total porosity for shale contents   50
    Allocation between shale morphology types   51
    Carbonate porosity   55
    Vug porosity evaluation from acoustic and resistivity logs   56
    NMR logging of vuggy porosity   62
3. Permeability Estimation   67
    Permeability is a vector   67
    Prediction of permeability from porosity   67
    Flow-zone indicator (*FZI*) discrimination of hydraulic units   69

Application of *FZI* to permeability prediction   73
Permeability predictions from porosity and "irreducible" water saturation   77
NMR estimation of permeability in clastic pore systems   79
Permeability estimation in carbonates dominated by interparticle porosity   81
Evaluation of permeability in dual- and triple-porosity systems   82
A Wilderness of mirrors   88

4. Compositional Analysis of Mineralogy   94
Some matrix algebra   94
Compositional-solution evaluation   97
Underdetermined systems   99
Overdetermined systems   102
Optimization models for compositional solutions   103
Multiple-model solutions of rock composition   105
Elucidation of clay minerals   106
Compositional analysis from geochemical logs   113
Inversion mapping of compositions   117

5. Petrophysical Rocks: Electrofacies and Lithofacies   122
Facies and electrofacies   122
Dunham textures and electrofacies   123
Petrophysical recognition of lithofacies   124
Zonation by cluster analysis   128
Theoretical, empirical, and interpretive electrofacies methods   131
Principal component analysis (PCA) of electrofacies   134
Classification by a parametric electrofacies database   146
Supervised electrofacies analysis methods   147
Electrofacies classification by discriminant function analysis (DFA)   148
Nonparametric discriminant analysis   152
Neural-network prediction of lithofacies from logs   157
Beyond product facies to the petrophysical prediction of process facies   162

6. Pore-System Facies: Pore Throats and Pore Bodies   171
The petrofacies concept   171
Equivalent hydraulic radius of tubes   172
Capillary pressure evaluation of pore-throat sizes   172
The Winland equation   176
The flow-unit concept   178
Petrofacies case-study applications of the Winland equation   179
Carbonate petrofacies pore-throat size distributions   185
Pore-body size distributions from NMR measurements   198
NMR facies in sandstones   200
NMR pore-size interpretation in carbonates   204
NMR-partitioned porosity and Dunham textural classes   205
NMR facies in carbonates   209

7. Saturation-Height Functions   217
    Integration: the saturation-height model   217
    The basics of reservoir saturation profiles   218
    Saturation-height modeling in sandstones from capillary pressure measurements   221
    Height functions for bulk-volume water   225
    Permeability-height functions   230
    Saturation-height modeling in carbonates   233
    Saturation-height modeling based on magnetic resonance logs   238
    Putting it all together: the static reservoir model   241

Index   245

FOREWORD TO THE SERIES

The **Studies in Mathematical Geology** (SMG) series was established 30 years ago to serve as an outlet for book-length contributions on topics of special interest not only to geomathematicians, but also to practitioners of various interdisciplinary scientific branches that looked to the IAMG—the International Association for Mathematical Geology—for leadership in the application and use of mathematics in geoscientific research and technology. In due course the full scope of the Association, founded in 1968, was recognized. Thus in 2009 it became the International Association for Mathematical Geosciences. The current monograph, *Principles of Mathematical Petrophysics*, is the first to appear in the "new" **Studies in Mathematical Geosciences** (SMG) series.

SMG No. 9 is designed to address all aspects of mathematical petrophysics. By combining petrophysics and mathematical modeling, it bridges the gap between existing mathematical knowledge and the use of modern methods for reservoir characterization, benefiting not only log analysts, petroleum engineers, petrophysicists, and geophysicists, but the field of geomathematics in general.

The Stanford IAMG conference mentioned in the author's Preface took place in August 2009. As SMG editor, I was an enthusiastic participant in the discussion about yet another IAMG monograph. Tentatively entitled "Mathematical Petrophysics," the author estimated an August 2011 date for delivery of the completed manuscript and illustrations. Perfect! SMG 11 would appear in 2011. The contract for what was to have been SMG 9 had been signed and sealed in January 2009. And OUP had in hand the manuscript proposed as SMG 10.

> Robert Burns summed it up nicely in 1785
> The best-laid schemes o' mice an' men
> Gang aft agley

The appearance of John Doveton's "SMG 11" (in 2011) was to have been my swan song. John has been a friend for over four decades, during three of which we were colleagues at the Kansas Geological Survey. No matter how awry various plans may have gone, I'm pleased to have seen this project through. John has succeeded in his effort to ensure that his book, SMG No. 9, incorporates every facet of modern petrophysics technology. Yes, I'm singing now!

Jo Anne DeGraffenreid, Editor
Baldwin City, Kansas, USA

PREFACE

Mathematical astrophysics and mathematical geophysics have their own journals and conferences, where practitioners discuss mathematical formulations that come to grips with the physical processes of the cosmos and planet Earth. Mathematical petrophysics is by no means new. It started in 1942 with the publication of an equation: the Archie equation. Strictly speaking, there are two Archie equations. The first describes the resistivity of a rock filled with salt water. The second equation is concerned with the resistivity of hydrocarbon-bearing formations and proposed a prediction of hydrocarbon saturation. This quantitative outcome moved the vagueness of "log interpretation" to "log analysis" and was finally dignified with the name of "petrophysics" by Archie in 1950. The Archie equations are still used today, and the huge economic value of these and further developments is the principal reason why mathematical petrophysics is not an academic speciality. The consequences of poor mathematical decisions raise more than academic passions, because of potential losses of millions of dollars. Furthermore, the traditional differentiation between reservoir rocks and seals has recently crumbled with the emergence of resource plays, typified by tight porosities and minimal permeabilities. Consequently, mathematical petrophysicists continue to be challenged to propose new algorithms that characterize nontraditional targets as well as refining established methodologies to manage large but aging fields.

The idea of writing this book started with a discussion at an open-air banquet on the Stanford golf course at an IAMG conference. Further encouragement came at an SPWLA topical conference on "Computational Petrophysics" held in Ashville, North Carolina. The extraordinary power of modern computer environments to actualize complex petrophysical models was not disputed, and the ability of young petrophysicists to navigate their way through a labyrinth of software options was considered to be admirable. However, a model is a model, and as the eminent statistician George Box famously said, "All models are wrong; some models are useful." Consequently, the ability to reflect on the limitations and strengths of any petrophysical model is an important consideration. Classic default equations implemented within large software packages are often based on small data sets, and their historical development will be reviewed in this book. As an additional consideration, an equation that gave a reasonable solution in the Texas Gulf Coast may not necessarily perform so well in the steppes of Kazakhstan. So, the purpose of this book is to review mathematical petrophysics ranging across microscopic to geographic scales from a perspective of strategic thinking rather than tactical cookbook recipes.

While reviewing a broad range of published petrophysical studies from around the world, a great proportion of the data sets analyzed in this book come from subsurface Kansas. In contrast with data from the Middle East, large numbers of these Midwest logs and core data sets are available in the public domain and can be downloaded from the Kansas Geological Survey website. The website also contains numerous petrophysical studies by Survey scientists, many of which I have cited. I have learned an extraordinary amount from my colleagues, past and present, in the Mathematical Geology and Energy Research sections, and my gratitude is reflected in my dedication of this book to them. My special thanks go to Jo Anne DeGraffenreid for her meticulous editing and John Davis for his technical critiques and advice over the period of this book project. However, any errors that remain are my responsibility. As a log petrophysicist, it has been my privilege to showcase the data supplied by the core petrophysicist maestro, Alan Byrnes, which have enriched and illuminated the wireline measurements. Finally, I owe special thanks to the Kansas Geological Survey as a stimulating and collegial institution for research, and to Survey directors, past and present, for their encouragement and support.

John H. Doveton
Kansas Geological Survey
Lawrence, Kansas
October 2013

ACKNOWLEDGMENTS

Acknowledgment is made for illustrations used or adapted from the following sources: Society of Petrophysicists and Well Log Analysts (Figures 1.5, 4.10, 5.9, 6.8, 6.9, 6.10, 7.9, 7.10); American Association of Petroleum Geologists (Figures 1.19, 2.6, 3.13, 5.8, 5.11, 5.24, 5.25, 6.7, 6.11, 6.23, 6.25, 7.2, 7.5, 7.6, 7.7); Society of Petroleum Engineers (Figures 4.1, 5.20, 6.6, 6.35, 7.15, 7.17); Society for Sedimentary Geology (Figure 5.27, Table 5.6); Canadian Well Logging Society (Figure 1.9); Geo-Arabia (Figure 7.18); Utah Geological Association (Figure 7.3); Elsevier (Figure 1.4); Springer-Verlag (Figure 7.14), the Geological Society of London (Figure 7.8).

Principles of Mathematical Petrophysics

CHAPTER 1
Fluid Saturation Evaluation

THE ARCHIE EQUATIONS

In his treatise on electricity and magnetism, Maxwell (1873) published an equation that described the conductivity of an electrolyte that contained nonconducting spheres as:

$$\Psi = \frac{C_o}{C_w} = \frac{2\Phi}{(3-\Phi)}$$

where the "meaning" of Ψ (*psi*) has been most commonly interpreted as some expression of tortuosity, C_o and C_w are the conductivity of the medium and the electrolyte, respectively, and Φ is the proportion of the medium that is occupied by the electrolyte. Since that time, considerable efforts have been devoted to elucidation of the electrical properties of porous materials, particularly with the advent of the first resistivity log in 1927, which founded an entire industry focused on estimating fluid saturations in hydrocarbon reservoirs from downhole measurements.

To some degree, spirited discussions in the literature reflect two schools of thought, one that considers the role of the resistive framework from a primarily empirical point of view, and the other that models the conductive fluid phase in terms of electrical efficiency. Clearly, the two concepts are intertwined because resistivity is the reciprocal of conductivity and the pore network is the complement of the rock framework. If the solid part of the rock is nonconductive, then the ability of a rock to conduct electricity is controlled by the conductive phase in the pore space, which should make the case for equations to be formulated from classical physical theory. This approach is typically developed using electrical flow through capillary tubes as a starting point. Unfortunately, the topological transformation of a capillary tube model to a satisfactory representation of a real pore network is a formidable challenge, so that mathematical solutions may not be acceptable, even though they are grounded in basic physics. The most successful model along these lines has been proposed by Herrick and Kennedy (1994), who maintain that while the Archie equation is a useful parametric function, it has no physical basis. Some of their conclusions are reviewed at the end of this chapter.

One of the appeals of the empirical resistivity approach of Archie has been that its interpretation has been made from both rock-framework and pore-space perspectives, rather than simply from the physics of the conductive pore network. An engineer may view a reservoir as simply a collection of holes in a rock container, but the rock framework is the primal cause, in that its fabric is created by depositional and diagenetic processes, with pore spaces and their connections as a consequence. In this book, we will pursue the empirical approach mainly because the Archie equation is so entrenched in published petrophysical theory, but also because it has stood the test of time and can provide useful insights into reservoir properties. However, we will also take into account observations from alternative approaches and variants that aid in understanding and provide the basis for improved models and equations when formations fail to behave as "Archie rocks."

From empirical observations of Gulf Coast sandstones, Archie (1942) established that the ratio of the resistivity of a completely brine-saturated rock (R_o) to the resistivity of its contained brine (R_w) was a constant for any given rock sample, and gave the name, resistivity formation factor (F) to this proportionality constant:

$$F = \frac{R_o}{R_w}$$

In a comparison with Maxwell's equation, we see that Maxwell's *psi* is equivalent to the reciprocal of Archie's formation factor, F. In the same paper, Archie (1942) further stated that "knowing the porosity of the sand in question, a fair estimate may be made of the proper value to be assigned to F, based upon the indicated empirical relationship $F = \Phi^{-m}$." This observation was based on many laboratory measurements, and Archie went on to suggest that "m has been found to range between 1.8 and 2.0 for consolidated sandstones...loosely or partly consolidated sands might have a value of m anywhere between 1.3 and 2." Although the origins of this first Archie equation are empirical, Bussian (1982) showed that the Archie equation was equivalent to the Hannai-Bruggeman equation at the low limit of electrical frequency and where the rock framework conductivity was effectively zero. This formulation is the first Archie equation, as distinct from the second Archie equation that considers the resistivity of partially water-saturated rocks.

Archie did not give a name to m, but Guyod (1944) introduced the term "cementation exponent," which is still widely used but is increasingly replaced by "porosity exponent," which more accurately acknowledges the multiplicity of processes that can affect m. The interpretation of the internal geometric elements that control m can be expressed either in terms of pore network or framework architecture, or both. One of the earliest interpretations considered the pore space in terms of a complex bundle of capillary tubes. If the tubes have a constant cross-sectional area, then resistivity variation is controlled by the length of the tubes. The porosity exponent, m, is then effectively a measure of pore channel tortuosity. However, capillary tubes are a restrictive representation of the elaborate branching of the pore network. Some authors, such as Perez-Rosales (1982), consider the pore space to be subdivided into channels and traps, where channel pores are the only elements that contribute to the electrical current flow. Under this model, the porosity exponent is a measure of the relative partition between conductive channel pores and nonconductive trap pores. Still other authors, such as Etris et al. (1989), have expanded the representation to a more realistic system of pore bodies connected by pore throats. In this case, it is argued that the major control on the formation factor is the distribution of pore-throat areas. In a later paper that expanded this model, Ehrlich and others (1991) suggested that the porosity exponent, m, was effectively a measure of the ratios of the logarithms of the pore-throat area to the pore-body area.

Turning from pore-network to framework models of explication, much has been learned from laboratory measurements of electrical current flow in grain packs with varying sizes, shapes, and degrees of sorting. Jackson et al. (1978) showed that size and sorting of grain-size had little effect on the porosity exponent, but that this was sensitive to grain shape. The porosity exponent increased as grains became less spherical, and it was highest in sediments dominated by platy grains. The conclusions from these empirical data were supported independently by a mathematical derivation of Sen et al. (1981). Mendelson and Cohen (1982) observed that the porosity exponent was lowest in rocks where the grains had the same shape and orientation. However, porosity exponent values were increased if the grains were aligned with a distribution of shapes or if they were randomly oriented. In summary, grain properties that control porosity exponent variation are shape, shape distribution, and orientation.

Archie (1942) went on to consider the resistivity of partially saturated rocks, where the pore space was partitioned between conductive formation water and nonconductive hydrocarbon. He proposed a second Archie equation of the form:

$$S_w = \left(\frac{R_o}{R_t}\right)^{\frac{1}{n}}$$

where S_w is the fractional water saturation, R_t is the formation resistivity, and n is now known as the "saturation exponent," and he concluded that "the value of n appears to be close to 2" for both consolidated and unconsolidated sands that were clean (shale-free). Although he was encouraged by his laboratory measurements, Archie (1942) was cautious in extrapolating the relationship into the subsurface, observing that "there is a possibility that the manner in which oil or gas is distributed in the pores may be so different that these relations derived in the laboratory might not apply underground." The second Archie equation has proven itself to be a viable predictor in the years following Archie's work, being universally applied to hydrocarbon reservoirs, although major developments in the understanding of wettability have made Archie's cautionary comments rather prophetic. The complex relationship between wettability and the saturation exponent will be discussed later in this chapter.

Archie (1942) combined his two equations to solve for water saturation in what is now known as "the" Archie equation:

$$S_w = \sqrt[n]{\frac{FR_w}{R_t}}$$

which remains the most widely applied equation in petrophysics, in spite of all attempts to replace this empirical relationship with a more rigorous formulation made up of terms identified with distinctive physical properties.

THE HUMBLE EQUATION AND ITS VARIANTS

Archie's original pioneering work was based on shale-free sandstones, so we shall first review how his equations were deployed from the laboratory to subsurface reservoir evaluation of sandstones. In order for the first Archie equation of:

$$F = \frac{1}{\Phi^m}$$

to be used in oilfield operations, some knowledge of the value of m would be required, which could further be complicated by changes in its value within a logged formation. From their study of a variety of sandstone core measurements, Winsauer et al. (1952) proposed the relationship:

$$F = \frac{0.62}{\Phi^{2.15}}$$

which became known as the "Humble equation," named for the company that they worked for (now Exxon). If valid, then this modified Archie equation could be applied to sandstones without requiring any knowledge of the value of the porosity exponent, m. The relationship of the Humble equation with respect to the Archie equations with differing values of m is shown in Figure 1.1.

The generalization of the Archie equation to the form:

$$F = \frac{a}{\Phi^m}$$

moved it from an empirical, but functional, equation to an approximating statistical function because the formation factor must be unity at 100 percent porosity, which will only occur when a is unity. A variety of names have been used for the parameter a, of which the commonest is "tortuosity factor," while its detractors generally do not give it a name, presumably so as to avoid conferring any credibility on the parameter. So, for example, Maute et al. (1992) concluded from core measurements that a was "a weak-fitting parameter, with no physical significance. In general, we recommend that a be fixed to unity." Glover (2009) was more forthright with his criticism, stating that "reports that contain values of the constant a that are anything other than exactly unity are the result of sloppy thinking and the mindless application of curve-fitting programs."

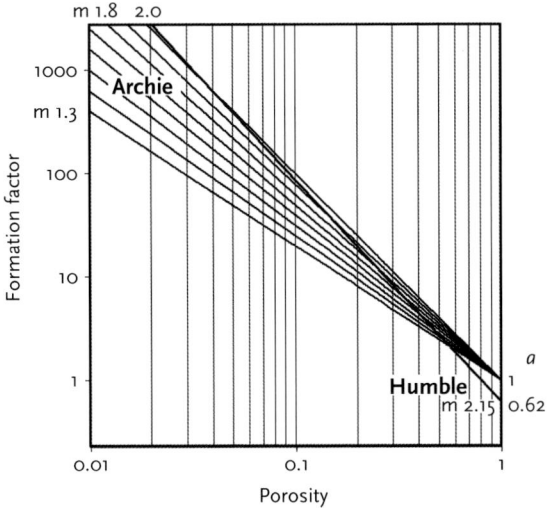

Figure 1.1: The first Archie equation for a range values of the porosity exponent, m, graphed together with the Humble equation on a double-logarithmic crossplot of the formation resistivity factor and porosity.

If the parameter a is significantly different from unity, then this implies that the porosity exponent, m, is correlative with pore volume. If a is less than unity, then the porosity exponent decreases with increasing porosity. This is the case with the Humble equation and at least makes it consistent with the observations of Archie (1942) that unconsolidated sands (with generally higher porosities) have lower porosity exponents than consolidated sandstones. However, the values of a and m are interdependent in practice (and demonstrated mathematically by Zinszner and Pellerin, 2007), so that once a takes a value different from unity, the porosity exponent is adjusted to accommodate the change, and all potential "meaning" of m is lost. If indeed there is a systematic trend of porosity exponent with pore volume, this is best modeled explicitly by the equation:

$$F = \frac{1}{\Phi^{c+d\Phi}}$$

where m has been replaced by a function of porosity with parameters c and d. In this case, the Archie model is driven by a variable-m porosity exponent, where the value of m is assigned by the porosity value. Not only is this variable porosity exponent a viable subject for interpretation, but the equation honors the physical constraint of a unit-value formation factor at total porosity.

In view of its almost universal status as the default equation for sandstones in log-analysis charts and software packages, it is sobering to reflect that the Humble equation is based on only thirty sandstone core samples. On the other hand, Winsauer et al. (1952) made a thorough and thoughtful analysis of their data by including mineralogy, clay content, porosity, permeability, packing index, grain size, sorting,

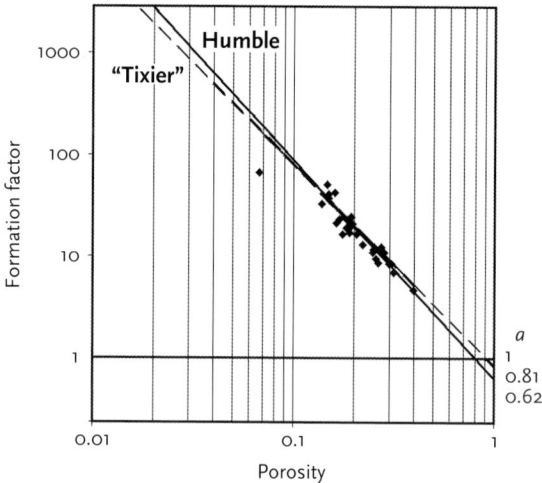

Figure 1.2: Default generalized first Archie equations for sandstones: the Humble equation and the Tixier equation. Data points of measured porosity and formation factor used to establish the Humble equation are also shown.

skewness, and roundness, in their considerations of properties that influenced the formation factor. Doveton (1986) reanalyzed their data statistically and concluded that there were significant correlations that collectively implied that better sorted sands with more rounded grains generated lower values for the porosity exponent. The Humble equation and the data points on which it is based are shown in Figure 1.2. Also shown is an alternative default equation that is widely used for sandstones, which is expressed by:

$$F = \frac{0.81}{\Phi^2}$$

Although it has no formal name, it is commonly known as the "Tixier" equation, in honor of the Schlumberger pioneer, possibly because he had devised it as a simpler alternative formulation to the Humble equation that was easier to calculate. The parameter a values of the Humble and Tixier equations can be seen as their intercepts at fractional porosity readings of unity (100 percent porosity), while m reflects their slope. The geometry of the plot also demonstrates the linked relationship between a considered as a variable and the porosity exponent, m, because changes in values of a to lower values results in higher values of porosity exponent slopes.

If the value of a is maintained at unity, then the data of Winsauer et al. (1952) can be analyzed from an orthodox Archie perspective (Figure 1.3). The broad scatter of porosity exponent values calculated for each core shows a generalized trend of decreasing values of m at higher porosities and is in general agreement with the ranges proposed by Archie (1942). However, the variability in the porosity exponent at any given porosity raises the question as to the effect on water saturation

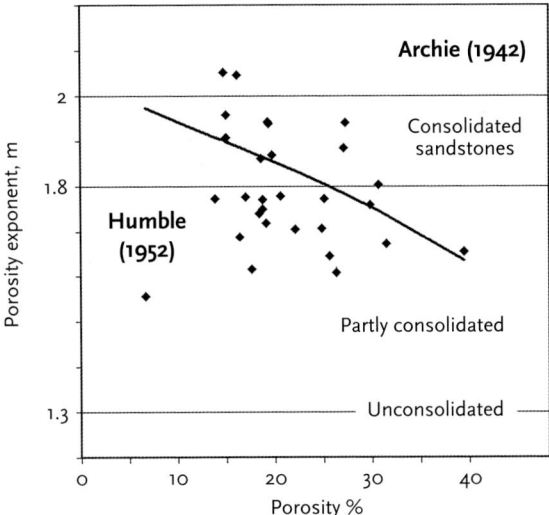

Figure 1.3: Crossplot of the porosity exponent, m, and the porosity for core samples used as the basis for the Humble equation compared with ranges suggested by Archie for sandstones at differing stages of consolidation.

calculated by the Archie equation. This will be discussed in some detail later in the chapter when considered in conjunction with the saturation exponent, n, as part of uncertainty analysis. In their discussion of the performance of the Humble equation, Winsauer et al. (1952) concluded that "resistivity factors can be approximated by means of a relation involving only resistivity factor and porosity," and that when applying the equation to data published by Archie (1950), the results were matched "with acceptable precision." As quoted earlier, Archie (1942) had stated that results from the Archie equation provided a "fair estimate" of the formation factor. Therefore, none of these authors had made extravagant claims as to the accuracy of the predictions, but a measured judgement that the methodology met the operational requirements of the mid-twentieth century when applied to logging measurements of that era. The Archie equation is still universally applied today for the calculation of water saturation because it works well, provided the acceptable error associated with the saturation prediction is considered explicitly, because this dictates the degree of accuracy required for Archie equation parameter values. The demands placed on water saturation estimations used in simulation modeling or reserve estimation are far more stringent than for a calculation used as the basis for a decision to run a drill-stem test.

Core measurements of formation factor and the elucidation of porosity exponent values are comparatively cheap procedures, so that reservoir-specific investigations are recommended, particularly in the low-porosity range, where variability in m has greater influence on water saturation estimates. Because the porosity exponent is sensitive to changes in fabric, values can be expected to cluster in associations that mirror aspects of lithofacies and porosity-permeability petrofacies. So,

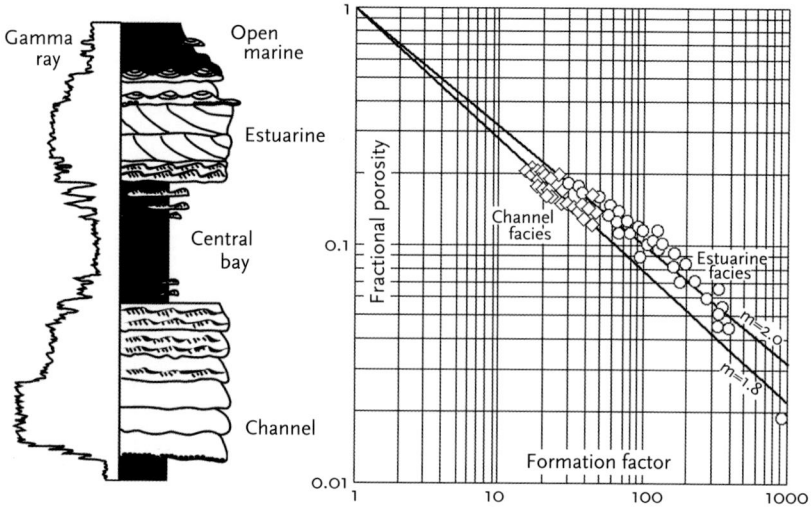

Figure 1.4: Type section of incised valley Pennsylvanian Morrow sandstone of southwest Kansas together with core measurements of the formation factor and porosity that differentiate distinctive porosity exponent values for channel and estuarine facies. Modified from Buatois et al. (2002), courtesy Elsevier.

for example, Bhattacharya et al. (2002) made core measurements of porosity exponents from Pennsylvanian Morrow sandstones from an incised valley, where channel facies cores were clearly distinguished from the overlying estuarine sand facies (Figure 1.4). While the channel facies had a porosity exponent of 1.8, the estuarine facies were characterized by a value of 2.0. This distinction was applied in selectively calculating water saturation from the Archie equations keyed to facies-specific porosity exponents.

SENSITIVITY ANALYSIS OF ARCHIE EQUATION PARAMETERS

In a letter to the editor of *The Log Analyst*, Brown (1997) reminded readers that regardless of the major improvements in resistivity tool technology reported extensively in the previous issue of the journal, accuracy in hydrocarbon estimation was still determined by the appropriate application of the Archie equation. In particular, he focused his attention on the Archie exponents of m and n, which he considered to be "frequently poorly characterized." Using simple graphs to make his point, Brown (ibid.) concluded that errors in the porosity exponent, m, are most critical at low porosities and high water saturations, while errors in the saturation exponent, n, have the greatest effect at intermediate saturations.

The role of each of the parameters in the Archie equation in their individual effects on water saturation estimation can be analyzed systematically through sensitivity analysis. Chen and Fang (1986) described the mathematical equations of error propagation necessary for error estimation. These use the partial derivative of each parameter with respect to water saturation applied to the partitioning of the water saturation variance between variances associated with each parameter. Now, when the Archie equations are combined and written as a solution for water saturation, then the equation is:

$$S_w = \left(\frac{aR_w}{R_t \Phi^m} \right)^{1/n}$$

which has six parameters. If these are assumed to be random variables, then the variance of the water saturation is given by:

$$\sigma_{S_w}^2 = \left(\frac{\partial S_w}{\partial a} \right) \sigma_a^2 + \left(\frac{\partial S_w}{\partial R_w} \right) \sigma_{Rw}^2 + \left(\frac{\partial S_w}{\partial R_t} \right) \sigma_{Rt}^2 + \left(\frac{\partial S_w}{\partial \Phi} \right) \sigma_\Phi^2 + \left(\frac{\partial S_w}{\partial m} \right) \sigma_m^2 + \left(\frac{\partial S_w}{\partial n} \right) \sigma_n^2$$

which is the sum of the squared partial derivative of each parameter multiplied by its variance. If C_x is the fractional error contribution of parameter x to the water saturation variance and σ_x is its standard deviation, then the effect of the parameter can be evaluated from the equation:

$$C_x = \left(\frac{n}{S_w} \right)^2 \left(\frac{\partial S_w}{\partial x} \right)^2 \sigma_x^2$$

A standard deviation must first be assigned to each parameter, either from experimental data or as hypothetical values to evaluate different scenarios. Also, the fractional error contributions of each parameter can either be made equal in the simplest model or assigned unequal uncertainties from petrophysical considerations. An example of the error analysis run in unequal uncertainties mode is shown in Figure 1.5, where Chen and Fang (1986) assigned low uncertainties to m and n, as contrasted with a higher uncertainty for porosity. In this case, while uncertainties in porosity become critical at low porosities, the relative importance of m and n changes with porosity. In allocating budgets for core measurements, investments should be focused on porosity exponent measurement in reservoirs with low porosity, while the saturation exponent should receive special attention at higher porosities. Because the error analysis is driven by simple equations, the procedure can be used to evaluate a variety of scenarios to evaluate the sensitivity of water saturation estimations to changes in Archie equation parameters.

An alternative methodology is to explore sensitivity effects of the Archie equation through the application of Monte Carlo simulation. The simulation can be based on normally-distributed Archie parameters taken as independent random variables applied to the generation of multiple water saturation estimates. In this case, the simulation results conform to the expectations of the partial derivative-variance solution without the uncertainty characterization. An advantage of the normal distribution is that results can be phrased in terms of probability, but the assumption of a normal distribution for the parameters may not be warranted. However, Monte

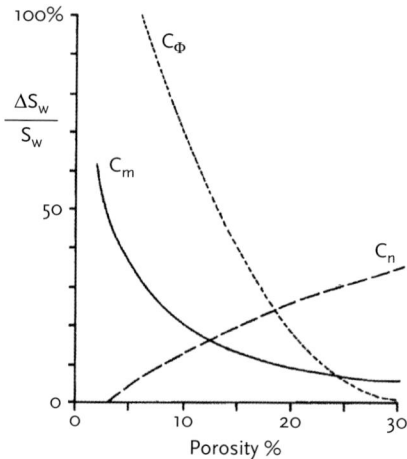

Figure 1.5: Example of an error analysis run in unequal uncertainties mode to evaluate the relative sensitivity of the water saturation estimation from the Archie equation with respect to porosity and the exponents, m and n, over a porosity range. From Chen and Fang (1986), courtesy SPWLA.

Carlo simulations can be designed to surmount nonnormality and other issues. A simple and pragmatic alternative to a normal distribution is a triangular distribution whose ends are anchored to minimum and maximum parameter values, with the apex set at the median. Also, the basic assumption that the parameters are independent is commonly not met. So, for a more realistic simulation, the independent first-order model should be expanded to accommodate parameter intercorrelation, as pointed out by Zeybek et al. (2009) who observed that inclusion of intercorrelation could either increase or decrease uncertainties in water saturation estimation.

NON-ARCHIE SANDSTONES

The range of sandstone reservoirs that qualify as viable exploration targets has expanded dramatically in recent years, so that tight sandstones that would have been dismissed as non-pay in the last century are now commercial gas producers. The Cretaceous Mesaverde sandstone in the western United States provides a classic example of a tight gas sand whose analysis by the Archie equation requires some judicious thought, as discussed by Cluff and Byrnes (2009). Formation factors measured for a large set of Mesaverde tight gas sandstone cores were converted to porosity exponents and plotted against porosity (Figure 1.6). Although the values of m at porosities greater than 10 percent show ranges that are fully consistent with the work of Archie and other investigators, at lower porosities there is a systematic plunge to markedly lower porosity exponents. As will be discussed later, a similar phenomenon is observed in carbonate rocks, which is usually attributed either to the influence of fractures or a change in pore geometry from pore bodies linked by throats

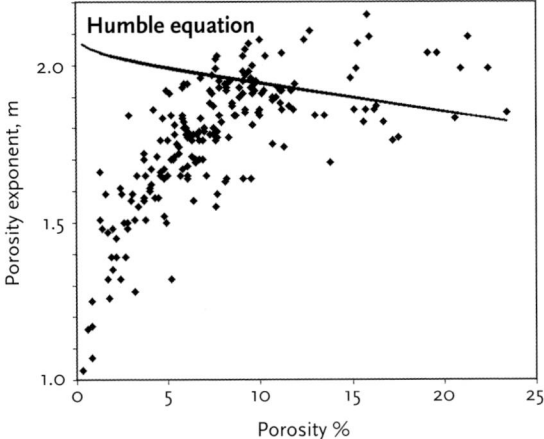

Figure 1.6: Porosity exponents from Cretaceous Mesaverde tight gas sandstones plotted against porosity and referenced to predictions from the Humble equation. Data from Cluff and Byrnes (2009).

to more slot-like shapes. These factors may apply to the Mesaverde sandstone but, more importantly, formation factors were observed by Cluff and Byrnes (2009) to vary systematically with water salinity. At these low porosities and low salinities, the role of conductivity introduced by the cation-exchange capacity of clays becomes increasingly important and contrasts with sandstones that have higher porosities and salinities, where the majority of the conduction is carried by the saltwater phase in conformance with the Archie equation model. A shaly sandstone resistivity equation model is therefore appropriate to replace an Archie equation formulation in the calculation of water saturations. The application of the Humble equation or an Archie equation with a typical porosity exponent would overestimate water saturation and so increase the possibility of overlooking significant gas reserves. In fairness, it should be noted that all but one of the core samples used by Winsauer et al. (1952) for the Humble equation exceeded 13 percent porosity. The tightest core sample had a porosity of 6.7 percent and a computed porosity exponent of 1.56 (see the outlier on Figure 1.3), a value which is consistent with the Mesaverde trend.

Difficulties in the estimation of water saturation over the expanding spectrum of commercial reservoir rock types has led to the concept of the "Archie rock" to provide a reference standard from which to develop strategies to analyze "non-Archie rocks." Quite simply, an Archie rock is one that meets the assumptions of the Archie model to an acceptable degree. These assumptions require a nonconductive matrix with a unimodal, connected pore system that is filled with water whose salinity is sufficiently high in order to provide electrical conductivity that dominates any potential surface effects. The Archie rock is somewhat analogous to an "ideal gas," in that Boyle's and Charles' laws of ideal gas behavior give generally acceptable results for real gases, which deviate from the ideal because of the neglect of small molecular interaction effects. As we have seen, Mesaverde sandstones deviate progressively from Archie rock at lower porosities and lower salinities to become increasingly more non-Archie in their behavior. In fact, the range between rocks that are acceptably Archie to non-Archie forms a continuum, as discussed by Worthington (1995), who suggested that this concept should be used as the basis for the petrophysical classification of reservoir rocks. Not only would criteria be developed to discriminate Archie from non-Archie rocks, but distinctions made between different variants of non-Archie rocks. By this means, appropriate water saturation algorithms could be applied to reservoirs with common styles of non-Archie properties. Worthington (1995) proposed the generic equation:

$$C_o = \left(\frac{C_w}{F}\right) + \left(\frac{x}{F}\right)$$

where C_o is the conductivity of the formation completely saturated with water, C_w is the conductivity of the formation water, F is the formation factor, and x is an extra conductivity term. When x is zero, the expression collapses to the first Archie equation, written in conductivity terms. The role of the extra conductivity term can be evaluated by comparing formation factors measured in core with pore water of different salinities. The extra conductivity term, x, is crossplotted against porosity from a sample of Mesaverde sandstone cores (Figure 1.7) calculated by a comparison of formation factors measured with pore waters of 40K and 200K ppm salinity. Worthington (1995) described a rule to discriminate Archie from non-Archie rocks, based on the ratio of Fa to F, where Fa is the measured formation factor and F is the expectation of the formation factor if electrical conduction was carried entirely by the formation water. For an ideal Archie rock, the ratio would be unity, and Worthington (ibid.) suggested that a

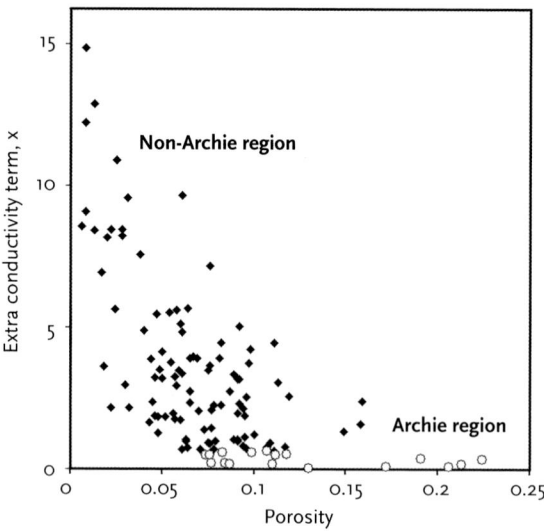

Figure 1.7: The extra conductivity term, x, crossplotted against porosity for Mesaverde sandstone cores, with discrimination of Archie from non-Archie rocks using the rule proposed by Worthington (1995) at a critical value of 0.9. Data from Cluff and Byrnes (2009).

ratio of 0.9 would be appropriate in recognition of the statistical variability of Archie rocks. Cluff and Byrnes (2009) observed that a salinity of 40K ppm was fairly typical for Mesaverde sandstone formation waters, so they concluded that a non-Archie model keyed to this salinity was appropriate to provide good estimates of water saturation in zones with low porosity. In a two-step procedure, they first estimated a modified porosity exponent matched with a formation water with 40K salinity as:

$$m_{K40} = 0.676 \log \Phi + 1.22$$

by using the demonstrated correlation of porosity exponent with porosity. They then generalized the model for varying Mesaverde sandstone salinities by further modification through the equation:

$$m_a = m_{K40} + \left[(0.0118\Phi - 0.355) * (\log R_w + 0.758)\right]$$

Mesaverde sandstones with porosities greater than 12 percent were considered to behave as Archie rocks and assigned a porosity exponent value of 1.95. A more conventional shaly sand model approach to the Mesaverde sandstones was described by Chisholm et al. (1987), who used clay cation-exchange data to evaluate water saturation using the Waxman-Smits equation.

It might be presumed that the term "non-Archie sandstones" is a synonym for "shaly sandstones," but this is not necessarily the case. The conductivity associated with shale is caused by the cation-exchange properties of the constituent clays. Shaly sandstones that do not contain clays with an appreciable cation-exchange capacity, especially in association with high-salinity formation waters, may be difficult to distinguish operationally from Archie rocks. In addition, shale-free clastics may show appreciable non-Archie characteristics if surface conductivity effects become important, such as in some siltstones or where low porosities severely reduce the formation water-conductivity component. In an extreme case, the sandstone matrix may contain metallic minerals whose conductivity is sufficient to violate the Archie-rock condition of a nonconductive framework. As an example, pyrite occurrence in the Sadlerochit sandstone of Prudhoe Bay required resistivity corrections in zones where the mineral occurred in clusters and exceeded a critical concentration (Clavier et al., 1976).

Consequently, the variety of potential reservoir sandstones with anomalously low resistivities favors a strategy that first classifies sandstones in the Archie/non-Archie continuum, as suggested by Worthington (1995), and then, by petrophysical comparison with other non-Archie reservoirs, determine the best method for water saturation estimation.

SHALY SANDSTONE ANALYSIS

The most common reason for low-resistivity non-Archie sandstone is the conductivity associated with clay minerals, which causes the Archie equation to overestimate the water saturation and therefore to be pessimistic in the search for oil or gas.

The economic implications of an ideal resistivity equation for shaly sandstones have stimulated considerable research and discussion by industrial petrophysicists over the years. In his review paper, Worthington (1985) documented the host of shaly sandstone equations that had developed up to that time and commented on their histories and interrelationships. Almost without exception, they revert to the Archie equation in the limit of no shale content, and are constructed with the concept of parallel resistances of the pore brine and conductive shale components. So, the generic equation takes the form of:

$$\frac{1}{R_t} = \frac{S_w^n}{FR_w} + X$$

where X is the conductivity contribution of the shale element, and the equation can be seen to be an expansion of the extra conductivity equation described earlier to accommodate partial water saturation.

Historically, two distinct model families of shaly sandstone equations have been developed. The older model considered the shale as a homogeneous conductive medium and developed resistivity equations keyed to V_{sh}, the volumetric fraction of shale in the rock. Although the physical basis of the model is incorrect, the equations often provide reasonable approximate solutions to water saturation, especially when the equation parameters are adjusted so that the results conform with local water saturation data measured from cores or production tests. Some of these equations are still widely used for this reason and also because of their relative simplicity and limited demands on additional input parameters. Probably the best known is that of Simandoux (1963), which originally was written as:

$$\frac{1}{R_t} = \frac{S_w^2}{FR_w} + \frac{\varepsilon V_{sh}}{R_{sh}}$$

where $\varepsilon = 1$ when $S_w = 1$, and $\varepsilon < 1$ when $S_w < 1$. The Simandoux equation was developed on the basis of laboratory measurements but was simplified by Bardon and Pied (1969) to a quadratic equation of water saturation to enable more practical field application:

$$\frac{1}{R_t} = \frac{S_w^2}{FR_w} + \frac{V_{sh} S_w}{R_{sh}}$$

The Simandoux equation was generally found to give reasonable estimates of water saturation in formations with higher salinity formation waters but was considered to be less satisfactory at lower salinities. In work with low-salinity sandstone reservoirs in Indonesia, Poupon and Levaux (1971) developed an equation that made a better match with field observations:

$$\frac{1}{R_t} = S_w^n \left[\sqrt{\frac{Vsh^{(2-Vsh)}}{R_{sh}}} + \sqrt{\frac{\Phi_e^m}{R_w}} \right]^2$$

The "Indonesia equation," as it is generally known, became widely used across the world, so, for example, Moss (1992) stated that in the North Sea Brent Group fields the commonest shaly sand equation used was the Indonesian, "although there is no theoretical basis for its applicability to these sands." Woodhouse and Warner (2004) pointed out that "the formula was empirically modeled in water-bearing shaly sand, but the detailed functionality for hydrocarbon-bearing sands is unsupported except by common sense and longstanding use." Successful empiricism in the field may trump poorly resolved laboratory models, as suggested by the dictum of Oscar Wilde that "Art is not to be taught in Academies. The real schools should be the streets." In particular, there are several alternative methods to estimate the volume of shale (V_{sh}), which gives some latitude for interpretation. In addition, the value of the resistivity of the shale component (R_{sh}) is questionable, because it is based on shales between the shaly sandstone units. The composition and morphology of clays within these external shales will probably be significantly different from clays within the shaly sandstones. However, as Worthington (1985) notes "disadvantages can be partially compensated by using R_{sh} as a tuning parameter to improve predictive performance in the water zone in the expectation that better estimates of S_w will thereby be obtained in the hydrocarbon zone."

Volumetric shaly sandstone equations continue to be widely used, but more realistic equations have been developed that reflect the physics of conductivity as a surface-area cation-exchange phenomenon. This second, and more recent, model is based on the ionic double-layer observed in shaly sandstones for which the classic equation is that of Waxman and Smits (1968), which is based on laboratory measurements.

In reality, the conductivity of the shale component is a function of the cation-exchange capacities of the various types and abundances of clay minerals that are present. Since the cations are exchanged primarily at broken bonds on the edges of flakes or by lattice substitutions on cleavage surfaces, the phenomenon tends to be surface-area dependent rather than controlled simply by the volume of clay minerals. This implies that a fine-grained clay has a higher exchange capacity than a coarser grained form of the same clay volume, and this observation is confirmed by experimental data. Since all the shale indicators estimate (at best) the volume of the shale component, no explicit assessment is made of the grain size or clay-mineralogical variation.

Although these factors are widely known among log analysts, it is difficult to design log-analysis procedures to accommodate them, in the absence of a tool that measures cation-exchange capacity directly. Consequently, the model equations that use cation-exchange data have been modified to variants that substitute quantities that can be measured on logs as surrogate variables, such as that of Juhasz (1901), which is suggested to be a "normalized" representation of the Waxman-Smits equation.

DOUBLE-LAYER SHALY SANDSTONE MODELS

Waxman and Smits (1968) formulated the cation-exchange mechanism that causes clay conductivity in shaly sandstones as a double-layer model equation for the solution of water saturation. As with other shaly sandstone models, it was built up from an Archie-equation base, but its terms are rooted in the causative physical phenomenon, rather than an empirical relationship. Other double-layer models were inspired by the theory of the Waxman-Smits formulation, but attempted to adapt the terms to quantities that could be measured directly from logs, because cation-exchange measurements from core samples are generally quite rare.

The Waxman-Smits model is based on laboratory observations of shaly sandstones. The resistivity behavior of shaly sandstones that are completely saturated with brine can be understood in terms of an Archie model modified by conductivity caused by cation exchange. If a clean sandstone was flooded with pure water, then both the conductivity of the water and the rock would be effectively zero. If the pore system was flooded with a sequence of successively more conductive brines, the conductivity of the rock would decrease progressively. When crossplotted on a graph of C_w (brine conductivity) and C_o (water-saturated rock conductivity), the points would form a straight line (Figure 1.8), whose slope is the reciprocal of the formation factor, because:

$$\frac{C_o}{C_w} = \frac{R_w}{R_o} = \frac{1}{F}$$

which is the first Archie equation.

If this procedure was repeated with a shaly sandstone, rock conductivity would take a different trajectory (Figure 1.8). When the pore fluid was pure water, the conductivity of the rock would be zero, because there would be no ions in the water for exchange at clay-mineral surfaces, and therefore no conductivity effects. With a small increase in salinity, some (but not all) sites would be activated, and a clay

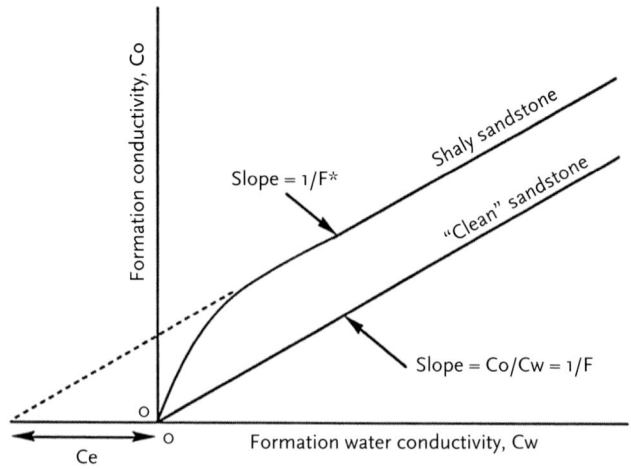

Figure 1.8: Crossplot of C_w (formation water conductivity) and C_o (water-saturated rock conductivity), with trends expected for "clean" and shaly sandstones, according to the Waxman-Smits (1968) model.

conductivity would be added to the conductivity caused by the ions in the pore space. As the salinity was increased further, there would come a point at which cation exchange was operating at full capacity, and beyond which no additional clay conduction effects would be added. The conductivity behavior of the water-saturated rock over a range of formation salinities would show a nonlinear increase at low-brine conductivities that would converge on a linear trend at higher conductivity and approximately parallel to the equivalent clean-sandstone line. The slope of the linear segment is, again, equal to the reciprocal of the formation factor, but this is the formation factor of both the open pore space and the clay-bound water and is symbolized as F^*. The Waxman-Smits model equations for water-saturated shaly sandstones are then:

$$C_o = \frac{1}{F^*}(C_w + C_e)$$

where C_e is the conductivity of the clay exchange ions and:

$$C_e = B \cdot Q_v$$

where B is the specific counterion activity (mho/m/equiv/liter) and Q_v is the concentration of exchange cations (meq/ml pore space).
So:

$$C_o = \frac{1}{F^*}(C_w + B \cdot Q_v) \text{ and } F^* = \frac{a^*}{\Phi^{m^*}}$$

In the case of a shaly sandstone that is partly saturated with hydrocarbon, the Waxman-Smits model expands in the equation development that follows.
Now:

$$C_t = C_o S_w^{n^*}$$

where C_t is the conductivity of the formation (the reciprocal of R_t) and n^* is the saturation exponent of the shaly sandstone.
So:

$$C_t = S_w^{n^*} \frac{(C_w + B \cdot Q_v')}{F^*} \text{ where: } Q_v' = \frac{Q_v}{S_w}$$

The modification of the exchange cations concentration (Q_v) occurs because the hydrocarbon phase concentrates the cations in a smaller volume of pore water.

Written in resistivity (rather than conductivity) terms, the Waxman-Smits shaly sandstone equation for water saturation becomes:

$$R_t = \frac{F^* R_w}{S_w^{n^*}\left(1 + \frac{BQ_v R_w}{S_w}\right)}$$

In order to apply the Waxman-Smits equation, values of Q_v are needed whose source are core measurements of the cation-exchange capacity (CEC) measured in units of meq/100 gm of sample:

$$Q_v = \frac{CEC \cdot (1-\Phi) \cdot \rho_{ma}}{100 \cdot \Phi}$$

Cores are generally taken from a limited sample of depths, so that some correlation with conventional logs must be established in order to evaluate water saturations over entire sections. Because of their sensitivity to clays, gamma-ray logs are an obvious choice as a surrogate measure of Q_v, but it is not always possible to establish a useable relationship because a gamma-ray value will be an expression of volume, while cation exchange is controlled by surface area. Commonly, there will be a correlation of Q_v with porosity so that, for example, Lavers, Smits, and Van Baaren (1975) proposed the empirical relationship:

$$Q_v = d\Phi^{-e}$$

where d and e are constants to be evaluated from a set of Q_v and porosity measurements from the core.

Johnson and Linke (1977) made measurements of the cation-exchange capacity from cores of Eocene sandstones in the Mackenzie Delta and were able to get a useful predictive relationship when related to gamma-ray values (Figure 1.9). They explained that the success of this relationship could be attributed to the radioactivity of the high-CEC illite, but low radioactivity of the low-CEC kaolinite content. By using their relationship, an instructive comparison can be made between water

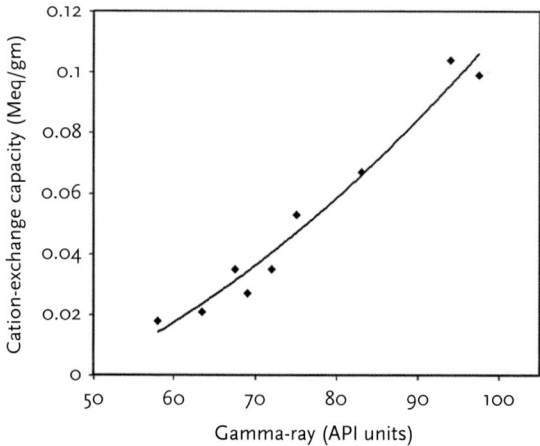

Figure 1.9: Crossplot of the cation-exchange capacity (CEC) measured from cores and gamma-ray values from logs of Eocene sandstones in the Mackenzie, fitted with a trend for use in prediction of the CEC in uncored wells. From Johnson and Linke (1977), courtesy Canadian Well Logging Society.

saturations calculated by the Archie, Simandoux, and Waxman-Smits equations in Imperial Oil Taglu G-33, which was the discovery well of the Mackenzie Delta Taglu field (Figure 1.10). As usual, the Archie equation overestimates water saturations in shaly zones by failing to accommodate the extra conductivity introduced by clays. In contrast, the Simandoux is characteristically overly optimistic on hydrocarbon saturation when related to the calibrated Waxman-Smits model prediction. All three models converge on the same water saturation estimates in shale-free zones, as the shale conductivity term is nullified, and the shaly sandstone equations become the Archie equation. Clearly, the Waxman-Smits model is the most credible prediction, not only because it is a better representation of the physics of the conductivity term, but because it is calibrated with respect to clays within the shaly sandstones rather than on the shale-log responses of the shales between the sandstones.

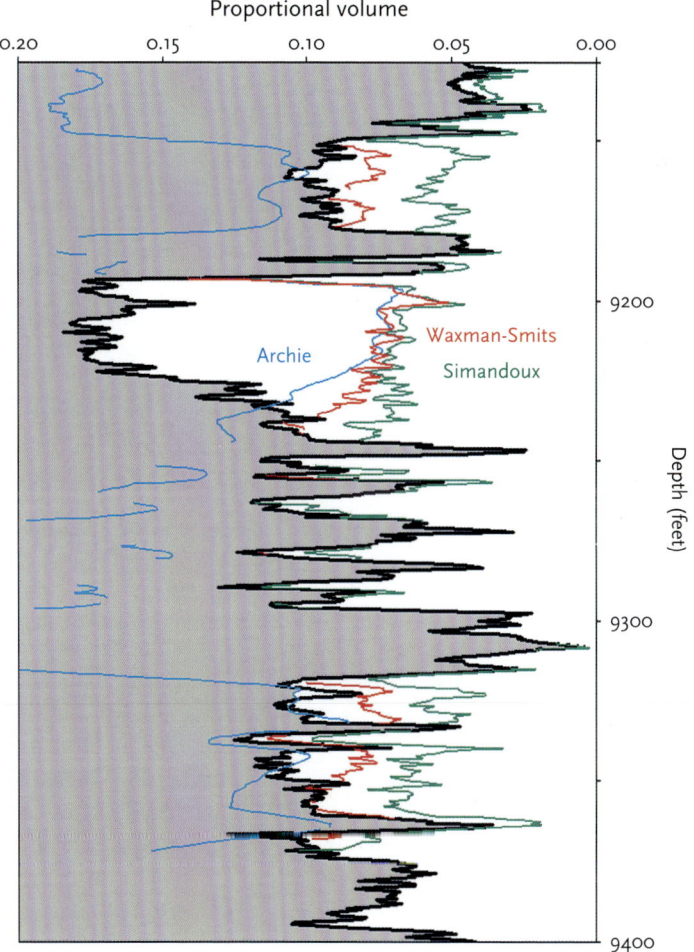

Figure 1.10: Comparison of water saturations calculated by the Archie, Simandoux, and Waxman-Smits equations in Eocene sandstones of the discovery well of the Mackenzie Delta Taglu field.

DUAL-WATER SHALY SANDSTONE MODELS

Core measurements of cation-exchange properties are not generally available, so that the derivation of these properties from the logs themselves would be preferable in most applications. These two considerations have inspired dual-water models that use the concepts of Waxman-Smits research work in coordination with properties that can be measured from the logs themselves. The most widely used variant is the "dual-water model" introduced by Clavier et al. (1977). As the name suggests, two kinds of water are postulated: formation water in the granular pore system (both "free water" and capillary-bound water) and clay-bound water, whose salinity and resistivity are different. While the resistivity of the formation water is symbolized by R_w, the clay-bound water resistivity is written as R_b.

The conventional application of the dual-water model requires the use of the neutron and density porosity logs. In common with other shaly sandstone methods, the volume of shale, V_{sh}, is first calculated and then the effective porosity, Φ_e, is computed from the porosity logs by correcting the apparent porosity for the shale effect. At this point, the dual-water model diverges, because rather than considering the shale to be a single medium, it subdivides the shale into dry matrix and a clay-bound water component. The volume of the clay-bound water added to the effective porosity is the total porosity, Φ_t. For any individual zone, the volume of clay-bound water is determined by the volume of shale multiplied by the "porosity" of the shale, Φ_{tsh}.

The volumetric aspects of the dual-water model may be clarified from the examination of the schematic neutron-density porosity crossplot (Figure 1.11). The shaly sandstone system consists of three components: quartz, fluid, and shale (the gray triangle). The quartz and fluid (mud filtrate) points are calibration points of the log; the shale point is chosen from "representative" shales in the section. In the dual-water model, the shale has a porosity of clay-bound water, and its value is a matter of local experience, but is computed from the equation:

$$\Phi_{tsh} = \delta \cdot \Phi_{dsh} + (1-\delta) \cdot \Phi_{nsh}$$

where δ takes a value between 0.5 and 1 (Asquith, 1990). The delta term is simply a pragmatic weighting factor to create a value that seems reasonable for the bound-water content of the shale, Φ_{tsh}.

We now have a new composition triangle of quartz, fluid, and "dry shale," where the fluid proportion contours are of total porosity, rather than the effective porosity contained within the triangle quartz-fluid-shale. Ideally, it would be preferable to work within the more fundamental quartz-fluid-clay system, because our aim is to correct for clay-mineral conductivity effects within the shaly sandstone. Shale consists of clay minerals and a silt-size fraction of quartz and other (mostly) nonconductive minerals, so that the "wet clay" and "dry clay" points are located on lines that are extrapolations of the shale-quartz lines. Unfortunately, it is highly unusual to see a clay zone in a section and shales are generally what we observe. Even if theoretical corrections are applied to transform shale volumes

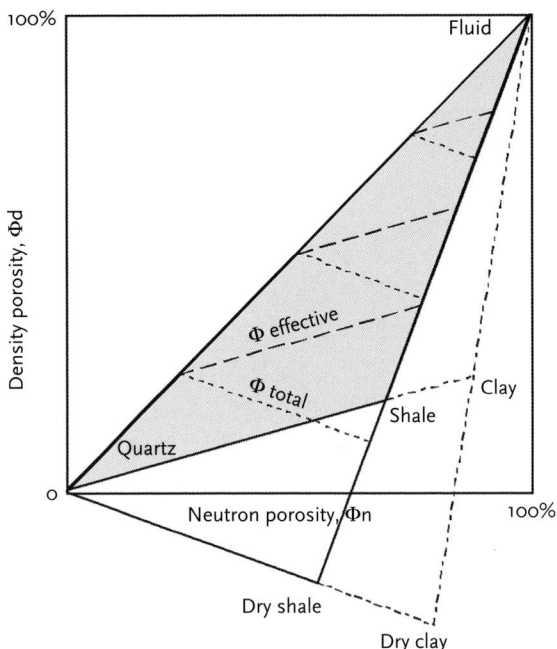

Figure 1.11: Schematic compositional system of the dual-water shaly sandstone analysis model on a neutron-density porosity crossplot.

to hypothetical clay volumes, we have no data on the resistivity of the clay, but are restricted to observations of the resistivity of shale, R_{sh}. In fact, this is all part of a larger problem in that we are using the properties of "external shales" (between the shaly sandstones) to represent "internal shales and clays," whose clay mineralogy and morphology may be radically different. Therefore, the external shale calibrators should be considered as initial model values that are selected to conform to shaly sandstone internal shales as closely as possible, but may be adjusted to give consistent solutions in the shaly sandstone. So, for example, a successful shaly sandstone model should generate values of water saturation close to 100 percent in fully water-saturated zones. If it does not, then the deviations can be used to fine-tune the model shale parameter values.

Once the total shale porosity value has been established, the total porosity of each zone can be calculated from:

$$\Phi_t = \Phi_e + Vsh \cdot \Phi_{tsh}$$

and the clay-bound water saturation computed by:

$$S_b = \frac{Vsh \cdot \Phi_{tsh}}{\Phi_t}$$

The bound-water resistivity is estimated by an Archie-style equation by combining the shale resistivity with the shale porosity:

$$R_b = R_{sh} \cdot \Phi_{tsh}^2$$

The dual-water model is a quadratic equation that can be solved for the *total* water saturation:

$$S_{wt}^2 - \left[S_b \cdot \left(1 - \frac{R_w}{R_b}\right)\right] \cdot S_{wt} - \frac{R_w}{\left(R_t \cdot \Phi_t^2\right)} = 0$$

which can be solved by:

$$S_{wt} = b + \sqrt{b^2 + \frac{R_w}{R_t \cdot \Phi_t^2}}$$

where:

$$b = S_b \cdot \left(1 - \frac{R_w}{R_b}\right)/2$$

Finally, the water saturation in the effective pore space can be calculated from:

$$S_w = \frac{(S_{wt} - S_b)}{(1 - S_b)}$$

The results of an initial run of the dual-water model should be evaluated carefully, particularly with respect to zones considered to be fully water-saturated. If the results appear to be at variance with what is known from other information sources, then a revised computation can be made with adjusted model parameter values. If the shale volume, V_{sh}, estimates are considered to be reasonable, then the parameter values that may need to be adjusted are the resistivity of the shale, R_{sh}, and the total porosity of the shale, Φ_{tsh}, as revised inputs in a second (or third) iteration.

Clearly, the ideal shaly sandstone model has not been resolved nor ever will be. Each variant has its own history of success and failure with respect to different formations and a key consideration prior to any analysis is the distribution of conductive clays within the shaly sandstone. Three endmembers are recognized: laminar shale, dispersed shale, and structural shale. Laminar shale consists of thin laminations of shale that separate stringers or beds of clean sandstone. Dispersed shale consists of pore-filling clay minerals, whose development leads to a progressive reduction in porosity. "Structural" describes sandstones in which some of the grains of the framework are shale fragments, diagenetically altered minerals, and other conductive grains. The Thomas and Stieber (1975) crossplot of shale volume versus porosity provides a widely used method to discriminate between these shale morphologies, which can be helpful both in the development of more complex models and the evaluation of operational results. However, this does not resolve the problem that the physics that underpins the volumetric shale-resistivity equations is unrealistic and the application of the ionic double-layer model is usually impractical.

Since none of these equations can possibly be "correct," we should evaluate their success, or lack of it, by their predictive power. From a statistical point of view, we should choose the model that minimizes the error between the estimates of the water saturation and their true values. At the same time, model equations should be favored that are simple (but not simplistic), with few parameters to estimate, thereby minimizing error terms and error propagation. For practical formation evaluation, the general similarity in structure of the many alternative equations that have been published provides a framework for model solution from a utilitarian viewpoint. Simple models can be used in an optimization procedure, where the terms are calibrated from water zones within the shaly sandstone reservoir. The optimization procedure also maintains internal consistency, because all shaly sandstone equations contain the Archie equation as an explicit component and are therefore already compromised by empiricism!

Some discussion has already been made about sensitivity analysis of the Archie equation by error propagation calculations and Monte Carlo simulations. With the inclusion of more parameters, shaly sandstone equations introduce additional sources of error, and their evaluation can be useful in attempting to evaluate the role of input variables. At the very least, a range of water saturations is a useful expression of the confidence that is associated with any prediction, particularly when it is expressed in percentile probability bounds. Freedman and Ausburn (1985) considered error propagation within the Waxman-Smits equation, while Bowers and Fitz (2000) evaluated the sensitivity of parameters within the dual-water model. In both cases, while the set of parameters is larger than those of the Archie equation, the method of partial derivatives expands to accommodate the extra terms. Characterizing the uncertainties associated with each parameter is particularly challenging, so that rather than presenting a clear-cut universal solution, the methodology should be considered as a toolbox for experimentation with individual reservoirs.

THE ARCHIE EQUATION IN CARBONATE ROCKS

Porosity in sandstones generally takes the form of intergranular pores between grains of quartz and other detrital minerals. Porosity in carbonate rocks (limestones and dolomites) can take a wide variety of forms because of the chemical nature of the framework that allows fracturing and dissolution at all scales, as well as compositional changes. Petrophysicists generally subdivide carbonate pore types between interparticle, fracture, and vugs. Interparticle porosity is an inclusive term for either intergranular or intercrystalline pore systems and is often referred to as "matrix porosity." Fracture porosity typically accounts for small pore volumes and so is difficult to differentiate by conventional log suites, although it plays a major role in fluid mobility. Vuggy porosity is created by a variety of dissolution processes and occurs over a wide spectrum of scales, with the smallest pores larger than grain size and ranging up to caverns. An important property of vugs is the degree to which they are connected, because this aspect influences both electrical and fluid flow through the pore system.

In the simplest case, a carbonate is characterized by a single porosity system, in which case the texture is broadly similar to a sandstone with interconnected pores between carbonate crystals or grains. A dual-porosity system couples interparticle porosity either with fractures or vugs. Finally, a triple-porosity system contains all three types of pores: interparticle, fracture, and vug.

The most widely used form of the first Archie equation for both limestones and dolomites is the basic:

$$F = \frac{1}{\Phi^2}$$

This choice is not intuitively obvious when contrasting the complex variability of carbonate pore types with the relatively simple pore structure of sandstones, whose porosity exponent takes a range of values. However, a porosity exponent value of two ($m = 2$) is a good choice for carbonate rocks, whose porosity is dominantly interparticle in nature, as shown by numerous core measurements and log evaluations (e.g. Focke and Munn, 1987).

In a dual-porosity system, the addition of either fractures or vugs causes changes in the formation factor that are reflected in the value of the porosity exponent. The volume of fracture porosity is typically very small, but provides additional conductive pathways through the rock. In theory, a rock that was composed entirely of planar fractures oriented parallel to the current flow would provide a purely linear electrical circuit and an expectation of a porosity exponent of unity. Consequently, the porosity exponent of dual-porosity systems with fractures should take a value somewhere between 1 and 2 (Figure 1.12). In contrast, the addition of vugs to a carbonate commonly adds appreciable volume to porosity, but its effects on the formation factor will vary according to the degree to which they are connected (adding conductivity) or unconnected (electrical dead space). Unconnected vugs do not contribute to the conductivity, so that the porosity exponent takes values higher than two. While high values of m in sandstones are often considered to reflect increased tortuosity of the pore network, the higher porosity exponent in carbonates caused by nonconnected vugs is simply an expression of pore space that does not participate in the flow of electrical current through the rock. The uncorrelated pore space will contain both electrically connected pore space and "dead-end" pore space that is bypassed by electrical flow. An increase in "dead-end" pore space (nontouching vugs) is matched by an increase in m (Figure 1.12).

In carbonate lithologies with variable amounts of vugs or fractures, the value of m will clearly be variable, rather than constant. This problem has been recognized in adaptations of the basic Archie equation. So, for example, the great variability of pore geometry in the Niagaran (Silurian) pinnacle-reef belt of the Michigan basin

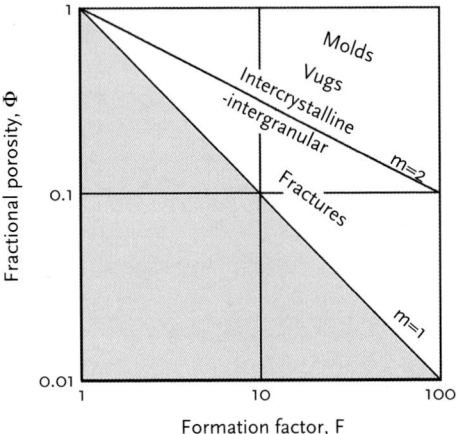

Figure 1.12: Schematic crossplot of formation factor versus porosity for carbonates with a subdivision between fields dominated by interparticle, fracture, and vug pore systems.

causes significant errors in water saturation calculations, if modeled with a cementation exponent of 2. Log analysts such as Labo (1977) used the relationship to accommodate reef vugginess:

$$F = \frac{1}{\Phi^{2.5}}$$

which works reasonably well over the normal range of reef porosities. However, Bigelow (1992) pointed out that the poor performance of this equation in low- and high-porosity ranges caused unacceptable errors in reserve estimations and proposed a "variable-m" modification which took the form:

$$F = \frac{1}{\Phi^{(a+b\Phi)}}$$

The equation was fitted empirically to core measurements of formation factor and porosity and makes m a simple function of porosity. As discussed earlier in this chapter, the structure of this equation is an improvement over the generalized Archie equation of:

$$F = \frac{a}{\Phi^m}$$

because the quantity a is simply a fitting parameter that holds the porosity exponent to an artificial constant, but the variable-m function produces values that can be related immediately to individual core-sample measurements.

Porosity Partitioning in Dual-porosity Systems

If the porosity exponent is considered to take a value of two in carbonates with interparticle porosity, then the variability of m can be used to partition the porosity between different pore types. Several authors proposed independently some simple expansions to the Archie equation that incorporate either fracture porosity or vug porosity for two types of dual-porosity models. So, for example, Aguilera (1976) evaluated changes in the value of m in the interparticle-fracture system. Watfa and Nurmi (1987) derived equations for both dual-porosity systems and reported that the application of these equations gave estimates of water saturations in a variety of Middle Eastern carbonate reservoirs that were an improvement over the simple Archie equation with cementation exponent, m, equal to two. If planar fractures have an m value of unity and the matrix has intergranular and intercrystalline porosity with $m = 2$, then the apparent m of a fractured carbonate can be solved by considering the fractures and matrix as resistances in parallel. Then, the expansion of the Archie equation leads to the result:

$$\Phi^m = \Phi_f + \Phi_{mx}^2$$

where Φ is the total porosity made up of Φ_f, the fracture porosity, and Φ_{mx}, the "matrix porosity" (interparticle porosity). If the vugs are not connected, then electrical current does not flow through them and they are nonconductive voids. As a consequence, the equation for this model is:

$$\Phi^m = (\Phi - \Phi_{nc})^2$$

where Φ_{nc} is the vug porosity.

There is insufficient information to solve for both fractures and vugs using these equations. However, if the apparent m of the carbonate is clearly higher than two, then unconnected or poorly connected "vugs" (either molds or vugs) are suggested. In this case, the vug equation can be used to solve for vuggy porosity. If the apparent m value is markedly less than two, then fracture porosity may be suspected. The two equations are graphed as contours of fracture porosity and vug porosity on the logarithmic chart of formation factor and porosity in Figure 1.13. The equations can be applied either to core measurements or downhole log evaluations of sections that are completely water-saturated, as demonstrated in the following examples.

Measurements of formation factor and porosity were made on core samples of the Permian Towanda limestone characterized either as unfossiliferous limestones or fossiliferous with the development of large dissolution molds. When the porosity exponent is calculated for each core and plotted against porosity (Figure 1.14), two distinctive electrical facies can be recognized immediately. Porosity exponents in the unfossiliferous limestones are matched closely by a value of two, as expected for an interparticle single-porosity system. In contrast, the limestones with fossil molds have higher porosities and a broad trend of increasing porosity exponent values that reflects the increase in the volume of unconnected pores. In a second example based

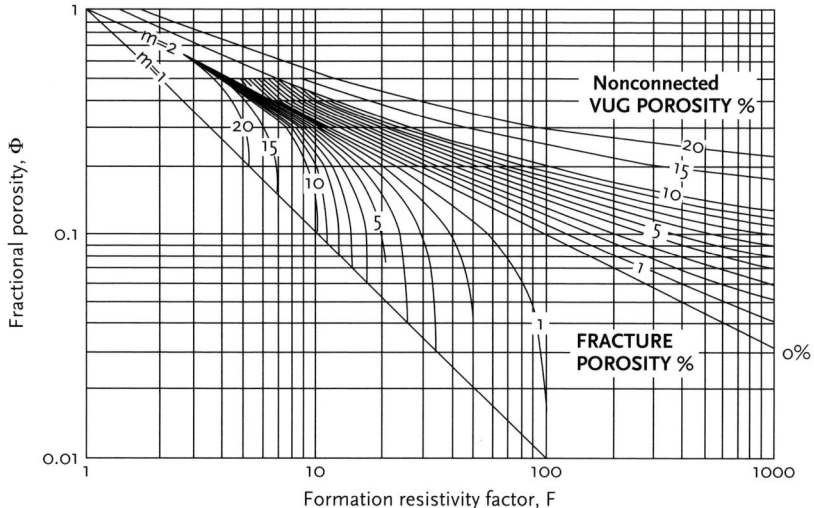

Figure 1.13: Subdivision of the formation-factor porosity crossplot by dual-porosity systems that incorporate either fractures or vugs with contours computed from expansions of the first Archie equation.

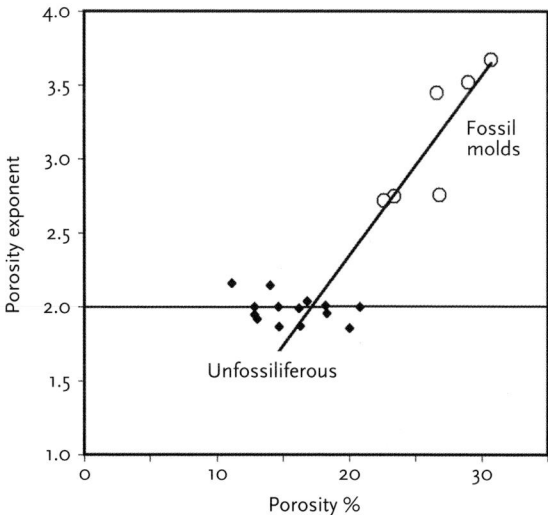

Figure 1.14: Crossplot of porosity exponent versus porosity from core measurements of Permian Towanda limestone, unfossiliferous limestones, and fossiliferous limestones with the development of large dissolution molds. The baseline at $m = 2$ matches the expectation for a single, interparticle pore system, while the trend of increasing apparent m with porosity is commonly observed in vuggy, dual-pore systems.

on cores, crossplots of formation factor versus porosity and porosity exponent, *m*, versus porosity are shown for Pennsylvanian Lansing "C" oomoldic limestones in Figure 1.15. Once again, alternative formulations of the Archie equation are presented that contrast the use of *a* as a fitting parameter and *m* as a mathematical artifact (left crossplot), with the variable *m* keyed to porosity (right crossplot). The variable-*m* trend offers some potential for the interpretation of the internal pore architecture of individual oolite bars, as discussed in the next example from the interpretation of downhole measurements.

Amoco #6 James was drilled as a development well in the Victory Field of Kansas with gas production from the Bethany Falls limestone. This Pennsylvanian limestone consists of a tight calcite wackestone deposited on an open shelf, which is succeeded by stacked oolite bars with high porosities from the dissolution of ooids and development of pervasive oomoldic porosity. A profile of the Bethany Falls limestone is shown in Figure 1.16 with a partition of the porosity between connected and unconnected pores (left) and computed porosity exponent, *m* (right). The subdivision between three successive bars (labeled informally as A, B, C) is obvious and matches outcrop observations of the Bethany Falls limestone to the east of the Victory Field (French and Watney, 1993). The anomalously high values of porosity exponent in the uppermost bar A are caused by gas saturations that increase the resistivity, as distinct from the lower bars, which are completely saturated with water. This well was perforated from 4,591 to 4,600 feet and gave an initial gas production of 380 MCF per day. A crossplot of porosity exponents, *m*, against porosity in bars B and C (Figure 1.17) show distinctive and separable trends. The bars in the outcrop are separated by unconformities, which have been interpreted as sequence boundaries, while the detailed examination of thin sections has revealed a complex history of dissolution, cementatation, and crushing with both improvement and decrease in pore connectivity (French and Watney, 1993). The implied better connectivity and relative homogeneity of the oomoldic

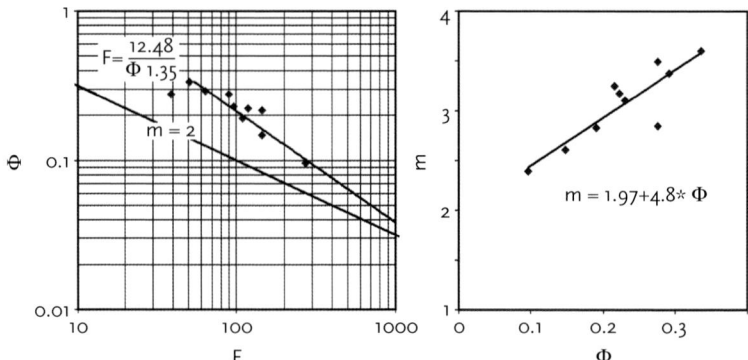

Figure 1.15: Crossplots of formation factor versus porosity (left) and porosity exponent, *m*, versus porosity (right) for Pennsylvanian Lansing "C" oomoldic limestones matched with their alternative formulations of the Archie equation.

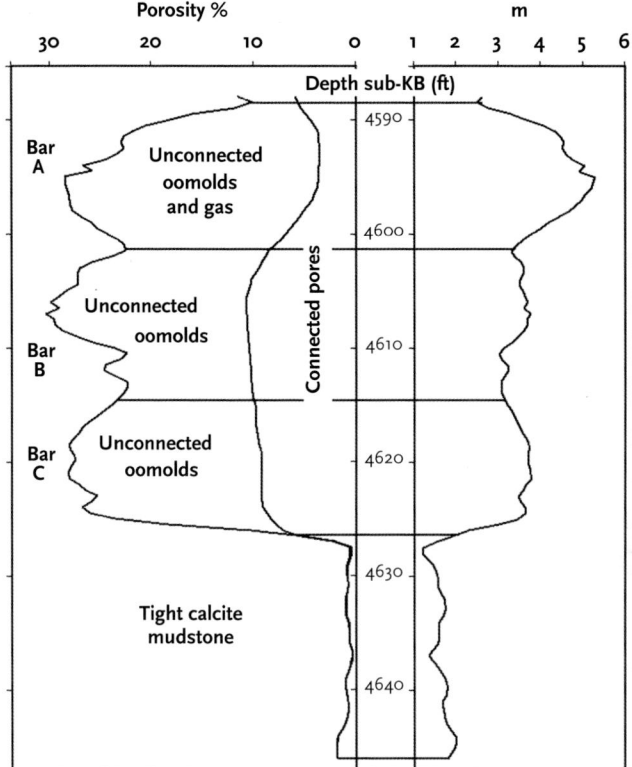

Figure 1.16: Partition of the porosity between connected and unconnected pores (left) and computed porosity exponent, m (right), of stacked oolite bars in a Pennsylvanian Bethany Falls limestone section in a gas development well in the Victory Field, Kansas.

limestones of Bar B are contrasted with those of the underlying Bar C, where the trend in porosity exponents is broader, and with values suggesting poorer connectivity (Figure 1.17). Clearly, these explicit patterns are tied to the separable diagenetic histories of the two bars and are of potential use in supplementing more traditional geological methods in subsurface interpretation. Most importantly, they demonstrate once again that although the porosity exponent, m, is empirical and a crude descriptor of pore geometry, it can often be interpreted in terms of mechanisms of deposition and diagenesis.

At another extreme, it is common to observe anomalously low values of the porosity exponent in carbonates with low porosities. From extensive core measurements, Focke and Munn (1987) observed that the porosity exponent in carbonates often shows a decline in value below 10 percent porosity, and becomes pronounced below 5 percent. This effect is shown clearly both in core measurements and in log analysis (Figure 1.18). In part, the effect reflects the increased sensitivity of small changes in porosity at these low levels, as can be seen on the plot of Figure 1.13.

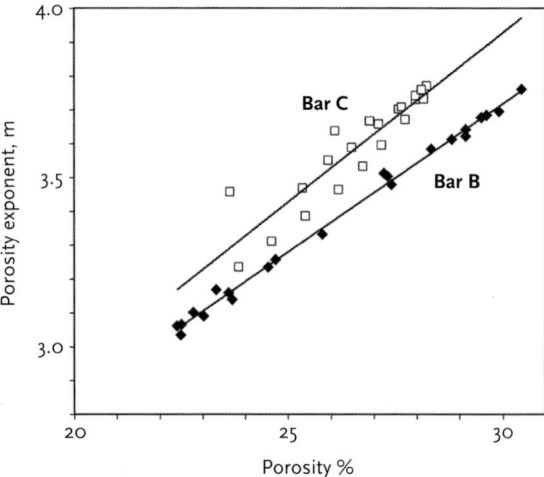

Figure 1.17: Crossplot of porosity exponents, m, against porosity in bars B and C in a Pennsylvanian Bethany Falls limestone section in a gas development well in the Victory Field, Kansas.

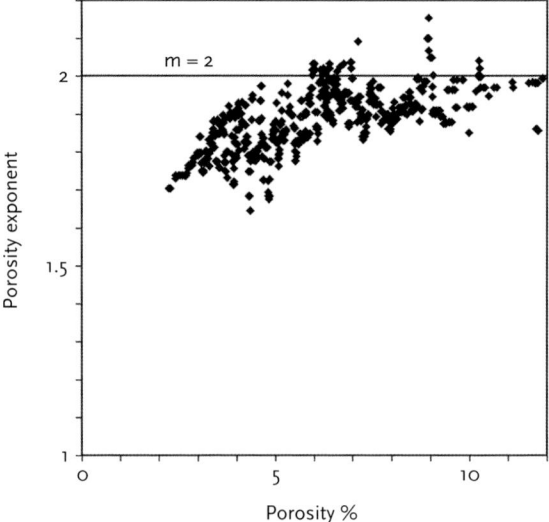

Figure 1.18: Porosity exponents computed from porosity and resistivity logs of a Cambro-Ordovician Arbuckle limestone aquifer, which shows the decline of the exponent value commonly observed in low-porosity zones from the general expectation of $m = 2$.

However, the bias towards lower values of m even in nonfractured rock appears to show a systematic effect with another physical explanation. One interpretation was offered by Wardlaw (1976), who attributed the effect to a tendency of the shape of carbonate pores to become more sheetlike and the pore throats to be more uniform in size.

THE POROSITY EXPONENT IN A TRIPLE-POROSITY SYSTEM

Dual-porosity carbonates can be classified either as interparticle-fracture or connected-pore–unconnected-vug systems. A pragmatic choice of the appropriate model is determined by whether the porosity exponent, m, takes a value of less than or greater than two. Clearly, there will be many instances where the carbonate pore system is triple in nature: interparticle, fracture, and unconnected vugs. The expansion causes a significant increase in complexity, but Aguilera and Aguilera (2004) proposed a useful methodology that capitalizes on experience with dual-porosity carbonates. They modeled the triple-porosity system as a parallel resistance network for interparticle and fracture porosity, coupled with nonconnected vugs in series. This model then also accommodates situations where the pore system is dual in nature. The porosity exponent, m, of a triple-porosity system expressed in the formulation of Aguilera and Aguilera (2004) is:

$$m = \frac{-\log\left[v_{nc}\varphi + \frac{(1-v_{nc}\varphi)}{v\varphi + (1-v\varphi)/\varphi_b^{-m_b}}\right]}{\log \varphi}$$

where ϕ is the total porosity, ϕ_b is the nonfracture porosity, v is the fraction of the fracture porosity divided by the total porosity, v_{nc} is the fraction of the nonconnected porosity divided by the total porosity, and m_b is the porosity exponent of the pore space when no fractures are present. The equation involves several unknowns, but its application is helped considerably by the different morphology of the pore types. So, the effect of fractures on m is most pronounced at low porosities because they typically occupy minimal volumes. This contrasts with nonconnected vugs, whose origin in dissolution creates greater volume, and so their effects are noticeable in carbonates with higher porosities. The complexity of the equation should not blind us to the fact that this triple-porosity formulation is only a model, grounded in the empirical Archie equation and built from concepts of electrical flow in parallel and series networks. Its utility lies in whether it helps us to better understand the electrical properties of carbonates and whether it can be applied to viable predictions in the subsurface. While older methods of log analysis were limited to the estimation of petrophysical unknowns by direct calculation, undetermined models of this kind can be the basis for forward modeling. This newer approach computes logs based on alternative reservoir models for comparison with actual log measurements and is easy to implement in today's computer environment.

DIELECTRIC LOGGING MEASUREMENT OF THE POROSITY EXPONENT

Clearly, variable-m methods are fully appropriate for carbonate rocks with any pore complexity. However, up to this point, the prediction of the porosity exponent has been keyed to the porosity, either by a fitted relationship with total pore volume or by porosity partitioning methods. As a first-order model, this is not unreasonable because the addition of vuggy porosity will both increase pore volume and the anticipated value of m. However, the prediction will be based on a trend rather than a specific value for each zone. The direct and continuous estimation of the porosity exponent, m, by a logging-tool measurement would clearly be a major improvement over an indirect inference of its value. This is possible by means of dielectric logging. The electrical properties of rocks are completely defined by three properties: magnetic permeability, electrical conductivity, and dielectric permittivity. The magnetic permeability of most reservoir rocks is minimal, because of the paucity of magnetic minerals. Electrical conductivity, or its reciprocal, resistivity, has already been discussed extensively as to its role in both Archie and non-Archie rock formulations. Finally, the dielectric permittivity reflects the polarization of a material when exposed to an electrical field at a given frequency. The dielectric constant of water is considerably higher than that of other rock and pore components, and so its measurement by dielectric logging gives the volume of water in the formation, with only minor effects due to salinity.

The operational principle of the dielectric tool is essentially the same as a microwave oven, but where the microwaves travel through a subsurface formation and measurements are made of their attenuation and phase shift. The earlier tools were strongly affected by borehole rugosity, which discouraged their acceptance, but more recent developments by the logging industry have resulted in major improvements. The measurement investigation is shallow, so that its evaluation is within the flushed zone, where formation water has effectively been replaced by mud filtrate. Sherman (1983) confirmed experimental work by other investigators that the internal geometry of the pore network has additional effects on the dielectric constant and suggested that this could be used to predict the porosity exponent in dielectric logging through use of the depolarization factor. Consequently, m can be solved for each zone and then used in the evaluation of saturations in the uninvaded formation. The improvements in dielectric logging and the challenge of estimating the porosity exponent in vuggy carbonate formations have encouraged the use of this methodology as a more routine application (Seleznev et al., 2006).

PETROGRAPHIC EVALUATIONS OF THE POROSITY EXPONENT IN CARBONATES

The "meaning" of the porosity exponent in carbonates is more difficult to grasp than is the case for sandstones, which is complex in itself. Sandstones are assemblages of grains with an interparticle pore system whose basic control on m can be considered in terms of grain and pore shape, together with the tortuosity of the pore network. Of course, at a deeper level, aspects of pore-throat aspect ratios, pore coordination numbers, and other morphological descriptors need to be considered. As discussed earlier, the chemical nature of carbonates results in a more complex and heterogeneous pore architecture through dissolution, mineral transformation, and other diagenetic processes.

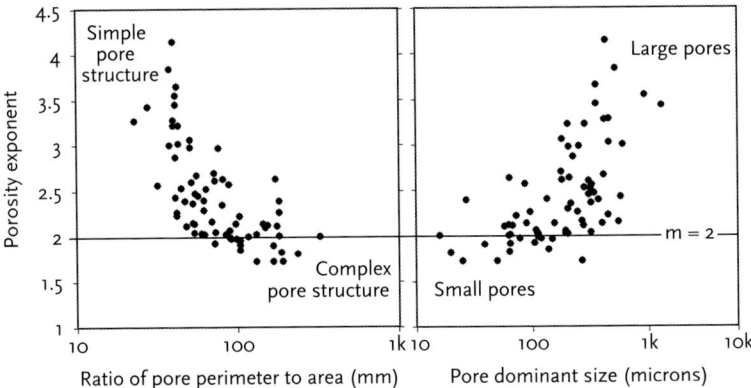

Figure 1.19: Crossplots of porosity exponent against pore perimeter to area ratio (left) and dominant pore size (right) from carbonate core measurements. Adapted from Verwer et al. (2011), © 2011 American Association of Petroleum Geologists (AAPG), reprinted by permission of the AAPG, whose permission is required for further use.

Because carbonate reservoirs are so important worldwide, substantial efforts continue to be directed to laboratory studies of core samples with the aim of improving the understanding of pore networks and isolating key predictors that can be applied to the subsurface. In a recent study by Verwer et al. (2011), seventy-one carbonate core samples with textures ranging from wackestones to grainstones were evaluated by a variety of methods. Measurements of resistivity, porosity, and permeability were related to pore morphology parameters computed from digital image analysis of thin sections. Verwer et al. (ibid) considered the relationship between the porosity exponent and the pore perimeter/area ratio and the dominant pore size. The pore perimeter/area ratio is the two-dimensional equivalent of the specific surface area, with smaller values suggesting a simple pore geometry and larger values indicating an intricate pore system. Crossplots of these two measures with porosity exponents (Figure 1.19) show distinctive trends that can be interpreted. Lower values of the porosity exponent are associated with small pores in a complex pore structure, while high values are linked with larger pores in simple pore structures. Additionally, an increase in the number of pores and pore connections, rather than the size of the pore throats, appeared to have a major influence on lowering the porosity exponent.

THE SATURATION EXPONENT, N

As mentioned at the beginning of this chapter, Archie (1942) proposed a second Archie equation to describe the resistivity of formations partially saturated with hydrocarbons, which took the form:

$$S_w = \left(\frac{R_o}{R_t}\right)^{\frac{1}{n}}$$

where R_t is the resistivity of the sample at S_w, which is the fractional water saturation, R_o is the resistivity of the completely water-saturated sample, and n is the saturation exponent. Although a considerable body of research has been devoted to the first Archie equation as applied to rocks that are completely saturated with brine, until recently, less work has been directed to the consideration of the saturation exponent, n, which controls the resistivity index—water-saturation relationship. In part, this discrepancy reflects the extra difficulties involved in both measurement and evaluation. Core estimations are based typically on a crossplot of the resistivity index measured over a range of water saturations, where the saturation exponent is determined by the slope (Figure 1.20).

Archie's original suggestion (Archie, 1942) that n takes an approximate value of two is still widely accepted as a viable default value, and it is not unreasonable for water-wet rocks. Adisoemarta et al. (2001) considered this value to be a representative number for laboratory measurements, but argued from a theoretical derivation that the exponent should actually be equal to one. They attribute the difference between theory and practice to the pore structure of bodies and throats, where the resistivity is controlled by pore-throat constrictions, but is used to predict the proportion of the water within the pore body. From this perspective, a saturation exponent of greater than unity is the necessary adjustment to scale a pore-throat cross-sectional area parameter to a pore-body volumetric.

As an example of laboratory measurements, saturation exponents from core samples of the Permian Towanda limestone are plotted against porosity exponents, in Figure 1.21. These are the same cores of unfossiliferous and vuggy, fossiliferous limestones for which the porosity exponents were plotted in Figure 1.14. Notice that the marked and systematic change in the porosity exponent with porosity does not

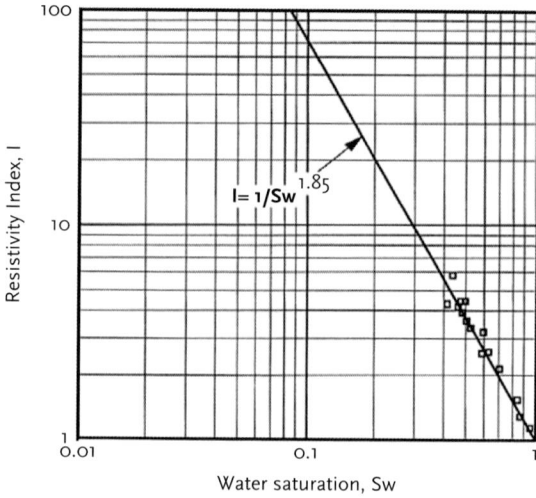

Figure 1.20: Example of the determination of the saturation exponent in an Arbuckle limestone core sample from the slope of the trend of resistivity index and water-saturation measurements.

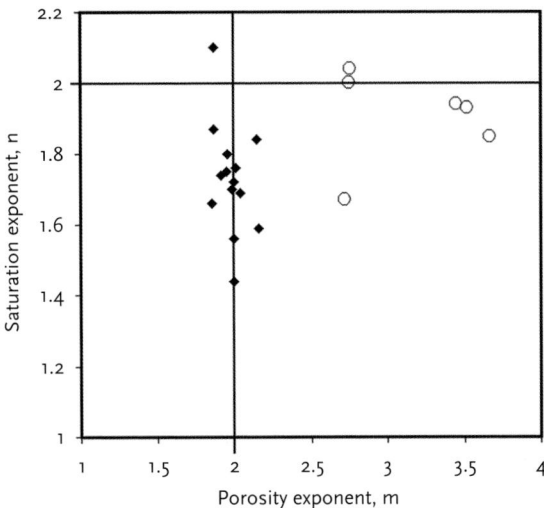

Figure 1.21: Crossplot of saturation exponents versus porosity exponents measured from core samples of the Permian Towanda limestone. Values for unfossiliferous limestones are contrasted with vuggy, fossiliferous limestones, and both referenced to the general expectation that $m = n = 2$.

appear to be matched with major differences in the range of saturation exponent values. However, there is a distinctive shift to higher values in the moldic pores, as compared to the interparticle porosity of the unfossiliferous limestones, and this could be interpreted in terms of pore-throat area to pore body ratios.

Laboratory measurements most commonly report a single value for the saturation exponent, which perpetuates the commonly held belief that it is always a constant. However, when the saturation exponent is evaluated in conjunction with capillary pressure measurements, then distinctive changes are often seen in the slope of the resistivity index—water saturation crossplot. The changes in slope, and so also the saturation exponent, are observed to occur at points that correspond to distinctive pore-throat sizes. Consequently, because the mechanism of oil emplacement within pores is controlled by capillary forces, the saturation exponent may vary with the height within the reservoir. At greater heights, the lowering of water saturation represents the progressive breaching of successively smaller pore throats, so that at partial saturations, larger pores contain oil, while smaller pores remain fully water saturated. If the pores are effectively filled by oil, such that the residual water saturation makes a minimal contribution to conductivity, then all the electrical current is carried preferentially by smaller pores. These conclusions have been supported by a number of laboratory studies of core resistivity and capillary pressure. So, for example, Swanson (1985) made simultaneous measurements of capillary pressure and electrical resistivity in sandstone samples. He observed a subdivision between macropores and micropores that was reflected in a change in the slope of the resistivity index and was commonly caused by the occurrence of microporous chert.

Brown (1999) proposed a "binary Archie model" (BAM) to fit resistivity-index–water-saturation trends where a distinctive break in slope suggested a separable micropore-macropore system. The equations of the BAM can be used to solve both the porosity exponent and saturation exponent values of the macropores and micropores. The binary expansion of the second Archie equation can be written as:

$$I = \frac{R_t}{R_o} = \frac{Co_{mic}V_{mic} + Co_{mac}(1-V_{mic})}{Ct_{mic}V_{mic} + Ct_{mac}(1-V_{mic})}$$

where:

$$C_o = C_w \cdot \Phi^m \quad \text{and} \quad C_t = C_w \cdot \Phi^m \cdot S_w^n$$

are the conductivity terms applied to each component and V_{mic} is the proportion of microporosity. Notice that the m and n values for the micropores and macropores in the conductivity terms will be different and can be solved by a least-squares nonlinear optimized fit to the resistivity index—water-saturation measurements. Brown (1999) provided an example of the binary Archie model applied to the Iveshak formation of Prudhoe Bay. McCoy and Grieves (1997) reported that the saturation exponent was found to change with height above the oil-water contact from the effect of microporous chert in the Iveshak. By applying the *BAM* model to the Iveshak resistivity index—water-saturation trend, Brown (1999) concluded that the Archie exponents for the sandstone macropores were $m = 1.30$ and $n = 1.49$, and those for the microporous chert were $m = 1.98$ and $n = 1.53$.

In using these insights for improved reservoir analysis, it is clear that the height of the hydrocarbon column in the reservoir coupled with pore-size distribution are the major considerations. Conventional analysis with Archie-equation exponents set as constants will often be sufficient to characterize limited sections and also longer columns with unimodal pore distributions. Large reservoirs with heterogeneous pore distributions may show distinctive changes in Archie exponents with height, but the effects can be evaluated from core analysis of capillary pressure and resistivity. These issues will apply to both clastic and carbonate reservoirs, but may prove to be of relatively minor concern if wettability is a factor, as discussed in the next section.

WETTABILITY EFFECTS ON THE SATURATION EXPONENT

Wettability was not generally thought to be a significant issue when Archie's pioneering paper was published in 1942. At that time, reservoirs were considered to have been formed from aquifers in which the wetting agent continued to be water, as hydrocarbons accumulated in the center of pore bodies. Once oil wettability was demonstrated in many reservoirs, its impact on the Archie equation was immediately recognized in its potential effect on the saturation exponent, *n*. One of the defining characteristics of an Archie rock is that its water phase should provide a continuous conductive path through the rock. In a water-wet situation, even at low saturations, the immobile water maintains a continuous film at grain surfaces for conduction. In an oil-wet situation, formation water becomes trapped in isolated globules and filaments that do not contribute to electrical flow, thereby creating a non-Archie rock. Because the formation resistivity is increased substantially, viable predictions of water saturation from the Archie equation would require unreasonably high and erratic values for the saturation exponent.

Rather than water-wet and oil-wet situations being discrete alternatives, they actually form endmembers of the range of wettability, with intermediate types variously known as Dalmatian, fractional, or mixed wetting. The degree to which a formation can be considered to be an Archie rock can be gauged by the departure of the saturation exponent values from their expectations under water-wet conditions, typically a value of approximately two. Two aspects need to be considered: the proportion of the grain surface that is oil wet and the water saturation of the pore space. Morgan and Pirson (1964) measured saturation exponents on cores with different partial wettabilities and found that the values stayed fairly constant and consistent with water-wet expectations when oil wettability was less than 40 percent. However, at higher oil wettabilities, the saturation exponent increased rapidly to extremely high values. With regard to the relationship with water saturation, Anderson (1986) summarized the conclusions of several experimentalists who were in fundamental agreement that under oil-wet conditions, the saturation exponent maintained essentially its water-wet expectation at water saturations exceeding about 35 percent. Below this value, there was a substantial increase in the saturation exponent, and microscopic observations showed the water phase to be increasingly isolated within the pore space.

In summary, Archie rock appears to be sufficiently robust to tolerate a certain degree of oil wettability, as indicated by experimental work with cores. However, in contrast to laboratory work, serious effects can often be seen from field analyses, particularly in reservoirs with high capillary pressures associated with long oil columns and carbonates whose surfaces often show a greater susceptibility to chemical reactions from humic acids than to clastics. Wettability may change as a function of height within the oil column, so that water-wet conditions can be expected to prevail at the base of the transition zone, but with a potential for increasing oil wettability above the transition zone. Fortunately, recent advances in logging technology can be brought to bear on the problem, particularly as the degree of wettability often varies zone by zone within a reservoir. So, for example, differences in the surface relaxivity of water and oil can be evaluated by magnetic resonance image logs and used to calculate a wettability index (Looyestijn, 2007).

In addition, because of its sensitivity to the water volume, dielectric logging has been linked with conductivity measurements in a connectivity equation that accommodates oil wettability effects and converges on the Archie equation under water-wet conditions (Montaron, 2007).

ARCHIE REDUX

Over seventy years have passed since Archie published his equations that have been the basis for water-saturation estimations ever since. In the twenty-first century, the debate concerning the meaning and viability of the equations continues in the pages of respected scholarly journals, and a variety of alternatives continues to be offered. One motivation is an instinctive dislike for empirical equations, and instead, there is a quest for true meaning that can be explained in terms of the fundamental physics of electrical flow. Another motivation is the increasing importance of hydrocarbon reservoirs that would have been dismissed as noncommercial until recently, and commonly occur in "non-Archie rocks." New models with proven predictive power for these challenging rock types would be welcomed, particularly if they were rooted in comprehensible physical theory rather than raw empiricism.

Kennedy and Herrick (2012) pointed out that the Archie equations were products of their time. Before computers, or even hand calculators, data analysis was driven by graph paper, and data was plotted on linear, semilogarithmic, or logarithmic axes. Even with these limitations, Archie had some choices available to him and elected to use a power law based on resistivity, rather than functions of conductivity, which are generally favored by his detractors. He probably chose to work with resistivity so that useful results could be applied immediately to the resistivity logs of his day. However, physicists tend to think in terms of electrical efficiency and focus on the conductive aspects of the medium.

Although conductivity is the reciprocal of resistivity, rewriting the formulation of the electrical properties of porous rocks gives potential new insights into the nature of the controlling parameters. The formation conductivity factor, G is given by:

$$G = \frac{\sigma_o}{\sigma_w}$$

which is the reciprocal of the formation resistivity factor, and where ρ_w and ρ_o are the conductivities of the pore water and the water-filled rock, respectively. The conductivity factor is most commonly related to porosity through a consideration of the connectivity of the pores using percolation theory, following the seminal work of Kirkpatrick (1973). For any porous medium, there will be a minimum critical porosity at which there are no connections between the pores and therefore no conductivity. This critical value is the percolation threshold porosity. A simple model of a reservoir can be constructed from a cubic lattice of cells in which all the cells initially are solid, and the pore cells are then added progressively and randomly. It is not until a porosity of 0.3116 is reached that a conductive path can be established through the cubic lattice. Clearly this high value marks one small step towards a useful reservoir model, because the connection of each cell with adjacent cells is limited to six cube faces. When additional connections are allowed at the cell corners, edges, and sides, then the threshold percolation porosity drops to about 0.1 (Montaron, 2009).

Real rocks have considerably lower thresholds, and measurable conductivity is maintained at very small pore volumes. So, for example, Bourbie and Zinszner (1985) estimated the threshold percolation porosity of the Fontainbleau sandstone to be 0.025. The Fontainebleau sandstone is widely recognized as the archtypical Archie rock, so that this estimate could be considered to be a default value.

The conductivity formation factor can then be related to porosity with the equation:

$$G = \left[\frac{\varphi - \varphi_{pt}}{1 - \varphi_{pt}}\right]^q$$

where φ_{pt} is threshold percolation porosity and q is an exponent whose value is typically taken to be close to two. With a percolation threshold porosity of zero, this expression becomes the first Archie equation, rewritten in terms of conductivity.

The resistivity index, I, can also be interpreted in terms of connectivity, by applying a "water connectivity index," X_w in a model proposed by Montaron (2009):

$$I = \left(\frac{1 - S_c}{S_w - S_c}\right)^\mu \quad \text{where} \quad S_c = \frac{X_w}{\varphi}$$

and S_c represents the critical water saturation. When the water connectivity index is zero and μ takes a value of two, then the formula becomes the second Archie equation. However, the second Archie equation assumes a constant value for the saturation exponent, whereas the introduction of the water connectivity index with nonzero values creates a variable slope when the resistivity index is plotted against water saturation. With positive values of the water conductivity index, the resistivity index takes on progressively higher values at lower water saturations, which emulates oil-wet rocks in this lower range. The physical interpretation of these models then recasts resistivity concepts of tortuosity, length, and area into the degree of connectedness of the conductive water within the reservoir pore space.

As a power law, the Archie equation might be a natural candidate for characterization by some kind of fractal model because the self-similarity functions of fractals are created by power laws keyed to size. Roy and Tarafdur (1997) demonstrated how the Archie equation can be created from a fractal model, but they cautioned that this relationship will only hold true provided that the size of the pores are also fractal. Detailed work by Krohn (1988) determined that the distributions of pore volumes in sandstones were subdivided between a fractal short-length regime and a Euclidean (nonfractal) long-length regime. She attributed the nonfractal behavior to a reflection of the original sandstone porosity, as contrasted with fractal behavior at smaller scales, caused by diagenesis, which could be fitted by power laws of scale. The creation of pore space within carbonates is generally the result of a more complex mix of processes that modify the pore architecture over a wide range of scales. However, the application of fractal models to carbonate reservoirs appears to be gaining ground from a time when they were considered to be a novelty with little practical application. Montaron (2005) concluded that each of the carbonate pore systems of interparticle, vugs, and fractures could be independently and successfully modeled by fractal distributions. Their characterization by fractals would be especially useful, because the daunting complexity of some carbonate pore systems could be described by power laws with relatively few parameters.

In spite of these newer developments, the Archie equation continues to be used successfully in conventional reservoirs on a daily basis. Critics of the Archie equation concede that the necessary textural information must indeed be contained within the porosity exponent, m, and the saturation exponent, n. Their complaint that the physical meaning of the exponents is obscured in the formulation is probably of less importance to the geologist than the physicist. Rocks are complicated. Facies that are differentiated by changes in m can take their place with geological facies defined by a host of qualitative criteria. Meanwhile, engineers can be satisfied by equations that are fit for purpose and deliver results at an acceptable level of accuracy. In the end, perhaps we can all agree with the widely quoted aphorism of the statistician, George Box, "All models are wrong; some models are useful."

REFERENCES

Adisoemarta, P., Anderson, G., Frailey, S., and Asquith, G., 2001, Saturation exponent n in well log interpretation: Another look at the permissible range: Society of Petroleum Engineers, SPE 70043-MS, 4 p.

Aguilera, R., 1976, Analysis of naturally fractured reservoirs from conventional well logs: Journal of Petroleum Technology, v. 28, no. 7, pp. 764–772.

Aguilera, R.F., and Aguilera, R., 2004, A triple porosity model for petrophysical analysis of naturally fractured reservoirs: Petrophysics, v. 45, no. 2, pp. 157–166.

Anderson, W., 1986, Wettability literature survey-part 3: The effects of wettability on the electrical properties of porous media: Journal of Petroleum Technology, v. 38, no. 12, pp. 1371–1378.

Archie, G.E., 1942, The electrical resistivity log as an aid in determining some reservoir characteristics: Transactions of the American Institute of Mechanical Engineers, v. 146, pp. 54–62.

Archie, G.E., 1950, Introduction to petrophysics of reservoir rocks: American Association of Petroleum Geologists Bulletin, v. 34, no.5, pp. 943–961.

Asquith, G.B., 1990, Log evaluation of shaly sandstone reservoirs: A practical guide: American Association of Petroleum Geologists, continuing education course notes, Tulsa, Oklahoma, 59 p.

Bardon, C., and Pied, B., 1969, Formation water saturation in shaly sands: Transactions of the Society of Professional Well Log Analysts, 10th Annual Logging Symposium, Paper Z, 19 p.

Bhattacharya, S., Byrnes, A.P., Gerlach, P.M., and Olea, R.A., 2002, Reservoir characterization to inexpensively evaluate exploitation potential of a small Morrow incised valley-fill field (Abs.): American Association of Petroleum Geologists, Search and Discovery Article #90007, 1 p. 3

Bigelow, E., 1992, Introduction to wireline log analysis: Western Atlas International, Houston, Texas, 312 p.

Bourbie, T., and Zinszner, B., 1985, Hydraulic and acoustic properties as a function of porosity in Fontainebleau sandstone: Journal of Geophysical Research, v. 90, no. B13, pp. 11524–11532.

Bowers, M.C., and Fitz, D.E., 2000, A probabilistic approach to determine uncertainty in calculated water saturation: Transactions of the Society of Professional Well Log Analysts, 41st Annual Logging Symposium, Paper QQ, 13 p.

Brown, G., 1997, Letter to the editor: The Log Analyst, v. 38, no. 2, p. 11.

Brown, G., 1999, Analysis of non-linear resistivity index vs. saturation data using a binary Archie model: Transactions of the Society of Professional Well Log Analysts, 40th Annual Logging Symposium, Paper R, 12 p.

Buatois, L.A., Mángano, M.G., Alissa, A., and Carr, T.R., 2002, Sequence stratigraphic and sedimentologic significance of biogenic structures from a late Paleozoic marginal-to open-marine reservoir, Morrow sandstone, subsurface of southwest Kansas, USA: Sedimentary Geology, v. 152, no. 1, pp. 99–132.

Bussian, A.E., 1982, A generalized Archie equation: Transactions of the Society of Professional Well Log Analysts, 23rd Annual Logging Symposium, Paper E, 12 p.

Chen, H.C., and Fang, J.H., 1986, Sensitivity analysis of the parameters in Archie's water saturation equation: The Log Analyst, v. 27, no. 5, pp. 39–44.

Chisholm, J.L., Schenewerk, P.A., and Donaldson, E.C., 1987, A comparison of shaly-sand interpretation techniques in the Mesaverde group of the Uinta Basin, Utah: Society of Petroleum Engineers, Formation Evaluation, v. 2, no. 4, pp. 478–486.

Clavier, C., Heim, A., and Scala, C., 1976, Effect of pyrite on resistivity and other logging measurements: Transactions of the Society of Professional Well Log Analysts, 17th Annual Logging Symposium, Paper HH, 11 p.

Clavier, C., Coates, G., and Dumanoir, J., 1977, The theoretical and experimental bases for the "dual water" model for the interpretation of shaly sands: Society of Petroleum Engineers, SPE 6859-MS, 10 p.

Cluff, R.M., and Byrnes, A.P., 2009, Evidence for a variable Archie porosity exponent "M" and impact from saturation calculations for Mesaverde tight gas sandstones; Piceance, Uinta, Green River, Wind River, and Powder River Basins (Abs.): American Association of Petroleum Geologists, Search and Discovery Article #90092, 11 p.

Doveton, J.H., 1986, Log analysis of subsurface geology: Wiley Interscience, New York, 273 p.

Ehrlich, R., Etris, E.L., Brumfield, D., Yuan, L.P., and Crabtree, S.J., 1991, Petrography and reservoir physics III: Physical models for permeability and formation factor: American Association of Petroleum Geologists Bulletin, v. 75, no. 7, pp. 1579–1592.

Etris, E.L., Brumfield, D.S., Ehrlich, R., and Crabtree, S.J., 1989, Petrographic insights into the relevance of Archie's equation: Formation factor without "m" and "a": Transactions of the Society of Professional Well Log Analysts, 30th Annual Logging Symposium, Paper F, 18 p.

Focke, J.W., and Munn, D., 1987, Cementation exponents in Middle Eastern carbonate reservoirs: Society of Petroleum Engineers, Formation Evaluation, v. 2, no. 2, pp. 155–167.

Freedman, R., and Ausburn, B.E., 1985, The Waxman-Smits equation for shaly sands: I. Simple methods of solution; 11. Error analysis: The Log Analyst, v. 26, no. 2, pp. 11–24.

French, J.A., and Watney, W.L., 1993, Stratigraphy and depositional setting of the lower Missourian (Pennsylvanian) Bethany Falls and Mound Valley limestones, analogues for age-equivalent ooid-grainstone reservoirs, Kansas: Kansas Geological Survey Bulletin 235, pp. 27–39.

Glover, P., 2009, What is the cementation exponent? A new interpretation: Leading Edge, v. 28, no.1, pp. 82–85.

Guyod, H., 1944, Fundamental data for the interpretation of electric logs: Oil Weekly, v. 115, no. 38, pp. 21–27.

Herrick, D.C., and Kennedy, W.D., 1994, Electrical efficiency—A pore geometric theory for interpreting the electrical properties of reservoir rocks: Geophysics, v. 59, pp. 918–927.

Jackson, P.D., Taylor-Smith, D., and Stanford, P.N., 1978, Resistivity-porosity-particle relationships for marine sands: Geophysics, v. 43, no. 6, pp. 1250–1262.

Johnson, W.L., and Linke, W.A., 1977, Some practical applications to improve formation evaluation of sandstones in the Mackenzie Delta: Canadian Well Logging Society, Transactions of the 6th Formation Evaluation Symposium, Paper R, pp. 1–19.

Juhasz, I., 1981, Normalised Qv—the key to shaly sand evaluation using the Waxman-Smits equation in the absence of core data: Transactions of the Society of Professional Well Log Analysts, 22nd Annual Logging Symposium, Paper Z, 36 p.

Kennedy, W.D., and Herrick, D.C., 2012, Conductivity models for Archie rocks: Geophysics, v. 77, no. 3, pp. WA109-WA128.

Kirkpatrick, S., 1973, Percolation and conduction: Reviews of Modern Physics, v. 45, no. 4, pp. 574–588.

Krohn, C.E., 1988, Sandstone fractal and Euclidean pore volume distributions. Journal of Geophysical Research, v. 93, no. B4, pp. 3286–3296.

Labo, J.A. 1977, Interpreting Silurian Niagaran reefs in the Michigan basin: Transactions of the Society of Professional Well Log Analysts, 18th Annual Logging Symposium, Paper I, 13 p.

Lavers, B.A., Smits, L.J.M., and Van Baaren, C., 1975, Some fundamental problems of formation evaluation in the North Sea: The Log Analyst, v. 16, no. 3, pp. 3–13.

Looyestijn, W.J., 2007, Wettability index determination from NMR logs: Transactions of the Society of Professional Well Log Analysts, 48th Annual Logging Symposium, Paper Q, 16 p.

Maute, R.E., Lyle, W.D., and Sprunt, E.S., 1992, Improved data-analysis method determines Archie parameters from core data: Journal of Petroleum Technology, v. 44, no. 1, pp. 103–107.

Maxwell, J.C., 1873, A Treatise on electricity and magnetism: Clarendon Press, Oxford, 554 p.

McCoy, D. D., and Grieves, W. A., 1997, Use of resistivity logs to calculate water saturation at Prudhoe Bay: Society of Petroleum Engineers, Reservoir Engineering, v. 12, no. 1, pp. 45–51.

Mendelson, K.S., and Cohen, M.H., 1982, The effect of grain anisotropy on the electrical properties of sedimentary rocks: Geophysics, v. 17, no. 2, pp. 257–263.

Montaron, B., 2005, Editorial: Schlumberger oilfield review, v. 17, no. 1, p.1.

Montaron, B., 2007, A quantitative model for the effect of wettability on the conductivity of porous rocks: Society of Petroleum Engineers, SPE105041-MS, 14 p.

Montaron, B., 2009, Connectivity theory–a new approach to modeling non-Archie rocks: Petrophysics, v. 50, no. 2, pp. 102–115.

Morgan, W.B., and Pirson, S.J., 1964, The effect of fractional wettability on the Archie saturation exponent: Transactions of the Society of Professional Well Log Analysts, 5th Annual Logging Symposium, Paper B, 13 p.

Moss, B., 1992, The petrophysical characteristics of the Brent sandstones: Geological Society, London, Special Publications, v. 61, no. 1, pp. 471–496.

Pérez-Rosales, C., 1982, On the relationship between formation resistivity factor and porosity: Society of Petroleum Engineers Journal, pp. 531–536.

Poupon, A., and Levaux, J., 1971, Evaluation of water saturation in shaly formations: Transactions of the Society of Professional Well Log Analysts, 12th Annual Logging Symposium, Paper O, 2 p.

Roy, S., and Tarafdar, S., 1997, Archie's law from a fractal model for porous rocks: Physical Review B, v. 55, no. 13, pp. 8038–8041.

Seleznev, N., Habashy, T., Boyd, A., and Hizem, M., 2006, Formation properties derived from a multi-frequency dielectric measurement: Transactions of the Society of Professional Well Log Analysts, 47th Annual Logging Symposium, Paper VVV, 11 p.

Sen, P.N., Scala, C., and Cohen, M.H., 1981, A self-similar model for sedimentary rocks with application to the dielectric constant of fused glass beads: Geophysics, v. 46, no. 5, pp. 781–796.

Sherman, M.M., 1983, The determination of cementation exponents using high frequency dielectric measurements: The Log Analyst, v. 14, no. 6, pp. 5–11.

Simandoux, P., 1963, Dielectric measurements on porous media application to the measurement of water saturations: study of the behavior of argillaceous formations: Revue de l'Institut Français du Pétrole, v. 18, Supplementary Issue, pp. 193–215.

Swanson, B.F., 1985, Microporosity in reservoir rocks: its measurement and influence on electrical resistivity: The Log Analyst, v. 26, no. 6, pp. 42–52.

Thomas, E.C., and Stieber, S.J., 1975, The distribution of shale in sandstones and its effect on porosity: Transactions of the Society of Professional Well Log Analysts, 16th Annual Logging Symposium, Paper T, 15 p.

Verwer, K., Eberli, G.P., and Weger, R.J., 2011, Effect of pore structure on electrical resistivity in carbonates: American Association of Petroleum Geologists Bulletin, v. 95, no. 2, pp. 175–190.

Watfa, M., and Nurmi, R., 1987, Calculation of saturation, secondary porosity, and producibility in complex Middle East carbonate reservoirs: Transactions of the Society of Professional Well Log Analysts, 28th Annual Logging Symposium, Paper CC, 24 p.

Wardlaw, N.C., 1976, Pore geometry of carbonate rocks as revealed by pore casts and capillary pressure: American Associaton of Petroleum Geologists Bulletin, v. 60, no. 2, pp. 245–257.

Waxman, M.H., and Smits, L.J.M., 1968, Electrical conductivities in oil-bearing shaly sands: Society of Petroleum Engineers Journal, v. 8, pp. 107–122.

Winsauer, W.O., Shearin, H.M., Jr., Masson, P.H., and Williams, M., 1952, Resistivity of brine-saturated sands in relation to pore geometry: American Association of Petroleum Geologists Bulletin, v. 36, no. 2, pp. 253–277.

Woodhouse, R., and Warner, H.R., 2004, Improved log analysis in shaly sandstones based on Sw and hydrocarbon pore volume routine measurements of preserved cores cut in oil-based mud: Petrophysics, v. 45, no. 3, pp. 281–295.

Worthington, P.F., 1985, The evolution of shaly-sand concepts in reservoir evaluation: The Log Analyst, v. 16, no. 1, pp. 23–40.

Worthington, P.F., 1995, A continuum approach to the petrophysical classification and evaluation of reservoir rocks: Petroleum Geoscience: v.1, no. 2, pp. 97–108.

Zeybek, A.D., Onur, M., Türeyen, Ö., Ma, M., Al-Shahri, A.M., and Kuchuk, F.J., 2009, Assessment of uncertainty in saturation estimated from Archie's equation: Society of Petroleum Engineers, SPE 120517, pp. 11–14.

Zinszner, B., and Pellerin, F.M., 2007, A Geoscientist's Guide to Petrophysics: IFP Publications, Editions Technip, Paris, 384 p.

CHAPTER 2

Porosity Volumetrics and Pore Typing

POROSITY OF SPHERICAL PACKS

The primary objective of porosity estimations based on measurements made either from petrophysical logs or core is the volume of pore space within the rock, given simply by the equation:

$$\Phi = \frac{V_p}{V_b}$$

The Greek letter, *phi*, is the standard symbol for porosity and is expressed in this equation as the ratio of the volume of void space (V_p) to the bulk volume of the rock (V_b). The simplest concepts of porosity are generally explained in terms of the packing of spheres as the sum of the pore volume of the space between the spheres. There are five basic arrangements of uniform-sized spheres that can be constructed: simple cubic, orthorhombic, double-nested, face-centered cubic, and rhombohedral packing (Hook, 2003). Each has a geometrically defined pore volume that represents an upper limit for granular rocks whose constituent grains have a variety of sizes and shapes and whose pore volumes have been reduced by compaction and diagenetic cements. This intergranular model is a useful starting point for the characterization of pores in clastic rocks and will be considered first, before reviewing the additional complexities of pore geometry introduced by dissolution in carbonate rocks.

CLASTIC "EFFECTIVE" POROSITY

The solid framework of a sandstone consists of a nonconductive "matrix" dominated by quartz, but commonly with accessory nonconductive minerals, and conductive clay minerals, whose electrical properties are caused by cation exchange with ions in saline formation water. It is important to distinguish between connected and unconnected pores, as well as larger pores that sustain fluid movement in contrast to smaller pores filled with capillary-bound water. A graphic presentation of these components (Figure 2.1) is widely used in the petrophysical literature as a reference basis to disentangle terminology that can be confusing and contradictory. In particular,

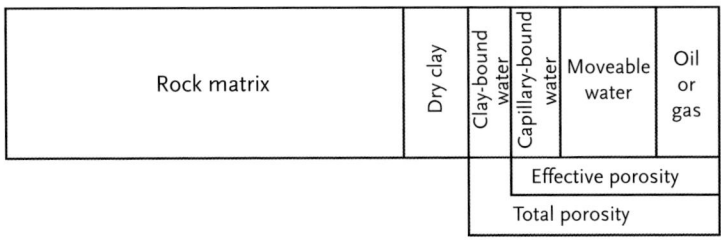

Figure 2.1: Descriptive model for the partitioning of a reservoir rock into its volumetric components. Effective and total porosities are marked as conventionally defined by petrophysicists.

the term "effective porosity" has different meanings that vary from one technical discipline to another. In their review of porosity terms, Wu and Berg (2003) concluded that many core analysts considered all porosity to be effective, log analysts excluded clay-bound water, while petroleum engineers excluded both clay-bound and capillary-bound from porosity consideration, thereby restricting effective porosity to pores occupied by mobile fluids. The increasing use of magnetic resonance logging has helped considerably to bridge this cognition gap through its ability to partition pore space between mobile and immobile fluids. Essentially it gives the petrophysicist the means to deliver what the petroleum engineer needs. However, when limited to porosity measurements from more traditional logging tools, the petrophysicist can only correct total porosity for clay-bound water and offer this as a shale-corrected "effective porosity."

Estimates of total porosity from borehole measurements are based most commonly on density, neutron, or sonic logs. These "porosity logs" must be scaled with reference to a specific lithology and pore fluid. The shallow depth of investigation by these tools generally restricts their measurement to the flushed zone, so the default pore fluid is mud filtrate. The conventional choices for the reservoir reference mineral are either calcite, quartz, or dolomite to represent zero porosity points for limestone, sandstone, and dolomite formations. Because of changes in lithology over logged sections, it is a standard practice to run them in combination so that variations in rock composition can be factored out by overlays or crossplots. Nuclear magnetic resonance (NMR) logs are also used to estimate porosity, but the measurement is free of lithology effects. The spectrum of NMR relaxation times partitions total porosity between clay-bound water, capillary-bound water, and free fluids (Figure 2.1). Consequently, effective porosity, as defined either by log analysts or petroleum engineers, can be evaluated explicitly.

NEUTRON-DENSITY SHALE VOLUMETRICS

In most situations, an NMR log is not available, in which case a combination of density and neutron logs is the best choice to evaluate porosity in clastic units. When referenced to a neutron-density crossplot (Figure 2.2), log readings of clastic zones can be located within a ternary system whose vertices are quartz, shale, and pore fluid. The "fluid point" corresponds to mud filtrate water because the shallow depth of

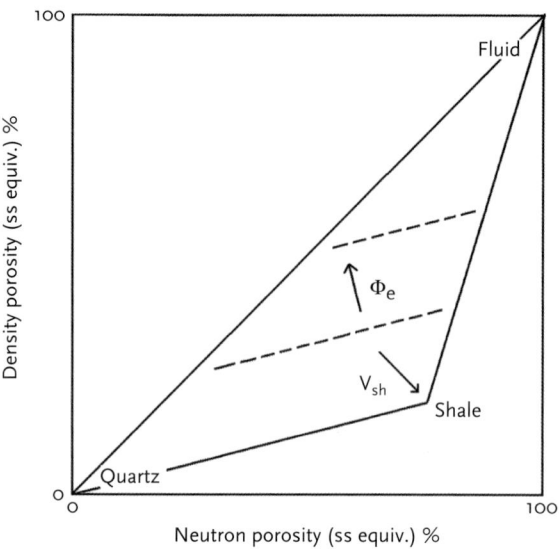

Figure 2.2: Shaly sandstone system on neutron-density porosity crossplot with vectors of effective porosity (Φ_e) and shale volume (V_{sh}).

investigation by these tools limits most of their response to the flushed zone. There are generally minimal perturbing effects by residual oil, because the density and hydrogen index of most oils are fairly similar to water. However, gas effects in shallow invasion zones are immediately recognized by points that plot outside the ternary composition space. Because shale properties are dictated not only by rock composition but also by their degree of compaction, the location of the "shale point" is empirical and must be based on representative shales in the section. As discussed in the first chapter, this shale point represents "external shales" between the sandstone units. Hopefully, these external shales will have a similar composition to shale laminae within the sandstones ("laminar shale"), but the neutron and density properties of the clays within the sandstone pores ("dispersed shale") may differ significantly. Once the ternary composition system is specified, contours of effective porosity in the traditional log-analysis sense are defined (Figure 2.2) and values are computed from the equation:

$$\Phi_e = \frac{(\Phi_n + \Phi_d)}{2} + V_{sh} * \frac{(\Phi_{nsh} + \Phi_{dsh})}{2}$$

This effective porosity contains both free fluids and capillary-bound water, and so is greater than the value of effective porosity considered by most petroleum engineers as the porosity containing mobile fluids.

In addition to porosity, the ternary system provides estimates of shale content in contours that parallel the sandstone line (Figure 2.2) and are given by the equation:

$$V_{sh} = \frac{(\Phi_n - \Phi_d)}{(\Phi_{nsh} - \Phi_{dsh})}$$

This is only one of several methods to estimate V_{sh}, and it is common practice to calculate a number of V_{sh} estimates, particularly if there is no petrophysical reason to favor one over another. The neutron-density estimate can be badly compromised in shaly sandstones that contain gas. Although strong gas effects are immediately recognizable when they plot outside the composition space, shale contents may mask smaller gas effects. Typically this results in an erratic profile of shale volume estimates caused by variability in flushed-zone gas saturation. Also, differences between shales within a sandstone and external shales may cause additional problems. If there are erroneous estimations of shale content, then effective porosity is also in error, so that V_{sh} values are a contributory factor in the porosity evaluation of clastic units. The most commonly used alternative method for volumetric shale estimation uses the gamma-ray log and is reviewed in the next section.

GAMMA-RAY ESTIMATIONS OF SHALE VOLUME

In most stratigraphic and petroleum geological applications, the gamma-ray log is used primarily as a "shale log," both to discriminate shales from "clean" formations and to estimate the proportion of shale in shaly reservoir units. Gamma rays counted by the logging tool from subsurface formations are emitted by radioactive isotopes in the decay families of thorium and uranium together with the potassium-40 isotope. The higher levels of radiation in shale are caused by thorium and potassium associated with clay minerals as well as uranium fixed by phosphatic or organic material. Older tools were recorded in "counts" or as equivalent weights of radium per ton, but all modern logs are scaled in terms of the American Petroleum Institute (API) gamma-ray unit. The API test pit at the University of Houston uses a concrete calibration standard that has a value of 200 API units, thought to be about double the gamma-ray emission of an "average" midcontinent shale (Belknap et al., 1959).

Expectations of the API unit log reading of a shale can be approximately reconstructed from the elemental abundances by applying multipliers of eight to the U (ppm), four to the Th (ppm), and sixteen to the K (%) estimates and summing their contributions (Luthi, 2001). This relationship provides a useful method to relate subsurface gamma-ray logging values of shale with samples from outcrop and core, based on laboratory geochemical measurements. Analyses of the North American shale composite (NASC) reference standard (Gromet et al., 1984) reported values of Th 12.3 ppm, U 2.66 ppm, K 3.2 percent, which converts to an equivalent gamma-ray log reading of 121.7 API units. Although higher than the vague assertion that a typical midcontinent shale should read about 100 API units, the hypothetical log value of the NASC standard is a good match with the actual values of gray shales logged in most midcontinent Pennsylvanian successions (Doveton, 1994).

The measurement scale of API units may be transformed into a volumetric estimate of shale proportion, V_{sh}, by interpolating between a minimum value, C, thought to represent zero shale and a maximum value, S, considered to be a typical gray shale.

This interpolation is expressed by the gamma-ray index (GRI) equation for a zone whose gamma-ray value is G:

$$V_{sh} = \frac{(G-C)}{(S-C)}$$

which attributes a linear relation between shale content and gamma radiation as a volumetric measure. Strictly speaking, the conversion to a proportion of radioactive material is by weight, but the volumetric estimation is a reasonable approximation because of the relatively minor differences in density within sandstones and shales. A more important consideration is that the equation attributes all sources of radioactivity to the shale component, but additional sources such as heavy minerals and potassium feldspar can be significant contributors. Finally, uranium sources have little relation to shale content, so that computed spectral gamma-ray logs that sum only potassium and thorium sources are commonly preferred for shale evaluation.

For these reasons, the shale proportions calculated from the linear interpolation equation may be overestimates, and this bias is recognized in many field studies. Remedial corrections can be attempted by using one of several nonlinear relationships that modify the linear estimate calculated by the "gamma-ray index" (*GRI*) to lower values. Two alternative equations for "Tertiary rocks":

$$V_{sh} = 0.083 * \left(2^{(3.7058*GRI)} - 1\right)$$

and for "older rocks":

$$V_{sh} = 0.33 * \left(2^{GRI} - 1\right)$$

were proposed by Larionov (1969), based on laboratory calibration of clay volumes from X-ray diffraction with gamma-ray values. The Larionov equations are still widely used, although their application encourages the notion that the volume of clay (V_{cl}) is equivalent to the volume of shale (V_{sh}). However, Yaalon (1962) reported that the average shale contains approximately 60 percent clay minerals, and this figure is supported by Bhuyan and Passey (1994), who suggested that if clay volume was the issue, then the simple relationship of:

$$V_{cl} = 0.6 \cdot * V_{sh}$$

was adequate and provided a reasonable fit to the Larionov predictions in the low-clay/shale range.

In contrast with these complex equations, Stieber (1970) used the simple formulation of:

$$V_{sh} = \frac{3*GRI}{(1+2*GRI)}$$

which he derived as a minimum-error fit to calculated pulsed-neutron sigma values of shaly sandstones in the Louisiana Gulf Coast. The precision that is implied by the terms and intricacy of the Larionov equations can be misleading when applied to most shale evaluations, so that the simple and pragmatic family of variants of the Stieber equations:

$$V_{sh} = \frac{GRI}{\left[GRI + c*(1-GRI)\right]}$$

is a popular nonlinear alternative, where c takes the value of either 1, 2, or 3. Interestingly, the alternative integer choices of the parameter c results in an emulation of the Larionov equations for "older rocks" ($c = 2$) and "Tertiary rocks" ($c = 3$) to a fair degree, while a value of $c = 1$ equates V_{sh} with the gamma ray index (GRI) (Figure 2.3).

In general, log practitioners such as Asquith and Krygowski (2004) recommend the use of the linear GRI estimate of shale as the model of first choice, but, if field evidence suggests the values to be overestimates, then a nonlinear model may be used to improve the estimation. As Ellis and Singer (2007) comment wryly, the estimation technique "attempts to determine too much from too little information." However, with that said, bad decisions on the choice of shale volumetric estimation will have distinctive consequences in poor estimations of effective porosity, water saturation, and other reservoir parameters. Of particular concern is that the coarse resolution of conventional logging tools causes thin beds and laminae of shale within sandstones

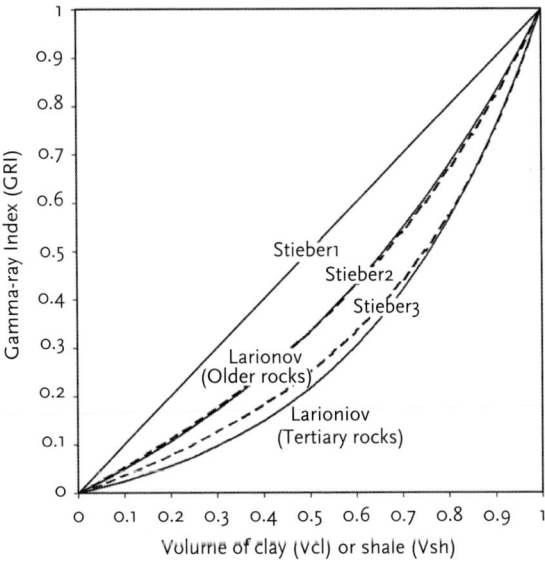

Figure 2.3: Alternative modification functions for the estimation of shale content from the gamma-ray index (GRI). Larionov equations (solid curves) are based on calibration of the GRI to clay content measured from X-ray diffraction (Larionov, 1969) for Tertiary formations and older rocks. The original Stieber equation (Stieber2, dashed curve) is mathematically simpler and was based on Gulf Coast log calibration (Stieber, 1970). The Stieber equation form can be generalized by changing the integer value of c (see text), with Stieber1 equal to GRI and higher values used as approximations to Larionov-style relationships.

to be averaged as shale content in what might appear to be a uniformly shaly sandstone, but are actually interbedded shale and sandstone layers. Consequently, feedback from reservoir models is necessary to either validate or provide guidance for corrective modifications to the shale estimation procedure. This will be discussed later when considering the distinctions between "dispersed shale" (clay) and "laminar shale" (shale) within shaly sandstones.

CORRECTION OF TOTAL POROSITY FOR SHALE CONTENTS

Estimating the volume of shale content is an important factor in evaluating hydrocarbon saturation because of clay-mineral conductivity effects, as discussed in the previous chapter. In addition, shale volumes also determine the reduction in total porosity estimated from logs to an effective porosity, equated here with both capillary-bound water and pores with mobile fluids. The response of the density log is a mass-balance relationship, so for a clean sandstone the porosity is given by:

$$\rho_b = \Phi \cdot \rho_f + (1-\Phi) \cdot \rho_{ma}$$

where ρ_b is the bulk density, ρ_{ma} is the matrix density, ρ_f is the pore fluid density and Φ is the porosity. The solution is an estimate of the total porosity, which is equivalent to the effective porosity because there is no perturbation by shale content. With shale content, the equation expands to:

$$\rho_b = \Phi \cdot \rho_f + V_{sh} \cdot \rho_{sh} + (1 - V_{sh} - \Phi) \cdot \rho_{ma}$$

where V_{sh} is the volume of shale and ρ_{sh} is the density of the shale component.

A common alternative and simpler equation corrects the total porosity, Φ_t, to the effective porosity, Φ_e by:

$$\Phi_e = \Phi_t - V_{sh} \cdot \Phi_{dsh}$$

where Φ_{dsh} is the density porosity reading of the shale component.

If both neutron and density porosity logs are available, then the effective porosity is computed from the relationship:

$$\Phi_e = \frac{(\Phi_n + \Phi_d)}{2} + V_{sh} * \frac{(\Phi_{nsh} + \Phi_{dsh})}{2}$$

In these equations, the pore fluid is assumed to be mud filtrate water because the shallow depths of investigation of both the density and neutron tools means their responses are primarily drawn from the flushed zone. However, the possible presence of gas close to the borehole wall must be included in the formulation. The equation commonly applied to estimate the porosity with a compensation for gas takes the form:

$$\Phi = \sqrt{\frac{\left(\Phi_n^2 + \Phi_d^2\right)}{2}}$$

The equation closely approximates the formula derived by Gaymard and Poupon (1968) from a petrophysical model of a gas-filled reservoir. An alternative and empirical equation that is also widely used is a simple weighted average of the neutron and density porosities, with a weighting of one-third applied to the neutron porosity and a weighting of two-thirds weighting applied to the density porosity (Asquith and Krygowski, 2004). This empirical equation closely emulates the gas correction vector shown on service company neutron-density crossplots. It also seems a good approximation to field observations, as suggested by the regression analysis of core porosities related to neutron and density fractional log porosities from limestones in the Chase Group of the Hugoton gas field (Dubois et al., 2006) which resulted in the equation:

$$\Phi = 0.62\Phi_d + 0.39\Phi_n$$

An important consideration in the estimation of effective porosity is whether the shaly content of the sandstone is dominated by either "dispersed shale" or "laminar shale." In the case of dispersed shale, clay content progressively occludes porosity as clay minerals develop on pore walls, so that the addition of clay results in a complementary reduction in porosity. However, if the shale content forms discrete laminae that separate layers of clean sandstone, then the effective porosity is averaged by the logging tool and appears as a low aggregate value, rather than layers of sandstone with potentially high effective porosities interbedded with shales.

ALLOCATION BETWEEN SHALE MORPHOLOGY TYPES

Geologists regard the shale content in sandstones to be distributed in forms that reflect their genesis as either autochthonous or allochthonous material. Autochthonous clays are developed in place, lining and filling pores and as diagenetic replacements for minerals such as feldspars. Allocthonous clays are introduced during sedimentation as components of rip-up mudstone clasts, biogenic pellets, burrows, and shale laminae. Wilson and Pittman (1977) applied this bipartite classification in a study of a large sandstone data set and found that the majority contained either authigenic or allogenic clays.

Petrophysicists have traditionally distinguished three types of shale as distinctive descriptors of the manner in which clay material is distributed within a shaly sandstone. They are:

(1) *Laminar shale,* which consists of thin laminations of shale that separate layers of clean sandstone. The occurrence of these laminations is not accompanied by a reduction in the porosities of the sandstone layers themselves, but there is an overall reduction in the bulk porosity of the total rock.

(2) *Structural shale*, which describes sandstones in which some of the grains of the framework are shale fragments or diagenetically altered minerals. The transition from a clean sandstone is not necessarily matched by any reduction in porosity.
(3) *Dispersed shale*, which consists of pore-filling clay minerals whose development has caused a progressive reduction in porosity.

So, the basis for description by petrophysicists is morphology, that is, the geometrical manner in which the shale is distributed through the sandstone. The three shale types are endmembers that encompass the spectrum of mixtures within shaly sandstones.

The three-shale-type system is commonly interpreted from a Thomas-Stieber plot of logs of sandstone-scaled density porosity and gamma rays (Figure 2.4). The line linking the clean sandstone point with a total shale point marks the trend of laminar shale and is the baseline for subdivision between shale types. The Thomas-Stieber model provides estimates of the volumes of both laminar shale and dispersed shale. In doing so, it presumes a constant sand porosity in which the pore space is filled with dispersed shale up to a maximum equal to the sand pore volume. Because the dispersed shale properties are based on the external shales, the dispersed shale is more representative of infiltration residues of detrital shale within the pore space rather than authigenic clay. Once again, the failure to distinguish clay from shale can have important implications with regard to the real effective porosity in sandstones with a high dispersed-shale component. In cases where the

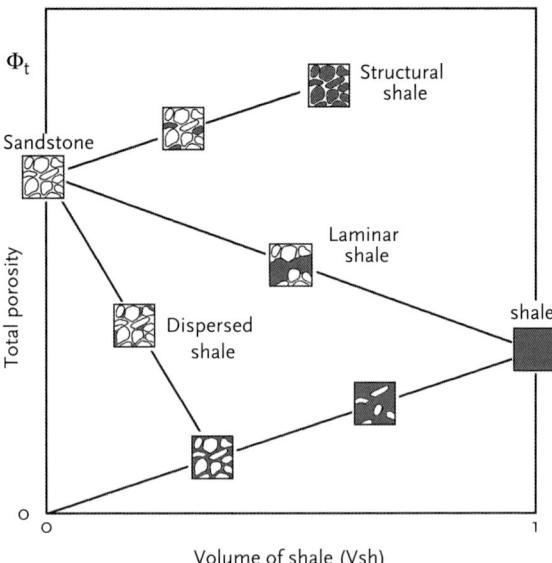

Figure 2.4: Framework of Thomas-Stieber plot with axes of volume of shale (V_{sh}) and total porosity (Φ_t) overlaid with trends of laminar, dispersed, and structural shale.

shaly sandstone is dominated by laminar shale, the model has good viability because the shale laminae are detrital shale rather than clay, so that volumetric estimates are more robust, provided that the shale laminae have the same properties as the shale beds used for calibration.

The Thomas-Stieber plot has important implications as a qualitative guide in petrophysical interpretation. The averaging of shale laminae with sandstone layers may result in low gross effective porosities, and thin sandstones with good porosity can be eliminated from pay consideration. However, the discrimination of laminar shale can be used to allocate the shaly sandstone between shale laminae and clean sandstones with higher effective porosities. The subdivision of shale types also has a bearing on the choice of resistivity model used in the evaluation of fluid saturations. If a shaly sandstone appears to be dominated by laminar shale, then older, volumetric shaly sandstone equations may be viable because the shaly sandstone is resolved as parallel resistance laminae of shale and sandstone. On the other hand, a strong dispersed-shale trend suggests a significant clay content within the pores, so that cation-exchange–driven models, such as the Waxman-Smits or dual-water models, may be more appropriate.

When used for quantitative analysis, the Thomas-Stieber model can be solved from the response equations for the density porosity and gamma-ray logs:

$$G = (1-V_L) \cdot (G_{SS} + V_D \cdot G_{SH}) + V_L \cdot G_{SH}$$

and

$$\Phi_t = (1-V_L) \cdot (\Phi_{SS} - V_D \cdot \Phi_{SH}) + V_L \cdot \Phi_{SH}$$

where V_L and V_D are the proportions of laminar and dispersed shale, G is the gamma-ray reading, and Φ_t is the density porosity, with the subscripts of SS and SH for sandstone and shale, respectively.

Although the Thomas-Stieber model had limited application following its introduction in 1975, the emergence of tensor resistivity logging has sparked greatly renewed interest in the model as a basis for quantitative analysis (Page et al., 2001). The ability to measure resistivity in three orthogonal directions has provided a novel means to resolve shale laminae in shaly sandstones that were below the resolution of conventional logging tools. An isotropic shaly sandstone would have a similar resistivity regardless of direction of measurement, as contrasted with anisotropic shaly sandstones of interbedded shale and sandstone layers. Conventional horizontal induction logging responds to shale/sandstone interbeds as resistances in parallel, while a vertical measurement reflects current transmission through a series of resistances. The two measurements are sufficient to differentiate thin shale/sandstone interbeds in the simplest scenario, where the borehole axis is orthogonal to the bedding (Figure 2.5); inclined strata require triaxial resistivity logging to resolve the more complex situation. Quirein et al. (2012) made a comparison between the Thomas-Stieber model and computations from tensor resistivity measurements and concluded that the results were equivalent when the beds were orthogonal to the

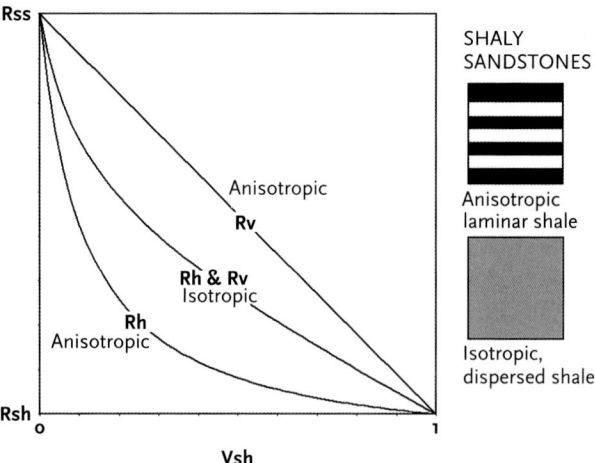

Figure 2.5: Diagrammatic representation of horizontal induction resistivity (R_h) and vertical induction resistivity (R_v) for anisotropic laminated shaly sandstones normal to the borehole and isotropic shaly sandstones with pores containing dispersed clay.

borehole. Where there was a relative dip, the two approaches could be reconciled if the dip and other parameters were incorporated. They also noted that more pay footage was located by applying a minimum effective-porosity cutoff to the sandstone effective porosity in sandstones with laminar shale, rather than the gross effective porosity.

Strategies for water saturation estimation in shaly sandstones become clearer once laminar shale and dispersed shale are evaluated as distinctive components. If the shaly sandstone is dominated by laminar shale with minimal dispersed shale, then the system is one of interbedded sandstones and shales with an apparent loss in porosity caused by the averaging of the logging tools. The averaged effective porosity is then a gross estimate, but can be corrected by eliminating the laminar shale from pay consideration and retaining the sandstone layers with their higher porosities. Water saturation within the sandstone layers is then most appropriately evaluated by a double-layer model (such as the Waxman-Smits or dual-water model), although a simple volumetric model (Simandoux) may be adequate. At the other extreme, a shaly sandstone dominated by dispersed shale implies an isotropic sandstone with no shale laminae, whose shale content consists primarily of clay minerals within the sandstone pores. The effective porosity is reduced by the clays within the pore space. Both the volumetric estimate of dispersed shale and water saturation are compromised by the use of external shales as reference points. Their gamma-ray, neutron, density, and electrical properties may differ significantly from pore-filling clays, particularly if they contain different clay minerals. However, these questions can be resolved by special core analysis (SCAL) and more complex evaluation strategies can be designed to accommodate these issues.

CARBONATE POROSITY

Limestones and dolomites are both chemical rocks in the sense that the carbonate mineral framework may dissolve to form vugs, a term for pore sizes that can range anywhere from a dissolved particle to a cavern. In common with siliceous clastic rocks, the pore space also contains interparticle porosity, which can form either between grains or between crystals. Finally, a third form of porosity can occur within fractures, which have a major influence on permeability but occupy a very small volume. The pore space of fractures is typically below the resolution of standard porosity logs, so that estimates of fracture porosity are speculative when based on density, neutron, or sonic logs. However, while both neutron and density measurements are responses to total porosity, it has been recognized for many years that the sonic log is primarily a measure of interparticle porosity. In one of the earlier papers to address this phenomenon, Nugent et al. (1978) pointed out that when the sonic porosity was less than either the density or neutron porosities, calculations of water saturation were overly optimistic. From this observation, they deduced that the apparent oil saturation was caused by isolated vuggy porosity, and that the sonic porosity was a useful measure of connected interparticle porosity. By utilizing neutron and density porosities in conjunction with sonic porosity, the total pore volume could be subdivided between "primary porosity" (sonic, interparticle) and "secondary porosity" (neutron-density minus sonic, vug).

The reasons that have been offered as to why the sonic log is "blind" to vug-sized pores have often been hazy, even though the phenomenon is widely exploited in petrophysical analysis. However, neutron and density measurements are based on the counting of radioactive particles from the volume of the entire formation, while the sonic tool selectively records the arrival of the fastest acoustic wave that travels to the receivers. Faster acoustic pathways in the borehole wall that do not intersect large vugs will result in shorter transit times. In addition, smaller vugs tend to have a more spherical shape and so have less influence on the acoustic wave than do typical elongated pore shapes (Ellis and Singer, 2007).

The measurement of the sonic or acoustic tool is the transit time of ultrasonic sound to traverse the formation of the borehole wall in units of microseconds per foot or meter. Conversion to porosity is most commonly made by linear interpolation between values for the matrix mineral transit time and the fluid in the pore space. The pore fluid is taken typically to be mud filtrate water because of the shallow depth of investigation by the fastest acoustic wave. The linear interpolation was first proposed by Wyllie et al. (1956) and based on laboratory measurements. In their experiments, they recognized nonlinearity at porosities that greatly exceeded those found in reservoir rocks, but proposed that for consolidated rocks, "the time-average formula is a good approximation" (Wyllie et al., 1958). The time-average relationship is universally known as the Wyllie equation and takes the form:

$$\Phi_s = \frac{(\Delta t - \Delta t_{ma})}{(\Delta t_f - \Delta t_{ma})}$$

where Φ_s is the porosity estimate, Δt is the acoustic travel-time log measurement, Δt_{ma} is the matrix transit time, and Δt_f is the pore fluid transit time.

An improvement on the Wyllie equation was proposed by Raymer et al. (1980) to accommodate the nonlinearity of the transit time to the porosity relationship, the Raymer-Hunt-Gardner (RHG) transform:

$$\frac{1}{\Delta t} = \frac{\Phi}{\Delta t_f} + \frac{(1-\Phi)^2}{\Delta t_{ma}}$$

The RHG transform is empirical, but based on a larger sample of measurements than previous studies and designed to accommodate the entire porosity range. This transform should be regarded as a second-order model to be used when improved accuracy is considered an important goal. In practice, the linear-interpolation method is still widely used as a reasonable approximation if the estimated porosity from logs is below 25 percent. In addition, calibrations based on core porosity are conventionally estimated by a linear relationship, which is equivalent to a Wyllie time-average function, but with the matrix travel time determined by statistical analysis.

VUG POROSITY EVALUATION FROM ACOUSTIC AND RESISTIVITY LOGS

Anselmetti and Eberli (1999) applied the Wyllie time-average equation to predict acoustic velocities in Miocene-Pliocene cores from the Great Bahama Bank and compared the expectations to measured velocities. When these were quantified as a velocity deviation, they were able to make a systematic corroboration of earlier qualitative observations that carbonates with separate-vug porosity had higher velocities than would be anticipated from the Wyllie equation. Digital image analysis was used by Weger et al. (2009) to relate measurements of pore size and shape to acoustic velocity deviations from the Wyllie time-average model. They demonstrated that the parameters of perimeter-to-area ratio and the dominant pore size explained much of the variability. Their findings from a grainstone and packstone subset of their data are reproduced in Figure 2.6 but are recast with reference to projected matrix transit times rather than velocity deviation. If the Wyllie time-average model was an adequate descriptor, then matrix transit times should match approximately the value measured for calcite in limestones whose porosity is entirely interparticle. However, some variability in this ideal value can be anticipated because of impurities and fabric differences of a limestone matrix with a pure calcite crystal. The presence of vugs causes the projected matrix transit times to be displaced to lower values that Wang and Lucia (1993) called pseudo-matrix transit times.

The acoustic velocity relationships with digital image parameters (Figure 2.6) and their interpretation mirrors the patterns observed between digital image parameters and porosity exponents (Figure 1.19), discussed in the previous chapter. Clearly, both acoustic and electrical properties are sensitive measures of the partitioning of

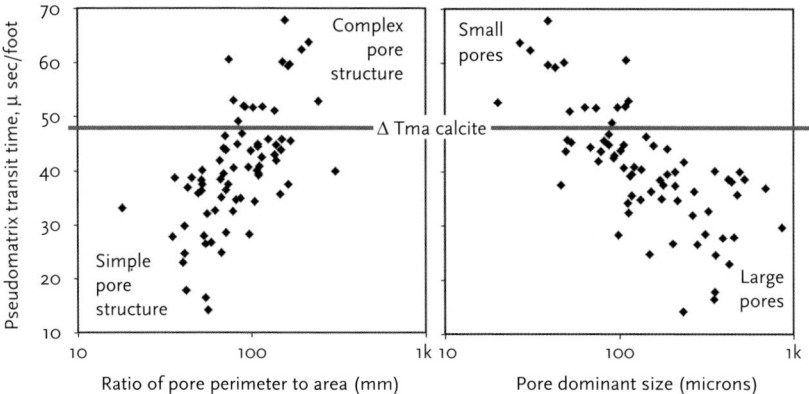

Figure 2.6: Pseudomatrix transit time against pore perimeter to area ratio (left) and dominant pore size (right) from carbonate core measurements. Adapted from Weger et al. (2009), © 2009 American Association of Petroleum Geologists (AAPG), reprinted by permission of the AAPG, whose permission is required for further use.

pore types in a dual-porosity system. However, the porosity exponent is a reflection of electrical efficiency, so that the discrimination of larger pores by resistivity measurements is also influenced by their degree of interconnection. In general, large dead-end pores will not participate in electrical flow, resulting in higher porosity exponent values. Consequently, vugs should be subdivided between separate and touching (connected) pores when comparing vug porosity evaluations based on core examination and resistivity properties (Lucia, 2007). On the other hand, the vug properties that primarily influence the acoustic log are the size and degree of sphericity.

Published research on acoustic and electrical properties of vuggy rocks have attempted to reconcile theoretical models with observations from core and logs. So, for example, Lucia and Conti (1987) empirically evaluated the separate-vug porosity (Φ_{sv}) in water-saturated carbonates from resistivity logs with an adaptation of the first Archie equation:

$$\Phi_{sv} = \left[\frac{(\log R_w - \log R_o)}{\log \Phi_t} - 1.76 \right] \cdot \Phi_t$$

Using the acoustic log, Wang and Lucia (1993) developed an empirical relationship between separate-vug porosity (Φ_{sv}) estimated by thin-section point counting and the acoustic transit time. Their relationship took the form of:

$$\Phi_{sv} = 10^{a-b(\Delta t - 141.5)}$$

where a and b are coefficients that vary with the carbonate rock type.

Oomoldic limestone reservoirs provide classic examples of vuggy porosity with increased resistivity reflecting poor connectivity between oomolds and higher acoustic velocities caused by oomold size and shape. However, the relationships are often complex because diagenetic cementation and oomold crushing may modify connectivity and oomold shape, with variation both within and between ooid bars (Byrnes et al., 2000). In a case-study, measurements from core and logs were integrated in an evaluation of oomoldic porosity in the Pennsylvanian Lansing-Kansas City "C" zone in a central Kansas well. Profiles of core, neutron-density, and sonic porosities are shown in Figure 2.7, together with dual-porosity partitions based on core measurements of formation factors and the resistivity log. The resistivity partition of porosity was computed from the Watfa and Nurmi (1987) formulation of:

$$\frac{1}{F} = \frac{R_w}{R_o} = \Phi_t^m = \left(\Phi_t - \Phi_{nc}\right)^2$$

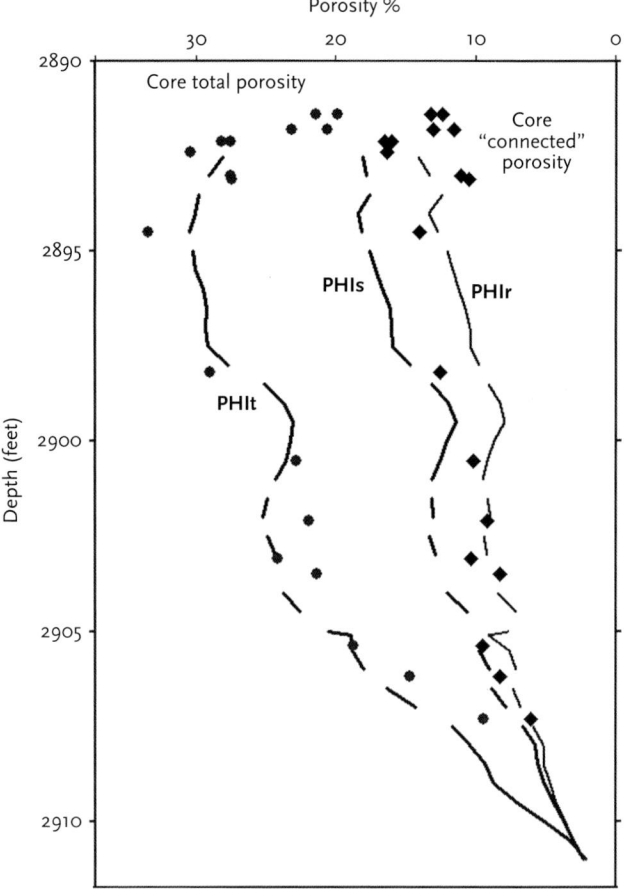

Figure 2.7: Comparison of core total porosity (circles) with neutron-density log porosity (*PHIt*), sonic log porosity (*PHIs*), and connected porosity estimated from the deep resistivity log (*PHIr*) and core measurements of formation factor (diamonds) in a Pennsylvanian oomoldic limestone from central Kansas.

Core measurements of the formation factor (F) and total porosity (Φ_t) can be used to calculate the apparent porosity exponent (m), which is then applied to the solution of the nonconnected porosity (Φ_{nc}). Alternatively, the resistivity log measurement in the fully water-saturated zones (R_o) divided by the formation water resistivity (R_w) provides a downhole estimate of the formation factor for the same purpose. The nonconnected porosity estimated from the resistivity model is systematically more than the "secondary porosity" implied by the difference between the sonic and total porosity from the neutron and density logs. This characteristic matches the observation by Brie et al. (1985) in their evaluation of oomoldic limestones in the Middle East that their estimates of acoustic spherical porosity were significantly larger than the secondary porosity.

The secondary porosity partition by the sonic log has been widely used to estimate the apparent porosity exponent by applying the Nugent equation (Nugent, 1984). In its original form, the Nugent equation is:

$$m_a = \frac{2 \cdot \log \Phi_s}{\log \Phi_t}$$

where m_a is the apparent porosity exponent, Φ_s is the sonic porosity, and Φ_t is the total porosity from the density or neutron log. Tiab and Donaldson (2011) considered that the Nugent equation gave acceptable porosity exponents for porosities below 10 percent, but tended to underestimate their true values at higher porosities. This opinion is consistent with the notion that the sonic-log secondary porosity is an underestimate of the vuggy porosity. In the case study of the Lansing-Kansas-City oomoldic "C" zone, it was found that the sonic porosity could be reconciled with the nonconnected porosity by the relationship:

$$\Phi_{nc} = \frac{(\Phi_t - \Phi_s)}{(1 - \Phi_t)}$$

which simply represents a rescaling of the secondary porosity to the matrix volume rather than to the bulk volume. The connected porosity is then:

$$\Phi_c = \frac{(\Phi_s - \Phi_t^2)}{(1 - \Phi_t)}$$

By replacing the sonic porosity with this estimate of connected porosity, the modified Nugent equation becomes:

$$m_a = \frac{2 \cdot \log \Phi_c}{\log \Phi_t}$$

Applications of Nugent-variant equations are a common approach to saturation estimations in vuggy carbonates when a sonic log is available and as an alternative to the resistivity methodology described in Chapter 1. In a case study of a Bethany Falls limestone, oomoldic-reservoir porosity exponents were estimated using the Archie equation applied to completely water-saturated zones. If a useable relationship can be developed between porosity exponents and porosity, then the porosity exponents can be estimated for the hydrocarbon zones, based on their porosity. In contrast, a Nugent-variant approach utilizes log data specific to each zone, rather than estimates based on a trend relationship.

Clearly, both resistivity and acoustic measurements play an important role in the reservoir evaluation of carbonate dual-porosity systems. However, rather than consider them by independent algorithms and then reconcile their results, it is more fruitful to combine them in a comprehensive model. The model should have a sound basis in physical theory but have the capability of delivering viable solutions in real rocks. An early attempt at this approach was described by Brie et al. (1985), who considered secondary pores as spherical inclusions in a homogeneous matrix. The primary porosity was represented by a Wyllie linear relationship, and its electrical properties were set by the Archie equation. An evaluation of the acoustic properties of the secondary pores was then determined by parameters of the Kuster–Toksöz model, and conductivity was determined by the Maxwell–Garnet equations. The coordinated methodology was then used to compute porosity exponent values on a zone-by-zone basis within test wells and to compare the predictions with measured values.

In a computer simulation study, Kazatchenko et al. (2006) created a double-porosity system with a homogeneous, isotropic matrix containing small pores and large-scale inclusions of secondary pores. The secondary pores were modeled by triaxial ellipsoids whose aspect ratios were varied in order to examine the influence of shape. From the simulation results, they concluded that the effects on both acoustic and electrical properties implied that there were four distinctive secondary pore types, which were: quasi-spherical vugs, oblate vugs, channels, and cracks. Each of these types could be identified when acoustic velocities were analyzed jointly with electrical conductivity. Kazatchenko et al. (2007) then applied their findings to the joint inversion of conventional well logs to evaluate double-porosity vuggy and fractured carbonate reservoirs. They reported that results from boreholes in southern Mexico showed good matches with core data, image logs, and geological descriptions, so that the methodology could be used to improve carbonate lithotype classification and permeability estimations by predicting secondary pore interconnections.

Oomoldic reservoirs provide classic examples of dual-porosity systems whose larger pores have distinctive effects on acoustic and resistivity logs, in part because the pore shapes are essentially spherical. In contrast, the brecciation and dissolution processes of karstic weathering can result in vugs with inchoate and variable shapes of all sizes. However, the degree to which a pore shape can be related to a sphere is by the ratio of the internal surface area to volume, which, for a sphere, is the minimum possible. As shown by both core studies (Weger et al., 2009; Verwer et al., 2011) and computer simulation (Kazatchenko et al., 2006), vugs with more equant dimensions are likely to have greater effects than more elongate shapes. Consequently, volumetric estimations of vuggy porosity volume must necessarily be semiquantitative because of the variety of vug shapes, unless shape is considered as a contributory parameter in the analysis. Even with this limitation, acoustic and resistivity evaluations of vuggy porosity are a valuable contribution to facies analysis of carbonates and the interpretation of dissolution fabrics. Interpretations can be refined when these analyses are linked with fabric observations from core and image logs.

A case-study example is provided from a well in central Kansas, whose trajectory was planned to cross a doline feature in the Cambro-Ordovician Arbuckle limestone, interpreted from volumetric curvature analysis of 3-D seismic data (Figure 2.8). The fault boundaries of the doline are clearly evident at the centre of the seismic record, with a dimension of about 1,000 feet, which was interpreted to be a collapsed paleocavern. The gamma-ray log from the horizontal section of this well (Figure 2.9) contrasts the shale and clastic content of the paleocavern breccias that disrupt the relatively low radioactivity of the stratiform carbonates lateral to the cavern. Rush (2013) applied the paleokarst classification system of Loucks (1999) to the subdivision of the section based on a combination of the image log and lithology logs. Rush (2013) distinguished two phases of vuggy-pore development with syndepositional evaporate dissolution of peritidal to supratidal carbonates contrasted with later dissolution and brecciation by glacioeustatic, vadose, and karstic processes. The image logs commonly showed bedding features in the stratiform carbonates adjacent to the collapsed section, but they were mostly absent within the paleocavern fill.

An assessment of vuggy porosity was made using the conventional acoustic and resistivity approaches applied to the sonic and resistivity logs. The profile of the secondary porosity from the sonic log shows a strong qualitative concordance with the nonconnected porosity estimated from the resistivity log within the stratiform margins and minimal occurrence of both within the paleocavern fill. These relationships

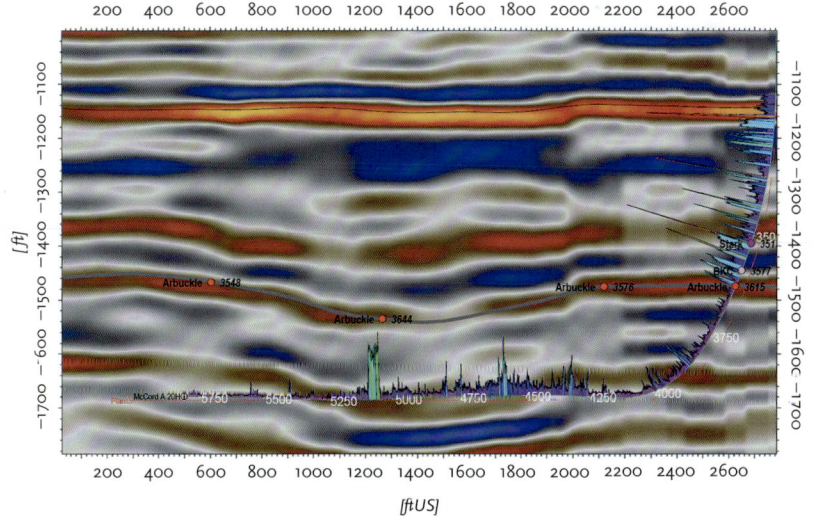

Figure 2.8: Seismic section from central Kansas indexed with trajectory of exploratory borehole to verify geophysical interpretation of Cambro-Ordovician Arbuckle limestone doline feature as a collapsed paleocavern. From Rush (2013).

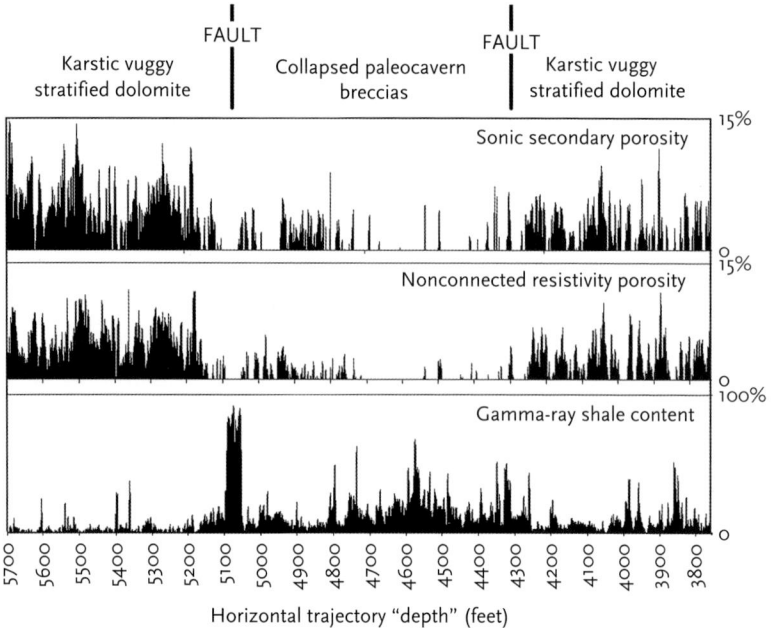

Figure 2.9: Log interpretation of sonic log secondary porosity, nonconnected porosity estimated from the deep resistivity log, and gamma-ray shale content in the horizontal section of the exploratory borehole shown in Figure 2.8. From Rush (2013).

suggest that nonconnected equant vugs are common within beds of the Arbuckle limestone outside the collapse zone. However, their relative absence within the chaotic breccia of the paleocavern fill may indicate that secondary pores are both connected and elongate. The distinctive but complex patterns make the case that the integration of secondary porosity transforms with image-log interpretation is an essential step to advance analysis from speculation to identification. However, once resolved, a useful prediction methodology could be developed for interpretation in wells that have resistivity and sonic logs but lack borehole images or core.

NMR LOGGING OF VUGGY POROSITY

The role of nuclear magnetic resonance (NMR) image logging data in examining the complete range of pore sizes will be discussed in great detail in Chapter 6. However, the immediate focus of this chapter section is to review petrophysical techniques that discriminate larger pores in a dual-porosity system. Longer relaxation times in the NMR T2-relaxation-time log correspond to larger pores, so that NMR logs should be an effective means for evaluating vugginess, provided that the logging speed is sufficiently slow to record relaxations associated with vugs. An exact correspondence is confounded by the increased importance of the bulk diffusion relaxation and diffusion coupling between small and large pores, as elaborated in Chapter 6.

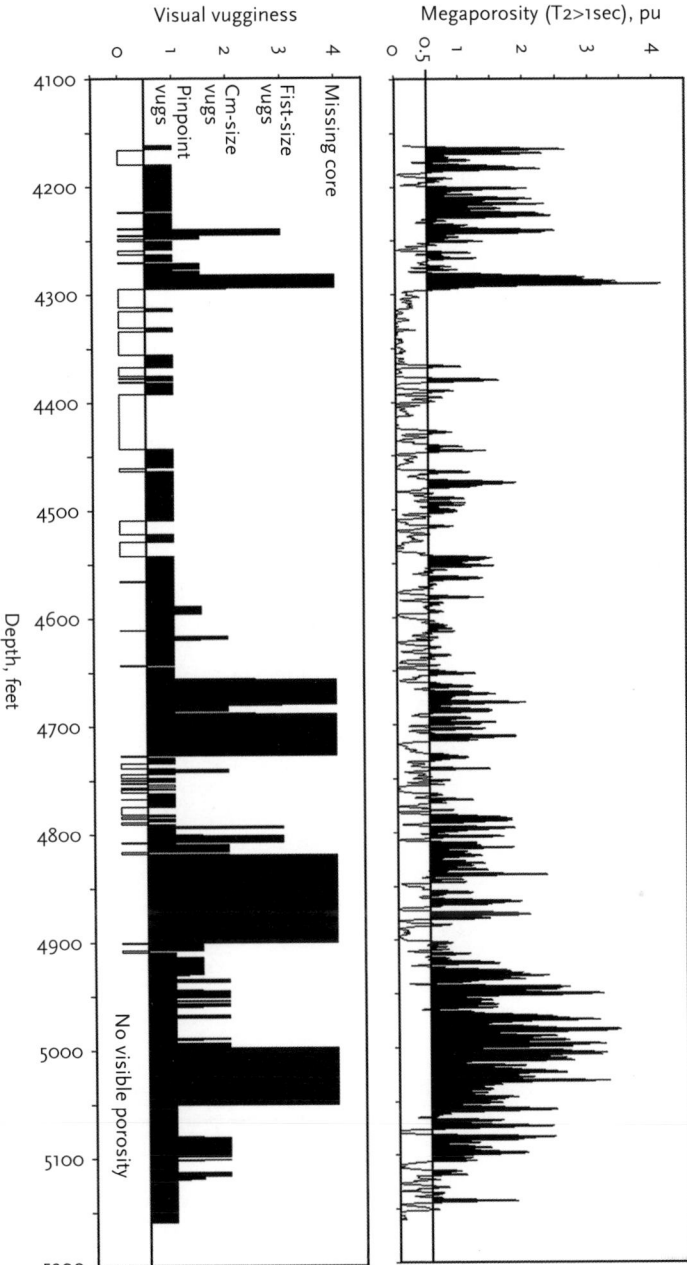

Figure 2.10: Visual observation of vugs from core in a section of Cambro-Ordovician Arbuckle limestone plotted against summed porosities of the NMR log for T2 relaxation times of greater than one second.

However, the occurrence of excessively slow times in the T2-relaxation-time spectrum are certainly a good indicator of vuggy porosity.

Chang et al. (1994) compared vugs observed in the core with T2 relaxation times and concluded that a threshold of 750 milliseconds corresponded to the appearance of vugs observed by image point counting. This critical value can be expected to show some variation in comparable studies of other carbonate reservoirs, due to the effects mentioned, as well as the variability of relaxivity in carbonate mineral surfaces. A generalized expectation of about 1,000 milliseconds (one second) might be a useful arbiter for vug recognition.

Logs of the NMR T2 distribution are recorded as pore counts within bins, so that the practical evaluation of vug occurrence is based on pore volumes in bins with the longest relaxation times. As a case-study example, the Arbuckle limestone in a well in southern Kansas was both continuously cored and logged by an NMR tool. Pore volumes were summed from T2-relaxation-time bins, whose times exceeded 1 second. It was postulated that the resulting vuggy "megaporosity" curve could be evaluated in conjunction with visual core assessment of the vugs. Qualitative descriptions of the occurrence of vugs in the Arbuckle limestone cores were allocated between categories and ordinal code numbers as: no vugs (0), pinpoint vugs (1), centimeter-sized vugs (2), and fist-sized vugs (3). Intervals with missing cores were assigned a code value of 4, because drilling through many of these intervals suggested pervasive vuggy porosity that destroyed the rock integrity. There is a good overall match between a profile of visual core vug assessment and megaporosity volume (Figure 2.10), and a cut-off level of 0.5 percent megaporosity was found to be the best discriminator for visual vugs within the Arbuckle limestone.

REFERENCES

Anselmetti, F.S., and Eberli, G.P., 1999, The velocity-deviation log: A tool to predict pore type and permeability trends in carbonate drill holes from sonic and porosity or density logs: American Association of Petroleum Geologists Bulletin, v. 83, no. 3, pp. 450–466.

Asquith, G., and Krygowski, D., 2004, Basic well log analysis (2nd Edition): American Association of Petroleum Geologists, Tulsa, Oklahoma, 244 pp.

Belknap, W.B., Dewan, J.T., Kirkpatrick, C.V., Mott, W.E., Pearson, A.J., and Rabson, W.R., 1959, API calibration facility for nuclear logs. Drilling and Production Practice: American Petroleum Institute, Document 59–289, pp. 289–317.

Bhuyan, K., and Passey, Q.R., 1994, Clay estimation from GR and neutron—density porosity logs: Transactions of the Society of Professional Well Log Analysts, 35th Annual Logging Symposium, Paper DDD, 15 p.

Brie, A., Johnson, D.L., and Nurmi, R.D., 1985, Effect of spherical pores on sonic and resistivity measurements: Transactions of the Society of Professional Well Log Analysts, 36th Annual Logging Symposium, Paper W, 20 p.

Byrnes, A.P., Watney, W.L., Guy, W.J., and Gerlach, P., 2000, Oomoldic reservoirs of central Kansas: controls on porosity, permeability, capillary pressure, and architecture: Proceedings of the American Association of Petroleum Geologists Annual Convention, New Orleans, Louisiana, p. A22.

Chang, D., Vinegar, H., Morriss, C., and Straley, C., 1994, Effective porosity, producible fluid and permeability in carbonates from NMR logging: Transactions of the Society of Professional Well Log Analysts, 35th Annual Logging Symposium, Paper A, 21 p. Later published in 1997: The Log Analyst, v. 38, no. 2, pp. 60–72.

Doveton, J.H., 1994, Geologic log interpretation: Society of Economic Paleontologists Mineralogists, short course notes, 169 p.
Dubois, M.K., Byrnes, A.P., Bohling, G.C., and Doveton, J.H., 2006, Multiscale geologic and petrophysical modeling of the giant Hugoton gas field (Permian), Kansas and Oklahoma, in P.M. Harris and L.J. Webber, eds., Giant hydrocarbon reservoirs of the world: From rocks to reservoir characterization and modeling: American Association of Petroleum Geologists Memoir 88, Tulsa, Oklahoma, p. 307–353.
Ellis, D.V., and Singer, J.M., 2007, Well logging for earth scientists: Springer, London, 692 p.
Gaymard, R., and Poupon, A., 1968, Response of neutron and formation density logs in hydrocarbon bearing formations: The Log Analyst, v. IX, no. 5, pp. 3–12.
Gromet, L.P., Dymek, R.F., Haskin, L.A., and Korotev, R.L., 1984. The "North American shale composite:" its compilation, major and trace element characteristics: Geochimica et Cosmochimica Acta, v. 48, no. 12, pp. 2469–2482.
Hook, J.R., 2003, An introduction to porosity: Petrophysics, v. 44, no. 3, pp. 205–212.
Kazatchenko, E., Markov, M., and Mousatov, A., 2006, Simulation of acoustic velocities, electrical and thermal conductivities using unified pore-structure model of double-porosity carbonate rocks: Journal of Applied Geophysics, v. 59, no. 1, pp. 16–35
Kazatchenko, E., Markov, M., Mousatov, A., and Pervago, E., 2007, Joint inversion of conventional well logs for evaluation of double-porosity carbonate formations: Journal of Petroleum Science and Engineering, v. 56, no. 4, pp. 252–266.
Larionov, V.V., 1969, Borehole radiometry: Nedra, Moscow, 238 p.
Loucks, R.G., 1999, Paleocave carbonate reservoirs; origins, burial-depth modifications, spatial complexity, and reservoir implications: American Association of Petroleum Geologists Bulletin, v.83, no. 11, pp. 1795–1834.
Lucia, F.J., 2007, Carbonate reservoir characterization: An integrated approach: Springer-Verlag, Berlin, 336 p.
Lucia, F.J., and Conti, R.D., 1987, Rock fabric, permeability, and log relationships in an upward-shoaling, vuggy carbonate sequence: The University of Texas at Austin, Bureau of Economic Geology, Geological Circular 87-5, 22 p.
Luthi, S.M., 2001, Geological well logs: Their use in reservoir modeling: Springer-Verlag, Berlin, 333 p.
Nugent, W.H., Coates, G.R., and Peebler, R.P., 1978, A new approach to carbonate analysis: Transactions of the Society of Professional Well Log Analysts, 19th Annual Logging Symposium, Paper O, 10 p.
Nugent, W.H., 1984, Letters to the editor: The Log Analyst, v. 25, no. 2, pp. 2–3.
Page, G., Fanini, O., Kriegshäuser, B., Mollison, R., Liming, Y., and Colley, N., 2001, Field example demonstrating a significant increase in calculated gas-in-place: An enhanced shaly sand reservoir characterisation model utilizing 3DEX multicomponent induction data: Society of Petroleum Engineers, SPE 71724-MS, 14 p.
Quirein, J., Donderici, B., Torres, D., Murphy, E., and Witkowsky, J., 2012, Evaluation of general resistivity density-based saturation in thin, laminated sand-shale sequences: American Association of Petroleum Geologists, Search and Discovery Article #41042, 21 p.
Raymer, L.L., Hunt, E.R., and Gardner, J.S., 1980, An improved sonic transit time-to-porosity transform. Transactions of the Society of Professional Well Log Analysts, 21st Annual Logging Symposium, Paper P, 12 p.
Rush, J., 2013, Prototyping and testing a new volumetric curvature tool for modeling reservoir compartments and leakage pathways in the Arbuckle saline aquifer: reducing uncertainty in CO2 storage and permanence: 9th Quarter Progress Report: DOE FE0004566, 34 p.
Stieber, S.J., 1970. Pulsed neutron capture log evaluation-Louisiana Gulf Coast: Society of Petroleum Engineers, SPE 2961-MS, 7 p.

Thomas, E.C., and Stieber, S.J., 1975, The distribution of shale in sandstones and its effect on porosity: Transactions of the Society of Professional Well Log Analysts, 16th Annual Logging Symposium, Paper T, 15 p.

Tiab, D., and Donaldson, E.C., 2011, Petrophysics: theory and practice of measuring reservoir rock and fluid transport properties: Gulf Professional Publishing, Houston, Texas, 688 p.

Verwer, K., Eberli, G.P., and Weger, R.J., 2011, Effect of pore structure on electrical resistivity in carbonates: American Association of Petroleum Geologists Bulletin, v. 95, no. 2, pp. 175–190.

Wang, F.P., and Lucia, F.J., 1993, Comparison of empirical models for calculating the vuggy porosity and cementation exponent of carbonates from log responses: Bureau of Economic Geology, University of Texas at Austin, Geological Circular 93-94, 27 p.

Watfa, M., and Nurmi, R., 1987, Calculation of saturation, secondary porosity and producibility in complex Middle East carbonate reservoirs: Transactions of the Society of Professional Well Log Analysts, 28th Annual Logging Symposium, Paper CC, 24 p.

Weger, R.J., Eberli, G.P., Baechle, G.T., Massaferro, J.L., and Sun, Y.F., 2009, Quantification of pore structure and its effect on sonic velocity and permeability in carbonates: American Association of Petroleum Geologists, Bulletin, v. 93, no. 10, pp. 1297–1317.

Wilson, M.D., and Pittman, E.D., 1977, Authigenic clays in sandstones: recognition and influence on reservoir properties and paleoenvironmental analysis: Journal of Sedimentary Petrology, v. 47, no. 1, pp. 3–31.

Wu, T., and Berg, R.R., 2003, Relationship of reservoir properties for shaly sandstones based on effective porosity: Petrophysics, v. 44, no. 5, pp. 328–341.

Wyllie, M.R.J., Gregory, A.R., and Gardner, G.H.F., 1958, An experimental investigation of factors affecting elastic wave velocities in porous media: Geophysics, v. 23, no. 3, pp. 459–493.

Wyllie, M.R.J., Gregory, A.R., & Gardner, L.W., 1956, Elastic wave velocities in heterogeneous and porous media: Geophysics, v. 21, no. 1, pp. 41–70.

Yaalon, D.H., 1962, Mineral composition of average shale: Clay Minerals Bulletin, v. 5, no. 27, pp. 31–36.

CHAPTER 3

Permeability Estimation

PERMEABILITY IS A VECTOR

Because it is a measure of flow, permeability is a vector quantity, as contrasted with conventional petrophysical log data, which are responses to static properties of the rock. In the absence of a direct measurement of permeability, predictions must be inferred from the rock framework characteristics that control the ability of fluids to move through the rock. In this chapter, we consider methods that predict *absolute* permeability, that is, permeability with respect to a single fluid. This is the most widely used meaning of the term and would be immediately applicable to aquifers. In engineering applications to reservoirs, a *relative* permeability is assigned to each fluid phase, so that relative fluid rates and volumes can be characterized explicitly. Although the fundamental physics of permeability in tubes has been understood for many years, reliable estimations are difficult to make in all but the simplest rock types. As we shall see, one approach attempts to adapt modifications to a tube model to accommodate the complexity of pore-system geometry. This model-driven methodology tends to be favored by engineers and contrasts with a data-driven geological approach that applies empirical relationships from core data from mercury porosimetry measurements.

PREDICTION OF PERMEABILITY FROM POROSITY

The most fundamental property used to predict permeability is that of pore volume. Both porosity and permeability are routine measurements from core analysis. If a useable relationship can be developed to predict permeability from porosity, then predictions of permeability can be made in wells that were logged with conventional measurements but not cored. The simplest quantitative methods used to predict permeability from logs have been keyed to empirical equations of the type:

$$\log k = P + Q \cdot \Phi$$

or

$$\log k = P + Q \cdot \log \Phi$$

where P and Q are constants determined from core measurements and applied to log measurements of porosity (Φ) to generate predictions of permeability (k). These equations are the basis for statistical predictions of permeability in regression analysis, where porosity is the independent variable and logarithmically scaled permeability is the dependent variable. The fitted function minimizes the sum of the squared deviations of the permeability about the trend line. In unimodal pore distributions such as those that typify sandstones, the pore volume is commonly observed to be matched by a normal distribution. Bimodal and multimodal pore systems in carbonates can create a positive skew to the distribution, although this effect may be small if there is a dominant primary mode. In either case, the scaling of porosity in arithmetic or logarithmic units is not a significant issue, because it is the independent variable. Distributions of permeabilities are almost invariably lognormal in form, probably resulting as the product of a multiplicative, rather than arithmetic, process. Consequently, the permeability is most commonly scaled logarithmically on porosity-permeability crossplots and is used within regression analysis in logarithmic form rather than in arithmetic darcy units. Because permeability is the dependent variable, the choice of scaling has implications, both with regard to the statistical fitting, but also with respect to the prediction results when applied to reservoir characterization. These implications will be discussed in more detail at the conclusion of this chapter, together with comparisons of other methodologies that compete with regression analysis in the prediction of permeability.

An example of fitting core measurements of permeability to porosity is shown for the Simpson sandstone (Middle Ordovician) from core measurements in Kansas reservoirs (Figure 3.1). As would be expected, there is a clear trend of increasing permeability with higher porosity, and this is fitted by a linear regression of log-scaled permeability on porosity. However, the prediction function has limited utility because of the broad scatter about the trend, with the result that a permeability at any given porosity can be estimated to an absurd degree of precision, but the observed permeability values at similar porosities range over several millidarcy decades. The degree of scatter will vary for different sandstones, but the major controlling factor for the dispersion is the grain size. Smaller grain sizes cause greater surface area, which decreases permeability; larger grain sizes in rocks of equivalent porosity causes a reduction in surface area and thus increased permeability (Nelson, 1994).

When the Simpson sandstone data set is subdivided by the grain size observed in the core (Figure 3.2), then the predictive error is reduced substantially by substituting functions for each of the facies with a different grain size, rather than a single global function. Notice that not only do the coarser-grained facies have higher permeabilities at any given porosity, but the scatter about their trends is relatively reduced, reflecting better sorting of coarser grains due to higher energy currents. The physical property that causes this permeability modulation is actually the internal surface area of the pore space, which is reflected both by the grain size and its complement, the pore size.

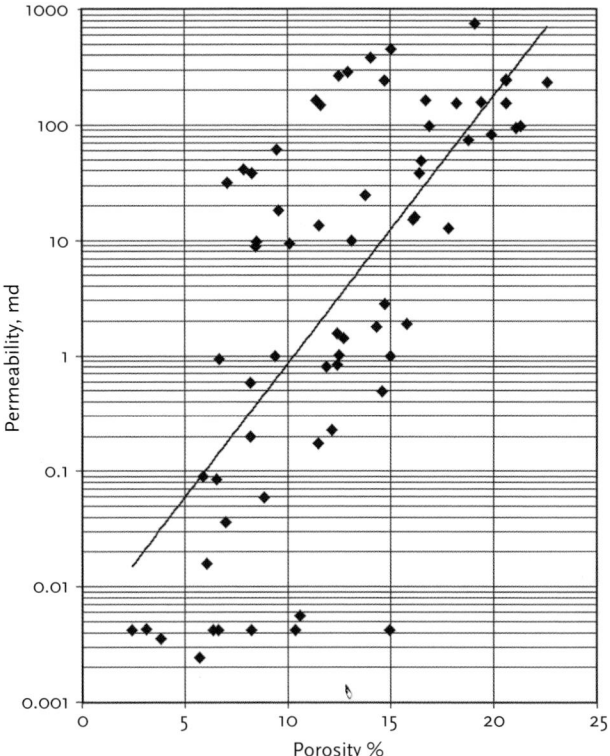

Figure 3.1: Crossplot of porosity and permeability for the Simpson sandstone (Middle Ordovician) from core measurements in Kansas reservoirs.

FLOW-ZONE INDICATOR (*FZI*) DISCRIMINATION OF HYDRAULIC UNITS

The control of permeability by pore volume and internal surface area has been known for many years, and the relationship for a rock modeled as a bundle of capillary tubes is given by the Kozeny-Carman equation:

$$k = \frac{\Phi^3}{(1-\Phi)^2}\left[\frac{1}{F_s\tau^2 S^2}\right]$$

derived by Kozeny (1927) and modified by Carman (1937). In this equation, the permeability (k) is in micrometer-squared units, effective porosity (Φ) is in fractional units, F_s is the shape factor of the tube, τ is the tortuosity of the tube, and S is the specific surface area (surface area per unit volume of solid). The collective term of $F_s\tau^2$ is widely known as the Kozeny constant, whose value (ironically for a constant) can vary anywhere between 5 and 100 in real reservoir rocks (Rose and Bruce, 1949). The issue of the variability of the Kozeny constant was addressed by Amafeule et al.

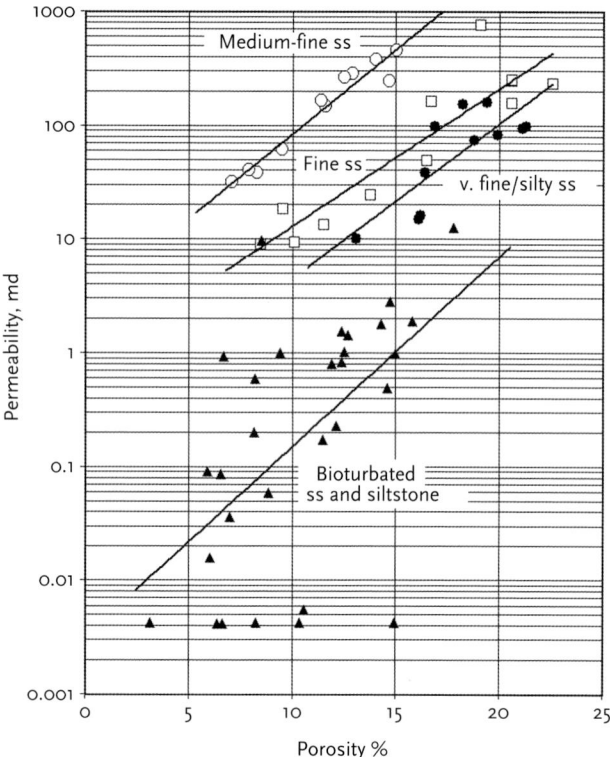

Figure 3.2: Crossplot of porosity and permeability for the Simpson sandstone (Middle Ordovician) from core measurements in Kansas reservoirs subdivided by grain-size observed in the core.

(1993) in a new formulation that first divided both sides of the Kozeny-Carman equation by the porosity and then took the square root:

$$\sqrt{\frac{k}{\Phi}} = \left[\frac{\Phi}{1-\Phi}\right]\left[\frac{1}{\sqrt{F_s \tau S}}\right]$$

When multiplied by 0.0314 to convert the permeability into millidarcy units, they designated the left-hand side of the equation as the reservoir quality index (RQI), so that:

$$RQI = 0.0314\sqrt{\frac{k}{\Phi}}$$

The pore volume-to-grain volume ratio (Φ_z) was then given by:

$$\Phi_z = \left(\frac{\Phi}{1-\Phi}\right)$$

and the flow-zone indicator (*FZI*) was introduced to encapsulate the terms of shape, tortuosity, and specific surface area, by the equation:

$$FZI = \frac{1}{\sqrt{F_s}\tau S}$$

The rearrangement and aggregation of the elements of the Kozeny-Carman equation by Amafeule et al. (1993) therefore gave the formulation of:

$$FZI = \frac{RQI}{\Phi_z}$$

In calculating the *FZI* from core measurements of porosity and permeability, samples with similar values have similar pore-throat characteristics and can be considered to be in a common flow unit. By assigning samples to a set of hydraulic units, predictions of permeability can then be made in any well, cored or uncored, based on effective porosity alone, providing that the appropriate hydraulic unit can be identified. This follows from the predictive relationship:

$$k = 1014(FZI)^2 \left(\frac{\Phi^3}{(1-\Phi)^2}\right)$$

with permeability in millidarcy units. The *FZI* values of Simpson sandstone core samples are shown in Figure 3.3, subdivided by the grain size reported from core observation. The coarser-grained facies show less variability in their *FZI* values, which

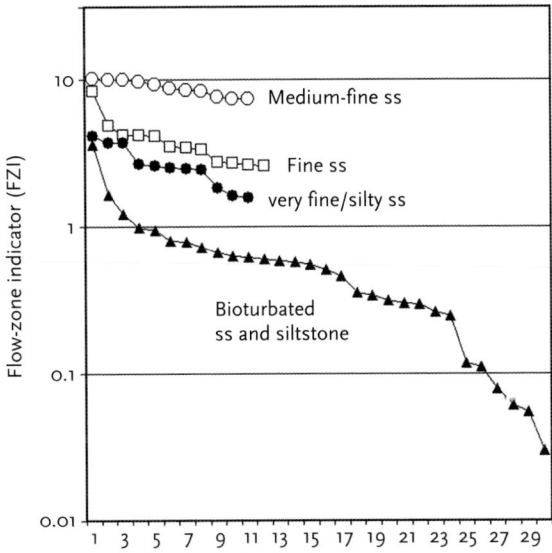

Figure 3.3: Ordered FZI values of Simpson sandstone core samples subdivided by the grain-size reported from core observation.

suggests more robust estimates of permeability, as contrasted with the FZI range for bioturbated sandstones and siltstones, and this is consistent with the visual impression of the relative line fitting of these facies in Figure 3.2. If permeability estimation in the finer-grained facies was an important goal, then a further subdivision of FZI within this facies could be pursued.

The flow-zone indicator methodology can be seen as a direct derivation of the Kozeny-Carman equation, where a formulation of tortuous capillary tubes has been adapted to accommodate the textural variability of real reservoir rocks. In this sense, it is model driven, but with the recognition that the FZI is controlled by the pore-throat size attributes, which ultimately determine the permeability. In an alternative approach, the principal pore-throat size measured by mercury porosimetry is commonly estimated by the Winland equation, published by Kolodzie (1980) as:

$$\log r_{35} = 0.732 + 0.588 \cdot \log k_{air} - 0.864 \cdot \log \Phi$$

where r_{35} is the pore-throat radius in microns (μ) at 35 percent mercury saturation, k is the absolute permeability to air (mD), and Φ is the porosity (%).

The application of the Winland equation will be described extensively in Chapter 6 as a method of defining distinctive petrofacies with common pore-throat sizes. These petrofacies then correspond broadly to the hydraulic flow units discussed here, with the former linked to pore-throat size explicitly and the latter, implicitly. This correspondence is demonstrated in the crossplot of Figure 3.4, where r_{35} estimates of principal pore-throat size show a strong trend with the FZI values calculated for the Simpson sandstone core samples. The result is not unexpected, either petrophysically or mathematically, because of the similar form of the equations and their common

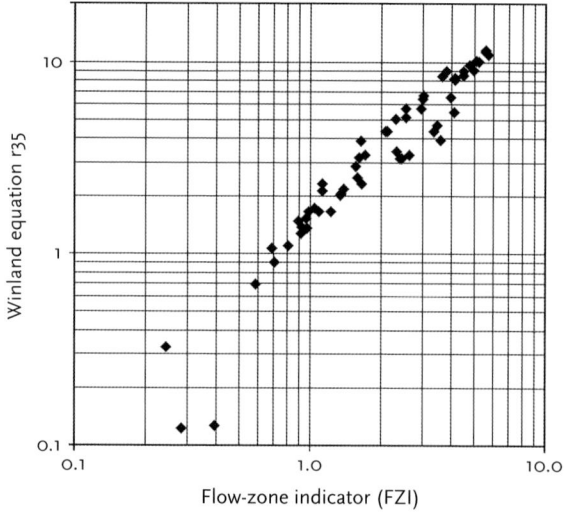

Figure 3.4: Crossplot of Winland equation r_{35} estimates of principal pore-throat size against FZI values calculated for the Simpson sandstone core samples.

input variables of porosity and permeability. Facies-sensitive geologists have often favored the Winland equation and its variants, while the model-driven *FZI* approach has generally been more popular with engineers. Ultimately, both methods recognize flow units (aka petrofacies) and use their discrimination methods of subdivision to strive for estimations of permeability that improve on global methods. This commonality was discussed by Corbett and Potter (2004), who noted the convergence of the methods as two closely related variants of permeability-driven rock-typing methods.

APPLICATION OF *FZI* TO PERMEABILITY PREDICTION

Bhattacharya et al. (2008) described flow-unit modeling and permeability prediction in Atokan sandstones of the Norcan East Field, Kansas, which provides an instructive case-study for reviewing the *FZI* methodology. This reservoir will be revisited in Chapter 6, when petrofacies will be discriminated by pore-throat sizes, rather than the present focus on permeability prediction.

The relationship between permeability and porosity from Atokan sandstone core measurements shows a weak positive trend, and the fitted linear function only accounts for an *R-squared* of 31 percent, and so it has a weak predictive power (Figure 3.5). However, the role of grain size appears to be captured when *FZI* values are calculated for the core samples and plotted against their gamma-ray values (Figure 3.6). The distinctive negative trend matches expectations that, as shale content increases and the grain size becomes finer, there is a systematic decline in flow-zone indicator

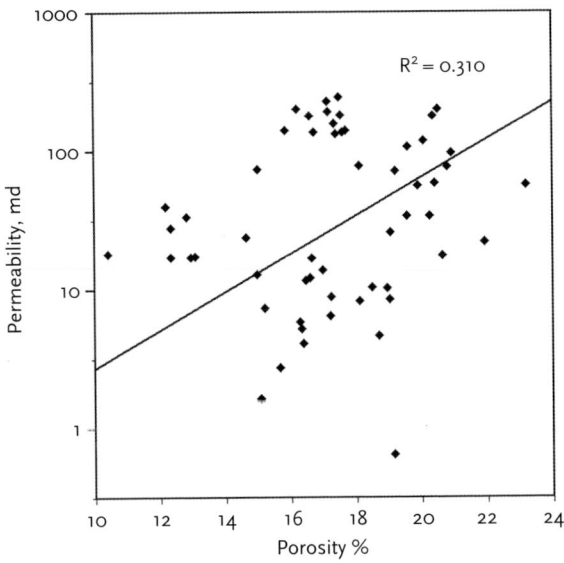

Figure 3.5: Crossplot of permeability with porosity from Atokan sandstone core measurements, together with a fitted linear regression trend.

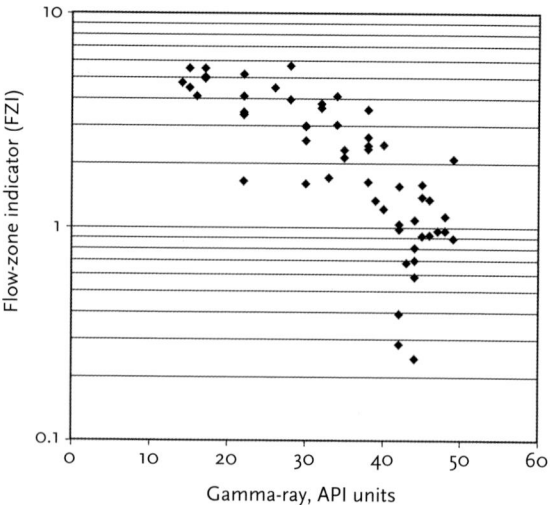

Figure 3.6: *FZI* values of Atokan sandstone core samples plotted against their gamma-ray values.

values. The next task is to subdivide the *FZI* range into distinctive hydraulic units that can be applied to permeability prediction in uncored wells. If successful, the gamma-ray log can then be used to assign any zone to the appropriate hydraulic unit, and then the permeability can be estimated by inserting the logged porosity into the prediction equation associated with the hydraulic unit.

A variety of methods have been applied to the hydraulic-unit subdivision of *FZI* values. The original approach described by Amafeule et al. (1993) was based on a visual inspection of a bilogarithmic plot of the rock quality index (*RQI*) versus the pore volume-to-grain volume ratio (Φ_z), where samples with the same *FZI* value will plot on a straight line with unit slope. Samples with similar *FZI* values are assigned to the same hydraulic unit. Clearly, some scatter is to be expected, even for distinctive hydraulic units, so that Amafeule et al. (1993) also suggested the examination of *FZI* histograms in a search for distinctive modes that would reflect hydraulic units identified by *FZI* values. Abbaszadeh et al. (1996) advocated subdividing by a variance-clustering method applied to logarithmically scaled *FZI* values, because they would be expected to conform more closely to a lognormal distribution, in common with their controlling variable of permeability. For this purpose, they applied Ward's algorithm of hierarchical cluster analysis (Ward, 1963), which minimizes variability within clusters while maximizing variability between clusters. This method generates a complete range of cluster possibilities, ranging from one cluster to the same number of clusters as data points. Consequently, a decision must be made about the appropriate number of clusters, from examination of the *R-squared* value, which expresses the proportion of variability within the clusters to the total variability. This decision will be conditioned by petrophysical judgment, because each cluster represents a distinctive hydraulic unit.

Ward's algorithm was applied to the *FZI* values of the Atokan sandstone core samples, ordered with respect to their gamma-ray reading. Examination of the *R-squared*

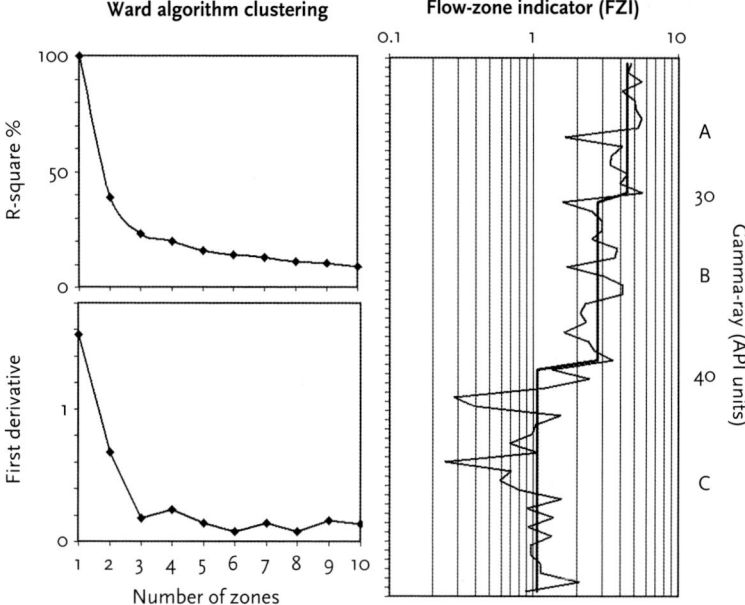

Figure 3.7: Ward's algorithm of hierarchical cluster analysis applied to the gamma-ray values of ordered *FZI* values of Atokan sandstone core samples to determine boundary values of potential hydraulic units.

scree slope and its first derivative indicated a basic discrimination between three clusters (Figure 3.7). The ranges of these hydraulic units are, successively, less than 30, 30 to 40, and greater than 40 API units. The (geometric) mean value of the *FZI* values within each hydraulic unit is then the value assigned to characterize that unit. Predictions of permeability generated by the corresponding three *FZI* equations are shown on Figure 3.8 and are related to the core measurements of porosity and permeability. The scatter of the observations about these trends clearly reflects the dispersion of sample *FZI* values about their hydraulic-unit mean. When applied to the prediction of permeability in uncored wells, the discontinuities of this tripartite model will be evident at gamma-ray curve values of 30 and 40 API units, where there will be abrupt changes in permeability as the model switches between hydraulic units.

Intelligent application of the *FZI* methodology requires some consideration of both the nature of the reservoir rock and the purpose for which the permeability predictions are to be applied. If the range of reservoir rock types constitutes a continuum, then their subdivision may be arbitrary, with the creation of artificial permeability discontinuities at their boundaries. On the other hand, if there are indeed distinctively different reservoir rock types, then they should be discriminated so that permeability relationships can be refined within each, rather than agglomerated within a heterogeneous global function. Geological knowledge concerning the depositional and/or diagenetic history of the rock types may already determine the appropriate model. However, an empirical perspective on the situation is given by the distribution of the *FZI* values themselves, as to whether they lend themselves

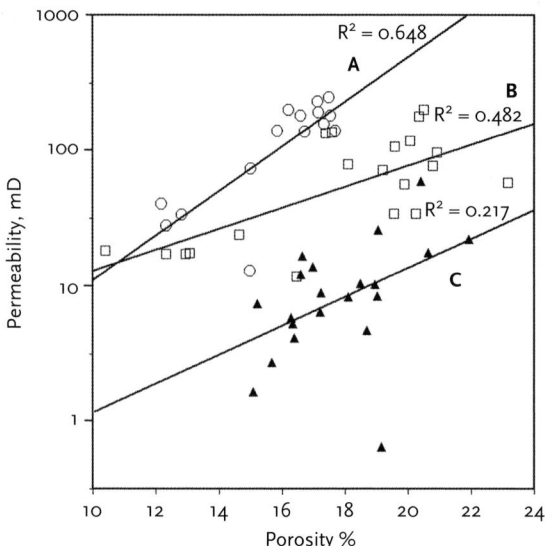

Figure 3.8: Predictions of permeability from *FZI* equations for Atokan sandstone hydraulic units located by Ward's clustering algorithm applied to gamma-ray log readings. The ranges of these hydraulic units are A: less than thirty, B: thirty to forty, and C: greater than forty API units.

to obvious partitioning or whether they appear to show generalized transitions. The second issue concerns purpose. Is the intent to characterize the permeability of individual zones within the reservoir, or will the results be used in an engineering simulator that uses discrete layers? If the permeability estimates are averaged by layer for a simulator, then hydraulic-unit zonation may be appropriate, so that the layer structure can be keyed to the actual variability observed in the *FZI* values.

Ambastha and Moynihan (1996) observed situations where *FZI* values formed a continuous spectrum with no reasonable expectation of a finite number of hydraulic units. They proposed that a regression analysis method should be substituted in these cases, bypassing the hydraulic-unit subdivision and treating *FZI* as a dependent variable to be predicted from petrophysical log responses. From a judicious choice of input logs or log transforms, an effective *FZI* predictor could be developed through analysis of fit and error statistics. Having sidestepped hydraulic-unit subdivision, it is tempting to continue this line of thought, by abandoning *FZI* calculations altogether and expanding the regression model so that it is keyed directly to permeability prediction, using multiple inputs recorded by petrophysical logs.

Returning to the Atokan sandstone data, the linear regression analysis of logarithmically scaled permeability on both gamma-ray value (γ) and porosity takes the form of:

$$\log k = P + Q \cdot \Phi + R \cdot \gamma$$

and there is an improvement in the *R-squared* fit to 72 percent. The regression prediction takes the form of a plane, whose gamma-ray contours are shown in Figure 3.9,

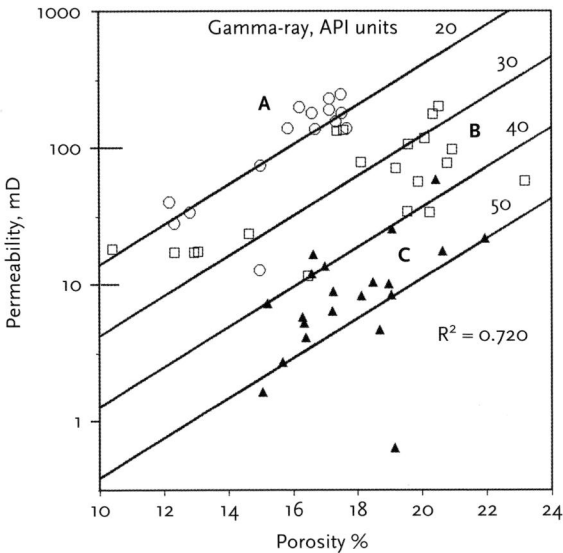

Figure 3.9: Multiple linear regression analysis of logarithmically scaled permeability on gamma-ray value and porosity for Atokan sandstone core data.

and maintains a good concordance with the hydraulic units (A, B, and C) discriminated by Ward's algorithm of variance partitioning. The *FZI* hydraulic units can be matched with facies seen in the core (Bhattacharya et al., 2008) so they are distinctive both statistically and geologically. However, it can be argued that the overlap in their properties favors modeling by a continuous function, which will still register abrupt changes caused by facies discontinuities when applied to the permeability prediction. A comparison between the multiple regression prediction of permeabilities and core measurements in an Atokan sandstone section is shown inFigure 3.10. This Atokan section will be revisited in Chapter 6, when the core data are transformed to pore-throat size estimates, which are ultimately the major control of permeability.

PERMEABILITY PREDICTIONS FROM POROSITY AND "IRREDUCIBLE" WATER SATURATION

In intergranular pore frameworks, the success of predictive methods that utilize estimators of particle size in characterizing permeability is keyed to their geometrically determined link with the internal surface area. By the same token, measures of pore-size distributions will be related to internal surface area because the pores are the complements of the grains. In a hydrocarbon reservoir, the water saturation at "irreducible" conditions (above the transition zone with no producible water) will be effectively controlled by pore size. Wyllie and Rose (1950) used this as the basis of a permeability prediction model in which they conjectured that irreducible water saturation was related to internal surface area, from their observations that the

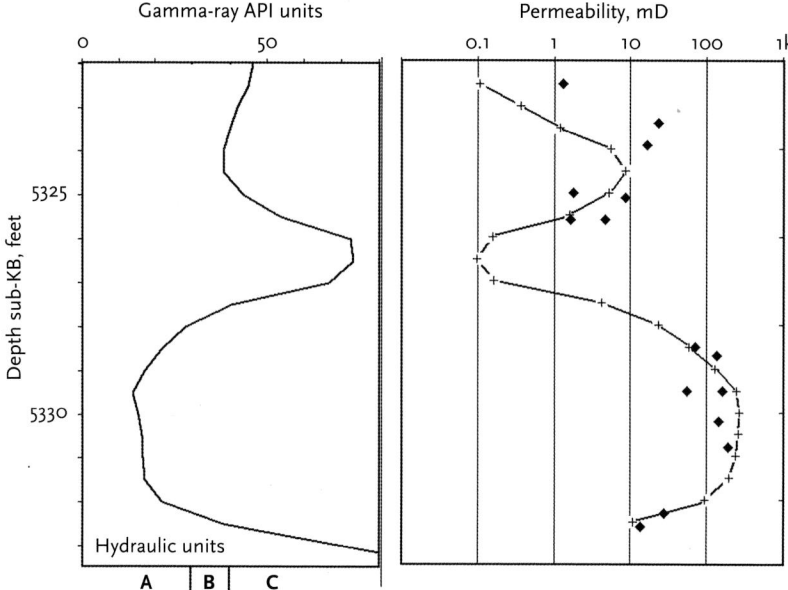

Figure 3.10: Regression analysis prediction of permeability (crosses) from gamma-ray and porosity logs in an Atokan sandstone section compared with core measurements (solid diamonds).

saturation increased both with decreasing grain size and poorer sorting. Their original model incorporated the formation factor as an expression of tortuosity, but it was simplified to the generic form:

$$k = \frac{P\Phi^Q}{S_{wi}^R}$$

which can be seen to be a Kozeny-Carman model, where the surface area term has been replaced by the irreducible water-saturation term (S_{wi}) as an ersatz surface-size measure.

Timur (1968) used maximum-likelihood statistics to determine the parameter values for the prediction of permeability in sandstones, based on porosity and irreducible water saturation from laboratory measurements of 155 core samples from US fields. The Timur equation is:

$$k = \frac{0.136 \cdot \Phi^{4.4}}{S_{wi}^2}$$

where both porosity and irreducible water saturation are in percentage units.

Timur's terminology for irreducible water saturation was "residual water saturation" (S_{wr}). Other workers use the term "immobile water saturation," while retaining the symbol S_{wi} or (S_{wirr}). The term "irreducible" is something of a misnomer, because

[78] *Principles of Mathematical Petrophysics*

mercury injection tests show that saturations at this level can ultimately be reduced to zero when subjected to sufficiently high pressures. However, the term "immobile" captures the essential meaning of S_{wi}. In rocks at these saturations, the remaining water is held by capillary forces at grain surfaces and within micropores. In Timur's laboratory measurements, the sandstone core samples were centrifuged at a differential pressure of 50 psi in an air/water system, which is approximately equal to 330 psi mercury injection pressure. This value represents the pressure at which pore throats larger than micropores have been penetrated by the nonwetting fluid. The remaining water is therefore "immobile" under conventional reservoir conditions. The water that has been expelled is equivalent to "free fluid," which exhibits normal Darcian flow.

The Timur equation is widely used as a default estimator of permeability in sandstones, either from a chart (e.g., Figure 3.11) or as an option within a log-analysis software package. Its predictions represent generalized values for a "typical" sandstone and are only valid in reservoir sections above the transition zone. At depths within the transition zone, water saturations will be higher than at "irreducible" conditions, with the result that the permeability estimates will be biased to lower values. An example of an application of the Timur equation is shown for a Morrow sandstone section in a western Kansas well (Figure 3.12). In the upper part of the sandstone, water-free production confirms that the water phase is immobile, and so the Timur-equation predictions are valid, although only approximate. In the transition zone below, there is a systematic decline of the predicted permeabilities to false values that result from water saturations that are increasingly larger than irreducible.

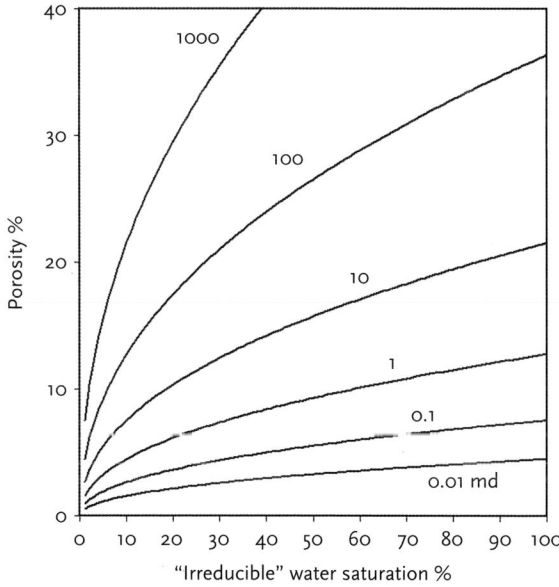

Figure 3.11: Chart for predicting permeability in sandstones from porosity and "irreducible" water saturation, using the Timur equation.

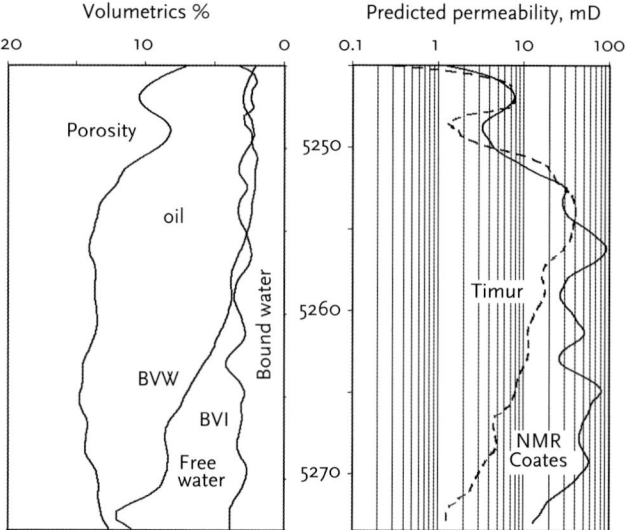

Figure 3.12: Comparison of predictions of permeability in a Morrow sandstone section using porosity and water saturation in the Timur equation and NMR partitioning of pore space in the Coates equation. The volumetrics profile shows porosity subdivided between oil and bulk volume water (*BVW*) calculated by the Archie equation overlaid by bulk volume irreducible (*BVI*) from the NMR T2-relaxation-time distribution.

These observations are substantiated by the nuclear magnetic resonance (NMR) log run in this well, where the partition between mobile and bound water identifies the reservoir, the transition zone, and a permeability estimate keyed to log measurements of pore-size distribution.

NMR ESTIMATION OF PERMEABILITY IN CLASTIC PORE SYSTEMS

The introduction of nuclear magnetic resonance (NMR) measurements as a laboratory procedure provided a reliable means to measure pore-size distributions in core samples. Seevers (1966) recognized the potential for permeability estimation based on NMR relaxation times, and the methodology was extended to borehole applications when NMR logging became a practical procedure. Because larger pore sizes would be expected to be correlated with higher permeabilities, Kenyon et al. (1988) proposed the predictive relationship, commonly known as the Schlumberger-Doll Research (SDR) equation:

$$k = a \cdot T2_{gm}^2 \cdot \Phi^4$$

where the principal pore radius is estimated by the geometric mean of the T2 distribution and the coefficient a is a formation scaling parameter. If the pore system is unimodal and the pore-body size is proportional to the pore-throat size, then the model is appropriate because permeability is controlled by pore-throat size. However, these ideal conditions are often poorly met, and the mode of the T2 distribution will be shifted in hydrocarbons, particularly gas, because of their low hydrogen index (HI).

In siliceous clastics, T2 relaxation times of 3 ms and 33 ms provide robust default partition values for subdividing the total pore space between clay-bound water, immobile capillary water (*BVI*), and free fluid (*FFI*). In carbonates, a T2 relaxation time of 92 ms is often used as the default value for the partition between free- and capillary-bound water. However, the more variable surface relaxivity of carbonate surfaces necessitates the choice of a formation-specific value for the cutoff time, when improved permeability estimates in carbonates are required. The Coates model (Coates et al., 1991) uses this pore-fluid partitioning in the equation:

$$k = \left(\frac{\Phi_{nmr}}{C}\right)^a \left(\frac{FFI}{BVI}\right)^b$$

where C is a formation scaling factor whose default value is conventionally set at 10, Φ_{nmr} is the NMR effective porosity, and a and b are exponents whose default values are four and two, respectively.

In applying the fluid partition ratio, the Coates model relates the permeability to the pore surface, which is in contrast with the SDR model, which is keyed to the pore radius. Commonly, both equations are computed as alternative models, because they have strengths and weaknesses with respect to different reservoirs. However, the Coates-equation estimation of permeability is generally considered to be more reliable, because the SDR equation is more adversely effected by hydrocarbons. With that said, it is common practice to modify the coefficients of the Coates equation through calibration with core measurements of permeability (e.g., Shafer et al., 2005).

PERMEABILITY ESTIMATION IN CARBONATES DOMINATED BY INTERPARTICLE POROSITY

The permeability prediction methods that have had some limited success in clastic rocks have fundamental limitations when applied to carbonates. The pore morphology of these chemical rocks is often highly variable in shape, size, and connectivity as a consequence of diagenesis and fracturing. However, in cases where the fabric of a limestone or dolomite is dominated by particle size (either grain or crystal), then the resulting pore network is not dissimilar to that of clastic textures, and predictions tied to pore volume and particle size can be viable. Lucia (1995) allocated Dunham carbonate fabrics between three petrophysical classes of grainstones, grain-dominated packstones, and mud-dominated packstones/wackestones, as defined by distinctive fields on a crossplot of permeability and interparticle porosity (Figure 3.13). Lucia (1995) recognized that the fabrics formed a continuum, so that the class subdivision functioned as a broad template for the development of predictive functions for permeability that incorporated crystal or grain size with interparticle pore volume.

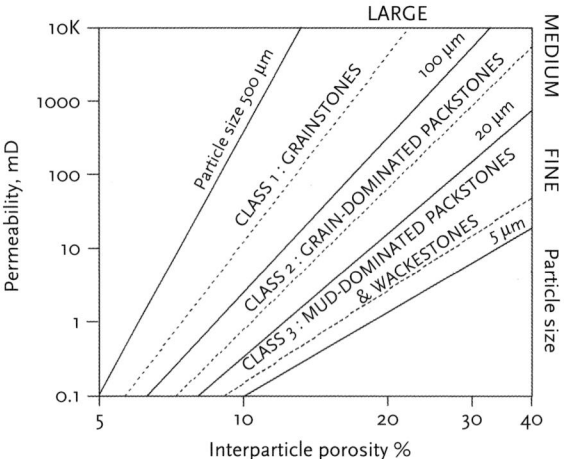

Figure 3.13: Allocation by Lucia (1995) of Dunham carbonate fabrics between three petrophysical classes of grainstones, grain-dominated packstones, and mud-dominated packstones/wackestones, as defined by distinctive fields on a crossplot of permeability and interparticle porosity. Figure simplified from original figure by Lucia (1995), © 1995 American Association of Petroleum Geologists (AAPG), reprinted by permission of the AAPG, whose permission is required for further use.

Even when dominated by interparticle porosity, carbonate rocks are sufficiently variable to require that permeability prediction functions be based on core measurements from a specific reservoir. Crossplots of porosity and permeability measurements from Lower Permian Chase and Council Grove limestone cores in the Hugoton gas field (Figure 3.14) show generalized trends and weak differentiation by Dunham texture. As would be expected, the flow-zone indicator (FZI) values for the Dunham classes are marked by a broad trend of values decreasing with particle size (Figure 3.15). The comparison is useful as a unifying theme that relates geological observations of texture from core with parameters developed from a modified Carman-Kozeny model. The link between the two approaches can yield expectations of FZI values from geomodels constructed from interpretations of depositional environments, and so it can yield potential improvements in permeability estimates.

EVALUATION OF PERMEABILITY IN DUAL- AND TRIPLE-POROSITY SYSTEMS

As discussed in the previous chapter, pore networks in carbonates can form single-, dual-, or triple-pore systems. Single-porosity systems are dominated by interparticle pores, and this is often termed "matrix porosity" by petrophysicists. While adding little to the pore volume, fracture porosity can radically increase permeability. By contrast, vugs commonly cause major increases in pore volume, but their effect on permeability is controlled by their degree of connectivity. Consequently, predictions

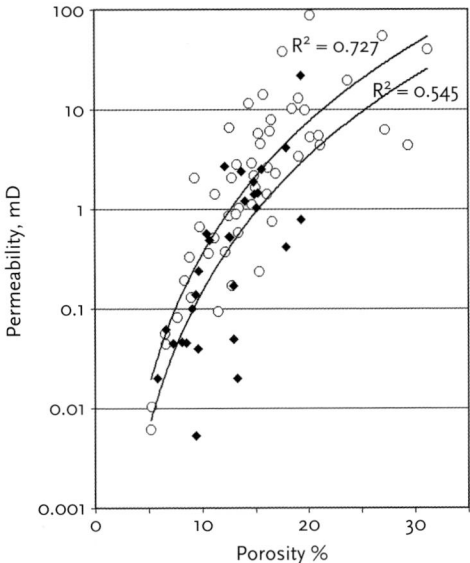

Figure 3.14: Crossplot of porosity and permeability measurements from Lower Permian Chase and Council Grove limestone cores in the Hugoton gas field, subdivided between grainstones and packstones (open circles) and wackestones and mudstones (solid diamonds).

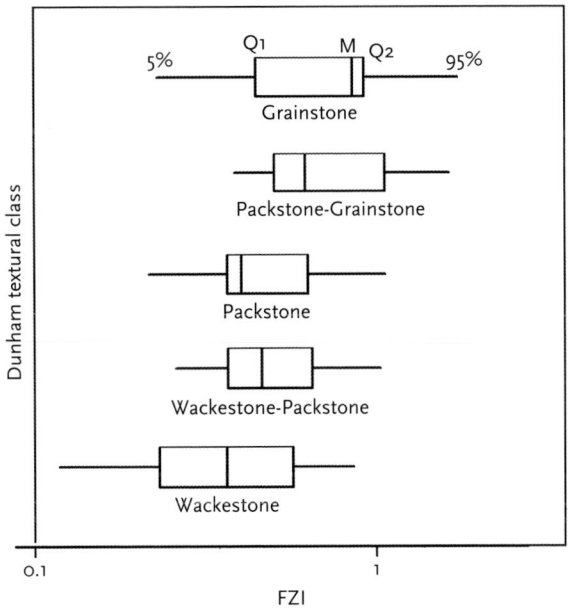

Figure 3.15: Box plots of ranges of flow-zone indicator (*FZI*) values for Dunham class groups summarized from Lower Permian Chase and Council Grove limestone cores in the Hugoton gas field.

PERMEABILITY ESTIMATION [83]

of permeability in dual- and triple-porosity systems are fraught with problems whose resolution must be attempted on a case-by-case basis.

While not presenting an easy solution, NMR logging has been a fruitful avenue to explore through its characterization of pore-body sizes within carbonates. The most commonly used prediction of permeability is based on the Coates equation:

$$k = \left(\frac{\Phi_{nmr}}{C}\right)^a \left(\frac{FFI}{BVI}\right)^b$$

but with an adjustment in the parameters that maximizes the fit with core measurements of permeability. Prior to adjusting the parameters, an initial step may be to calibrate the T2 cutoff value if core measurements of irreducible water saturation are available. Most commonly, the formation scaling factor is varied to find the best permeability match, while holding the exponents a and b at their default values of four and two. In a more exhaustive treatment, regression analysis can be applied to all the parameters as a nonlinear optimization problem (Shafer et al., 2005), although constraints are typically imposed on the exponents to avoid unrealistic results.

In fitting a modified-Coates-equation prediction of permeability to a section of Arbuckle limestone, the formation scaling factor, C, was adjusted from its default value of ten to twelve, which provided the best fit of the permeability predictions to the whole-core measurements of permeability. The resulting prediction curve and core permeabilities are shown in Figure 3.16, where extreme fluctuations in permeability reflect pore sizes over a range of scales. Lower permeabilities are matched with single-porosity systems of mud-supported and grain-supported dolomites, while high permeabilities are linked with vuggy intervals. This differentiation is demonstrated by the variance partitioning of Arbuckle core permeabilities by Ward's algorithm, where low permeabilities are matched broadly with mudstones, intermediate permeabilities with grain-supported textures, and high permeabilities with vuggy intervals (Figure 3.17). The role of the larger pores in increasing permeability can be seen in Figure 3.18, where averaged pore volumes with T2 relaxation times greater than 1,000 ms are shown plotted against core permeability. Although the relationship has some limited predictive power, it fails to take into account the degree of connectivity between the bigger pores, and so predictions have large associated error terms.

If carbonate pore frameworks can be classified in terms of single-, double-, and triple-systems, it seems reasonable that the permeability of an individual sample is the sum of contributory permeabilities from the different elements of matrix permeability, fracture permeability, and connected-vug permeability. As already discussed, variations within the matrix-permeability component can be related to the Dunham textural classes of particle size, which reflect their complementary pore size. At the larger scale, the contribution of vug permeability is dictated both by the volume of vuggy porosity and its degree of interconnection. To the degree that the T2-relaxation-time spectrum characterizes the pore-size distribution, and which in turn may mirror the pore-throat distribution, then the permeability should

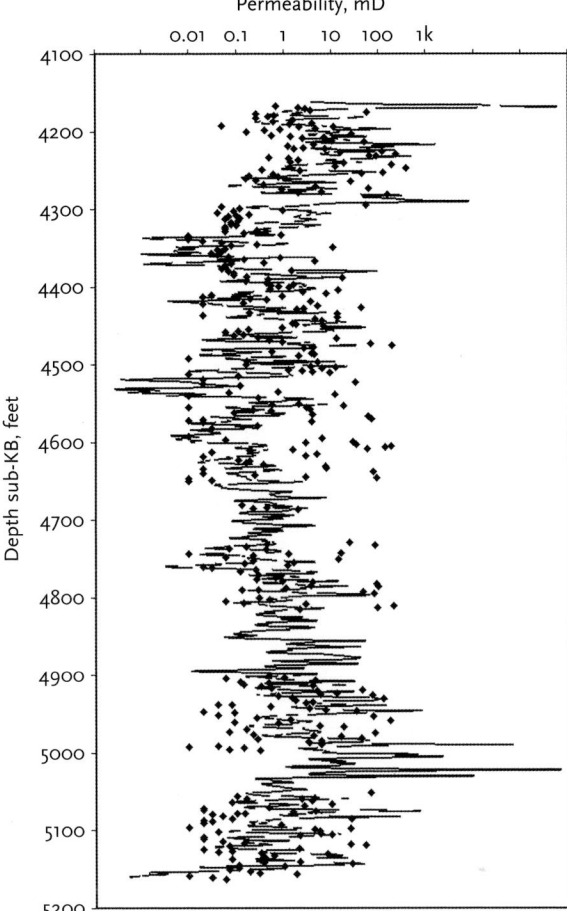

Figure 3.16: Prediction of permeability computed from a calibrated Coates equation applied to an NMR log compared with whole-core measurements of permeability in a Cambro-Ordovician Arbuckle limestone interval.

be based on the entire T2 spectrum in complex pore systems. In a single system of interparticle porosity, a dominant pore throat is commonly observed (as described in Chapter 6). The Kenyon equation then captures this model by keying the predictor to the geometric mean of the T2 distribution. However, if the T2 distribution is more dispersed, with the development of secondary modes, then it can be argued that the permeability prediction should be based on a weighted function applied to the binned porosities with the generalized form:

$$k = \Phi_{nmr}^{a} * \left(\sum_{i=1}^{n} w_i \cdot \Phi_i \right)^{b}$$

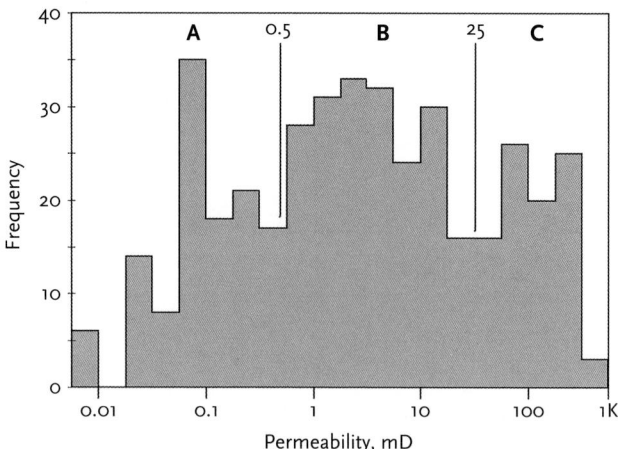

Figure 3.17: Histogram of Arbuckle limestone core permeabilities with variance partitioning by Ward's algorithm into groups A (mostly mud-supported fabrics), B (primarily grain-supported fabrics), and C (vuggy intervals).

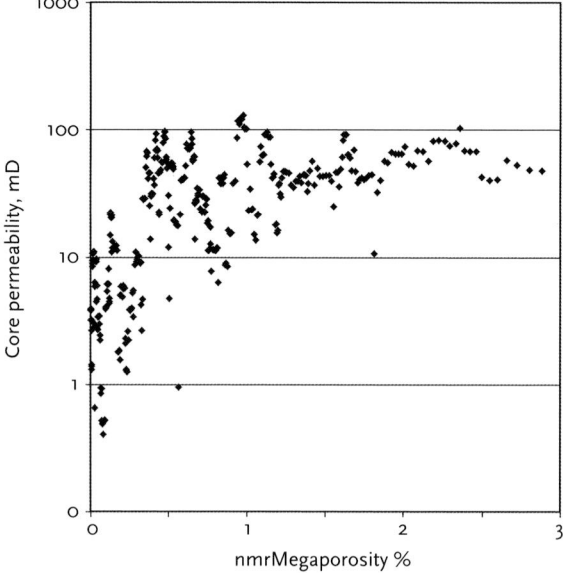

Figure 3.18: Averaged pore volumes with T2 relaxation times greater than 1,000 ms plotted against core permeability in Arbuckle limestone cores.

A case study of a T2-distribution binned-weighting procedure for predicting permeability in the Montney formation was described by Curwen and Molaro (1995), and the results were compared with those generated by the Coates equation. They concluded that there was no substantial difference between the two methods in some zones, while the binned approach performed much better in other zones. They cautioned that no individual method was the ultimate panacea and that the binned method had the intuitive appeal of providing a potential relation between pore-body size and pore-throat size that might or might not be substantiated in different field studies.

The statistical partitioning of Arbuckle limestone core permeabilities into three distinctive groups was evaluated in terms of their average T2-relaxation-time distributions (Figure 3.19). The total porosity is almost the same in all three groups, which is fortuitous, because it follows that differences in permeability must be interpreted with respect to the binned porosities. Notice that the predominantly grain-supported and vuggy groups, B and C, respectively, have very similar distributions, with a primary mode at 256 ms and a weak secondary mode at 16 ms. Vug dissolution occurred within the grainstones and is reflected by increased porosities at the longest relaxation times. In order to predict the higher permeabilities associated with the vuggy group, preferential weighting of the bins that exceed 1,000 ms would be appropriate and confirms the relationship shown in Figure 3.18. By contrast, the bulk of pore development in the lowest permeability group, A, is concentrated within the faster times of the T2-relaxation-time spectrum, as would be expected for the mudstones that dominate this group.

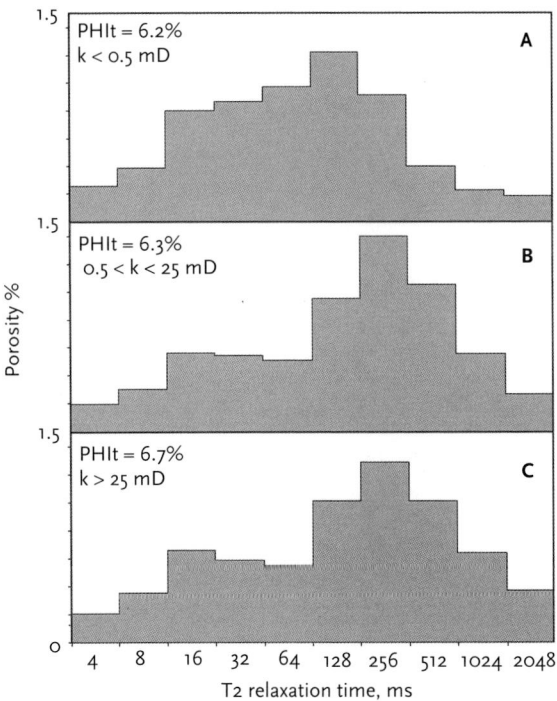

Figure 3.19: Average T2-relaxation-time distributions of Arbuckle limestone groups differentiated by Ward's algorithm: A (mostly mud-supported fabrics), B (primarily grain-supported fabrics), and C (vuggy intervals).

When the Coates equation is applied to carbonates, a default cutoff value of 92 ms is commonly used. It is interesting to note that the Coates partitioning of the distribution could be considered to be a selective weighting of grouped bins. Consequently, although the physical model of the Coates equation is generally conceived to be surface-area dependent, it might also be thought of in terms of the bin weighting of coarser pores versus finer pores. If the sizes of the pore throats are related to the pore-body size information in the T2 distribution, then in an alternative formulation, the Coates model is driven by the pore-throat size distribution.

A WILDERNESS OF MIRRORS

The lodestar of all permeability estimation procedures from logs should be the principle of "fit-for-purpose." The first question then follows as, "what is the purpose?," which leads to the second question, "What is the estimation method that will best achieve this purpose?" The most fundamental issue to be surmounted is that traditional logs are measurements of static properties, but permeability is a vector. Consequently, the logs must be calibrated to a permeability measurement, and core data remains the benchmark standard (Worthington, 2004). Permeability predictions will then be constrained to the scale of the core, which could either be a small plug or a whole core sample. Multiple scaling issues are introduced both in the analysis phase of reconciling core and log measurements with incompatible depth resolutions and when determining whether the predicted permeability is intended for small-scale characterization or application within a reservoir simulator. Various methods of averaging have been proposed to reconcile core and logs in permeability estimations that can be validated at the scale of the core, as reviewed by Worthington (2004), but the results require a careful strategy if they are scaled up to grid cells within a simulation, so that static and dynamic predictions of permeability match to within an acceptable tolerance.

Prior to the emergence of reservoir simulators as a routine industry procedure, the focus of permeability prediction was aimed at the scale of the typical log measurement. By this means, zones of high permeability could be identified and values predicted. However, this aim is thwarted when standard regression analysis is applied to permeability and scaled logarithmically. Without correction, the regression will underestimate the high permeabilities, and the effect becomes progressively greater at higher values. At the same time, the lower permeabilities will tend to be overestimated, but not to the degree of the values in the high range. Compensation can be made by applying a weighted regression (Davis, 2002), in which the weights are determined by the equation:

$$w_i = \frac{nk_i^2}{\sum k_i^2}$$

where w_i is the weight of the ith observation, k_i is its permeability, and n is the number of permeability measurements.

Both standard regression and weighted regression are parametric techniques, so that improvements in the methodology may not be matched by improvements in prediction if there is significant departure from normality in the logarithmically transformed permeability distribution. Also, the importance of permeability estimation may not be equally assigned to every permeability value. While agreeing with the observation that standard regression results in biased weighting, Wendt et al. (1986) chose more pragmatic criteria for their corrective strategy. They experimented with different weighting schemes and evaluated the results from the perspective that the high-permeability zones were of the greatest importance and the low-permeability zones, the least. Ironically, this is the reverse of the outcome of standard regression. Wendt et al. (1986) identified "high-permeability streaks" in their core database and compared it to the performance of an optimally weighted regression analysis in their prediction. In their case study, they achieved a success rate of 60 percent, with the additional interesting result that the number of high-permeability streaks that were missed approximately matched the number of false positives. Therefore, the total number of high-permeability streaks was a good estimate, but 40 percent were located at the wrong depths.

The application of neural networks to permeability prediction could be considered as a logical extension of this approach, but one in which the parametric model is abandoned and the weightings are determined by the neural network itself. Furthermore, no structural form of the model is assumed, such as linear, polynomial, or any other function that could be described as a global equation. Instead, the weights of the neural network are adjusted by training with log measurement inputs to match the permeability measurements that are output by back-propagation through hidden network layers. The methodology is a supervised learning technique in that the inputs and outputs are monitored, but the richly interconnected layers of the neural network result in a black-box operation, as contrasted with the white-box (or glass-box) design of formal regression models. As a result, judgments must be made by the experienced user as to the structure of the neural network that appears to be the best for the problem at hand, which in this case is the permeability.

One of the greatest dangers is the problem of overfitting, where extremely successful results can reflect the fact that the neural network is learning by rote, rather than from generalized associations. When an overfitted network is applied to prediction outside the calibration wells, the results are typically erratic. This characteristic is well known to all neural-network practitioners and is best accommodated through a disciplined strategy of calibration followed by validation, before moving on to prediction. A good introductory case study of neural-network prediction of permeability from logs is provided by Rogers et al. (1995), with an application to a Smackover field in southern Alabama.

Initial experimentation using simple neural networks to predict permeability has been followed by progressively more complex procedures that have attempted to provide realistic improvements. In particular, a single neural network applied to the same input patterns may generate different sets of weightings as its training converges on different minima of the objective function. To accommodate this problem, many alternative networks are trained, and the best network or combination of networks are then selected based on their performance. This procedure is the basis for the committee neural-network approach described by Bhatt and Helle (2002), in its application to permeability prediction from logs. They also concluded from their work that a three-layer model is the most efficient and reliable, consisting of input, hidden, and output layers, that improves on the overgeneralization of a simpler model, while avoiding being compromised by the overfitting that can result from a more complex network. The case study conducted by Bhatt and Helle (2002) indicated that when four input logs are used, then 150 samples would be sufficient for training, and the number of neurons in the hidden layer should be between eight and ten.

In other developments, neural networks have been subsumed as one of several methods collectively termed "soft computing." The most widely known of these is fuzzy logic. Fuzzy logic was first introduced by Zadeh (1965) and represents yet another step away from the parametric models of classical statistics. The comparative features are summarized well by Fang and Chen (1997) in their paper on fuzzy modeling and prediction of sandstone permeability. They point out that fuzzy modeling is assumption-free, not tied to a mathematical model, and can incorporate linguistic information as well as numerical data. Because the method is encoded in fuzzy rules, it is robust with respect to outliers and can deal with conflicting data. A central concept of fuzzy logic is the notion of "possibility" as distinct from the probability used in standard statistical methods.

Cuddy (2000) described the application of fuzzy logic to the prediction of permeability in Jurassic sandstones of the North Sea Ula field. In his introduction to the technique, he explained how core permeabilities are first allocated between bins whose ranges are bounded by equal percentile subdivisions on a logarithmic scale. The mean and standard deviation of each well log used for the permeability prediction is then determined for each bin to ascertain both the most likely value and its associated uncertainty, or fuzziness. Using the mean and standard deviation of each bin, the fuzzy possibilities of each log are combined at each depth to predict a permeability and its associated fuzzy possibility. Each of the permeability bins has an associated fuzzy possibility, and the highest fuzzy possibility is then taken as the most likely binned permeability. A numerical value for the predicted permeability can then be calculated from the weighted average of the two most likely bins. In the design of the fuzzy prediction model, Cuddy (2000) recommends that the number of bins should be selected with respect to the total sample size, so that a minimum of thirty core permeability measurements are allocated to each bin. The predictive ability of the technique is then evaluated by blind-test validation in cored wells not were not used for the calibration.

Most publications that describe yet another novel technique to predict permeability make comparisons of the supposedly improved performance over earlier methods, while itemizing the benefits and limitations of all approaches. The competitive nature of these discussions is appropriate, both to justify the publication in the first place and to encourage the reader to try the new technique. The designers of the eclectic techniques that make up soft computing appear to embrace collaboration rather than competition in their development of hybrid methods. So, for example,

Huang et al. (2001) introduced a neural-fuzzy technique combined with genetic algorithms in the prediction of permeability in petroleum reservoirs. Neural networks were first used to create membership functions and to estimate permeability automatically from log data. The trained networks were then used as fuzzy rules and hypersurface membership functions. Defuzzification operators applied to the results from these rules were optimized by genetic algorithms. Huang et al. (2001) applied the integrated methodology of the neural-fuzzy-genetic algorithm to a petroleum reservoir in offshore Western Australia and concluded that validation tests showed smaller errors than those associated with more conventional techniques.

Regardless of the complexity of the techniques described, almost all have been applied to sandstone reservoirs that are dominated by intergranular porosity. The addition of vugs and or fractures in carbonates introduces a level of complexity that is extremely difficult to accommodate by a general methodology. Useful solutions, if they exist, must be evaluated on a reservoir-by-reservoir basis to accommodate variable and multiple histories of diagenesis. Although fuzzy-logic models are nontraditional, the rules of a practical application would have some transparency and could be understood, at least intuitively, by a petrophysicist. It is easier to conduct an audit trail that explains the poor predictions of a fuzzy system than it is to remedy the unseen and convoluted results of a neural network. Prior to permeability prediction, it would be appropriate to apply fuzzy logic as a classifier applied to subdivisions of the carbonate reservoir in terms of distinctive permeability facies. So, for example, three subdivisions of Arbuckle core permeabilities were recognized and related to carbonate textures of mudstones, grainstones, and vuggy facies (Figure 3.17). So, the initial step would be to establish the best logs to apply in a fuzzy-logic model for classifying permeability textural facies, and then within each of these facies, develop a fuzzy prediction for the permeability value for each zone.

The role of the larger pores and their degree of connectivity can be assessed from borehole image logs based on the conductive connections between conductive spots. When evaluated through a connectivity coefficient, Russell et al. (2002) were able to improve permeability estimates when used in conjunction with other log data. In addition to its high resolution, imaging information has directional properties, unlike the nondirectional volumes that are measured by standard logs. Consequently, the azimuthal resistivity data can be applied to an anisotropic permeability evaluation, as described by Anxionnaz et al. (1999), when integrated with other logs. It is common practice to measure core permeability in two directions horizontally (maximum permeability and orthogonal permeability), as well as the vertical component, but it has been less common to consider the directional anisotropy of the permeability within a single coherent prediction model. However, as Delhomme (2007) pointed out, the increasing proportions of deviated and horizontal wells have made considerations of permeability anisotropy a major factor in both directional drilling and productivity.

The increasing and varied demands of production from both traditional and newer resource plays can make the goal of permeability prediction from logs seem like a trail through a wilderness of mirrors, particularly in view of the bewildering variety of mathematical methods that have been brought to bear on the task. The truism that common sense is a virtue, is the key to success. First, each reservoir must be considered on a case-by-case basis, particularly with regard to its degree of heterogeneity and the nature and interconnectedness of its pore systems. Secondly, the purpose of the permeability prediction must be stated clearly, so that permeability predictions of specific zones are differentiated from estimates scaled up for simulation, and considered with other factors such as anisotropy. Finally, the appropriate numerical strategy must be evaluated carefully in terms of whether the final result matches the expectations of the user. Does the prediction conform to the desired scale? Does apparent precision overwhelm accuracy? Are the uncertainties associated with the result within the acceptable tolerance range? Does the methodology introduce bias, and, if so, is the bias to high or low permeabilities? In the end, permeability is such a crucial component of reservoir development that methodologies for its prediction from logs will continue to evolve, as evidenced by the extensive literature on the subject that continues to grow every year.

REFERENCES

Abbaszadeh, M., Fuji, H., and Fujimoto, F., 1996, Permeability prediction by hydraulic flow units—theory and applications: Formation Evaluation, v. 11, no. 4, pp. 263–271.

Amafeule, J.O., Altunbay, M., Tiab, D., Kersey, D.G., and Keelan, D.K., 1993, Enhanced reservoir description: Using core and log data to identify hydraulic (flow) units and predict permeability in uncored intervals/wells: Society of Petroleum Engineers, SPE 26436-MS, 16 p.

Ambastha, A., and Moynihan, T., 1996, A simple and accurate method for an integrated analysis of core and log data to describe reservoir heterogeneity: Journal of Canadian Petroleum Technology, v. 35, no. 1, pp. 40–46.

Anxionnaz, H., Delhomme, J., and De Haan, S., 1999, Reconstructing petrophysical borehole images: Their potential for evaluating permeability distribution in heterogeneous formations: Society of Petroleum Engineers, SPE 56786-MS, 11 p.

Bhatt, A., and Helle, H.B., 2002, Committee neural networks for porosity and permeability prediction from well logs: Geophysical Prospecting, v. 50, no. 6, pp. 645–660.

Bhattacharya, S., Byrnes, A.P., Watney, W.L., and Doveton, J.H., 2008, Flow unit modeling and fine-scale predicted permeability validation in Atokan sandstones: Norcan East Field, Kansas: American Association of Petroleum Geologists Bulletin, v. 92, no. 6, pp. 709–732.

Carman, P.C., 1937, Fluid flow through granular beds: Transactions of the Institution of Chemical Engineers, v. 15, pp. 150–166.

Coates, G.F., Miller, M., Gillen, M., Henderson, C., 1991, The MRIL in Conoco 33-1: An investigation of a new magnetic resonance imaging log: Transactions of the Society of Professional Well Log Analysts, 32nd Annual Logging Symposium, Paper DD, 24 p.

Corbett, P.W.M., and Potter, D.K., 2004, Petrotyping: A base map and atlas for navigating through permeability and porosity data for reservoir comparison and permeability prediction: International Symposium of the Society of Core Analysts, SCA 2004-30, pp. 385–396.

Cuddy, S. J., 2000, Litho-facies and permeability prediction from electrical logs using fuzzy logic: Reservoir Evaluation & Engineering, v. 3, no. 4, pp. 319–324.

Curwen, D.W., and Molaro, C., 1995, Permeability from magnetic resonance imaging logs: Transactions of the Society of Professional Well Log Analysts, 36th Annual Logging Symposium, Paper GG, 11 p.

Davis, J.C., 2002, Statistics and data analysis in geology: Wiley, New York, 638 p.

Delhomme, J.P., 2007, The quest for permeability evaluation in wireline logging, in L. Chery and G. de Marsily, eds., Aquifer systems management: Darcy's legacy in a world of impending water shortage: Selected Papers on Hydrogeology 10, Taylor and Francis, Leiden, 12 pp.

Fang, J.H., and Chen, H.C., 1997, Fuzzy modelling and the prediction of porosity and permeability from the compositional and textural attributes of sandstone. Journal of Petroleum Geology, v. 20, no. 2, pp. 185-204.

Huang, Y., Gedeon, T.D., and Wong, P.M., 2001, An integrated neural-fuzzy-genetic-algorithm using hyper-surface membership functions to predict permeability in petroleum reservoirs: Engineering Applications of Artificial Intelligence, v. 14, no. 1, pp. 15-21.

Kenyon, W.E., Day, P.I., Straley, C., and Willemsen, C., 1988, A three-part study of NMR longitudinal relaxation properties of water-saturated sandstones: Journal fo Formation Evaluation, v. 3, no. 3, pp. 622-636

Kolodzie, S., Jr., 1980, Analysis of pore-throat size and use of the Waxman-Smits equation to determine OOIP in Spindle Field, Colorado: Society of Petroleum Engineers, SPE-9382, 10 p.

Kozeny, J., 1927, Uber kapillare leitung des wasser im boden sitzungberichte: Sitzungberichte das Akademie der Wissenschaften, Vienna v. 136, no. 11a, pp. 271-306.

Lucia, F.J., 1995, Rock-fabric/petrophysical classification of carbonate pore space for reservoir characterization: American Association of Petroleum Geologists Bulletin, v.79, no. 9, pp. 1275-1300.

Nelson, P.H., 1994, Permeability-porosity relationships in sedimentary rocks: Log Analyst, v. 35, no. 3, pp. 38-62.

Rogers, S.J., Chen, H.C., Kopaska-Merkel, D.C., and Fang, J.H., 1995, Predicting permeability from porosity using artificial neural networks: American Association of Petroleum Geologists Bulletin, v. 79, no. 12, pp. 1786-1796.

Rose,W., and Bruce,W.A.,1949, Evaluation of capillary character in petroleum reservoir rock: Journal of Petroleum Technology, v. 1, no. 5, p. 127-142.

Russell, S.D., Akbar, M., Vissapragada, B., and Walkden, G.M., 2002, Rock types and permeability prediction from dipmeter and image logs: Shuaiba reservoir (Aptian), Abu Dhabi: American Association of Petroleum Geologists Bulletin, v. 86, no. 10, pp. 1709-1732.

Seevers, D. O., 1966, A nuclear magnetic method for determining the permeability of sandstones: Transactions of the Society of Professional Well Log Analysts, 7th Annual Logging Symposium, Paper L, 12 p.

Shafer, J.L., Chen, S., and Georgi, D.T., 2005, Protocols for calibrating NMR log-derived permeabilities: International Symposium of the Society of Core Analysts, SCA2005-37, 15 p.

Timur, A., 1968, An investigation of permeability, porosity, and residual water saturation relationships for sandstone reservoirs: Log Analyst, v. 9, no. 4, pp. 8-17.

Ward, J.H., Jr., 1963, Hierarchical grouping to optimize an objective function: Journal of the American Statistical Association, v. 38, no. 301, pp. 236-244.

Wendt, W.A., Sakurai, S., and Nelson, P.H., 1986, Permeability prediction from well logs using multiple regression, in L.W. Lake and H.B. Caroll Jr., eds., Reservoir characterization: Academic Press, New York, pp. 181-221.

Worthington, P.F., 2004, The effect of scale on the petrophysical estimation of intergranular permeability: Petrophysics, v. 45, no. 1, pp. 59-72.

Wyllie, M.R.J., and Rose, W., 1950, Some theoretical considerations related to the quantitative evaluation of the physical characteristics of reservoir rock from electrical log data: Journal of Petroleum Technology, v. 2, no. 4, pp. 105-118.

Zadeh, L.A., 1965, Fuzzy sets: Information and control, v. 8, no. 3, pp. 338-353.

CHAPTER 4
Compositional Analysis of Mineralogy

SOME MATRIX ALGEBRA

Formation lithologies that are composed of several minerals require multiple porosity logs to be run in combination in order to evaluate volumetric porosity. In the most simple solution model, the proportions of multiple components together with porosity can be estimated from a set of simultaneous equations for the measured log responses. These equations can be written in matrix algebra form as:

$$CV = L$$

where C is a matrix of the component petrophysical properties, V is a vector of the component unknown proportions, and L is a vector of the log responses of the evaluated zone. The equation set describes a linear model that links the log measurements with the component mineral properties. Although porosity represents the proportion of voids within the rock, the pore space is filled with a fluid whose physical properties make it a "mineral" component. If the minerals, their petrophysical properties, and their proportions are either known or hypothesized, then log responses can be computed. In this case, the procedure is one of forward-modeling and is useful in situations of highly complex formations, where geological models are used to generate alternative log-response scenarios that can be matched with actual logging measurements in a search for the best reconciliation between composition and logs. However, more commonly, the set of equations is solved as an "inverse problem," in which the rock composition is deduced from the logging measurements.

Probably the earliest application of the compositional analysis of a formation by the inverse procedure applied to logs was by petrophysicists working in Permian carbonates of West Texas, who were frustrated by complex mineralogy in their attempts to obtain reliable porosity estimates from logs, as described by Savre (1963). Up to that time, porosities had been commonly evaluated from neutron logs, but the values were excessively high in zones that contained gypsum, caused by the hydrogen within the water of crystallization. The substitution of the density log for the porosity estimation was compromised by the occurrence of anhydrite as well as gypsum. Collectively, the mix of three minerals and porosity put the solution beyond the reach of graphical methods, such as crossplots and nomograms, which were the standard procedures of that time.

The log-analysis solution to this problem required the estimation of the true volumetric porosity that simultaneously accommodated the effects of the variable contents of gypsum, anhydrite, and dolomite. By using neutron, density, and sonic logs together, the log-response equations were:

$$\text{Neutron:} \quad \Phi_n = n_g \cdot G + n_a \cdot A + n_d \cdot D + n_f \cdot \Phi$$

$$\text{Sonic:} \quad \Delta t = \Delta t_g \cdot G + \Delta t_a \cdot A + \Delta t_d \cdot D + \Delta t_f \cdot \Phi$$

$$\text{Density:} \quad \rho_b = \rho_g \cdot G + \rho_a \cdot A + \rho_d \cdot D + \rho_f \cdot \Phi$$

which contain four unknowns within three equations. The model is completed by a fourth "unity" equation that combines the proportions of gypsum (G), anhydrite (A), dolomite (D), and the true fractional porosity (Φ), as a closed system:

$$\text{Unity:} \quad 1 = G + A + D + \Phi$$

The equations can be rewritten in a matrix algebra formulation as:

$$CV = L$$

where C is the matrix of neutron, transit times, and grain densities of gypsum, anhydrite, dolomite, and pore fluid, supplemented by a line of unit values; V is the vector of their unknown proportions in the zone; L is a vector of the zone log readings of neutron porosity, transit time, and bulk density, together with a unit value. The equation set describes a determined system, and the solution for the unknown vector, V is:

$$V = C^{-1}L$$

where C^{-1} is the inverse of the C matrix.

Savre (1963) described how this procedure was coded in a computer program, as a pioneer application of computers to petrophysics. An example of the graphical output drafted from one of the earliest computer runs is shown in Figure 4.1 (Alger et al., 1963), where profiles of porosity, dolomite, anhydrite, and gypsum are shown from a Permian San Andres formation section in West Texas. Savre (1963) compared the porosity estimations computed by this method with core measurements of porosity. The initial results seemed disappointing, but it was discovered that the core porosities had been overestimated because of dehydration of the gypsum in the core sample treatment. Once appropriate precautions were applied to the core analysis, there was an improvement to a much more reasonable match between the log estimates and the core measurements of porosity. The problem caused by the presence of gypsum in the accurate assessment of porosity from logs in the San Andres formation is now widely recognized and routinely resolved by compositional-analysis software (Hedberg and May, 1990).

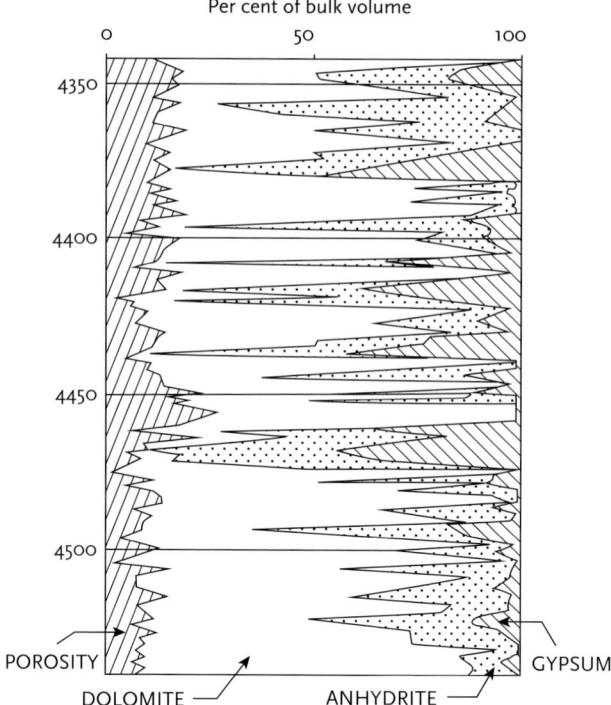

Figure 4.1: Graphical output profiles of porosity, dolomite, anhydrite, and gypsum from one of the earliest computer runs that processed neutron, sonic, and density logs of a Permian San Andres formation section in West Texas. From Savre (1963).

At the time that this early application was made, computing power was typically provided by a single mainframe computer in the company or university, which had extended computing times and limited memory, while programming code was a specialized and time-consuming task. The same application is very easy to implement today as a spreadsheet procedure, using standard matrix functions and graphical outputs. As an example, sonic, density, and neutron porosity logs are shown of a San Andres formation section of Waddell Field, Texas, and their divergent curves show the marked influence of both anhydrite and gypsum (Figure 4.2). The introduction of the photoelectric-factor log in the 1980's provided an important new measure of mineralogy because of its direct response to the aggregate atomic numbers of the formation. Consequently, the composition can be resolved using neutron, density, and photoelectric-factor logs, and the sonic log can be used to partition the pore space between interparticle and vuggy/moldic porosity. Integrated core and log studies by Lucia (1999) show this pore-partition method to be an effective measure of vug porosity in these dolomites. The compositional solution is shown in Figure 4.3, where log estimates are compared with core evaluations (Nissen et al., 2008). There is a close match between log-estimated total porosity and core porosity, as contrasted with the raw porosity logs (Figure 4.2). The depth range for estimated

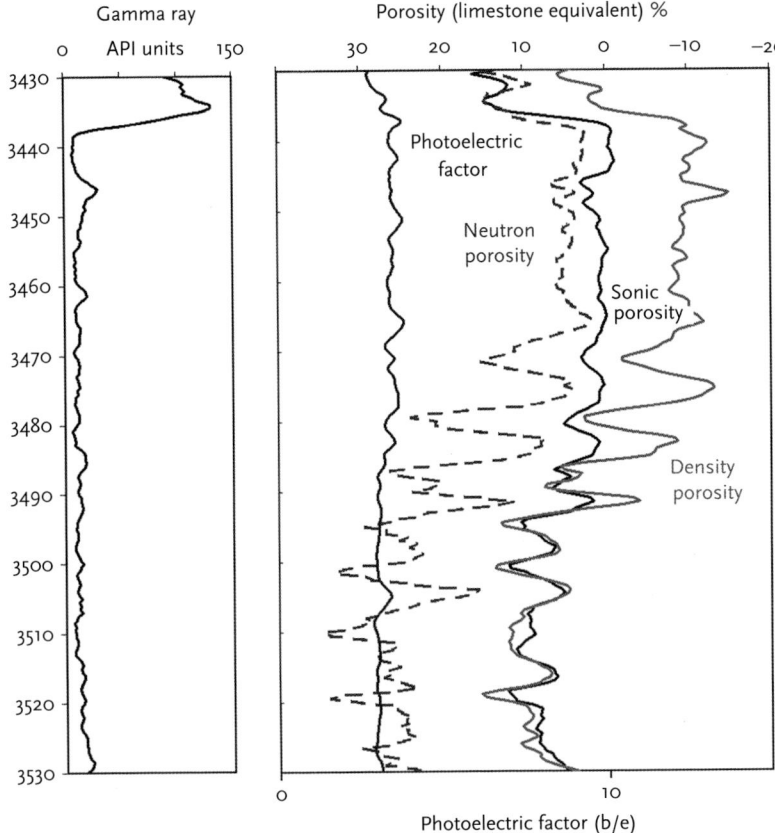

Figure 4.2: Gamma-ray, photoelectric-factor, sonic, density, and neutron porosity logs of a San Andres formation section of the Waddell Field, Texas.

anhydrite and gypsum content is closely matched by the observed occurrences of these minerals in the core, although the quantitative estimates are probably only approximate. However, better estimates can be made with the incorporation of the geochemical log whose measure of sulfur is tied directly to sulfate content (Cannon and Horkowitz, 1997).

COMPOSITIONAL-SOLUTION EVALUATION

The inverse solution is a simple and powerful procedure for compositional analysis, but its simplicity carries certain assumptions that must be considered carefully. In particular, the basic model contains no intrinsic constraint to preclude negative estimates of compositional proportions. The unity equation dictates the closure of the system so that the proportions collectively sum to unity. However, individual proportions can have a negative value or one that exceeds unity. Rather than representing

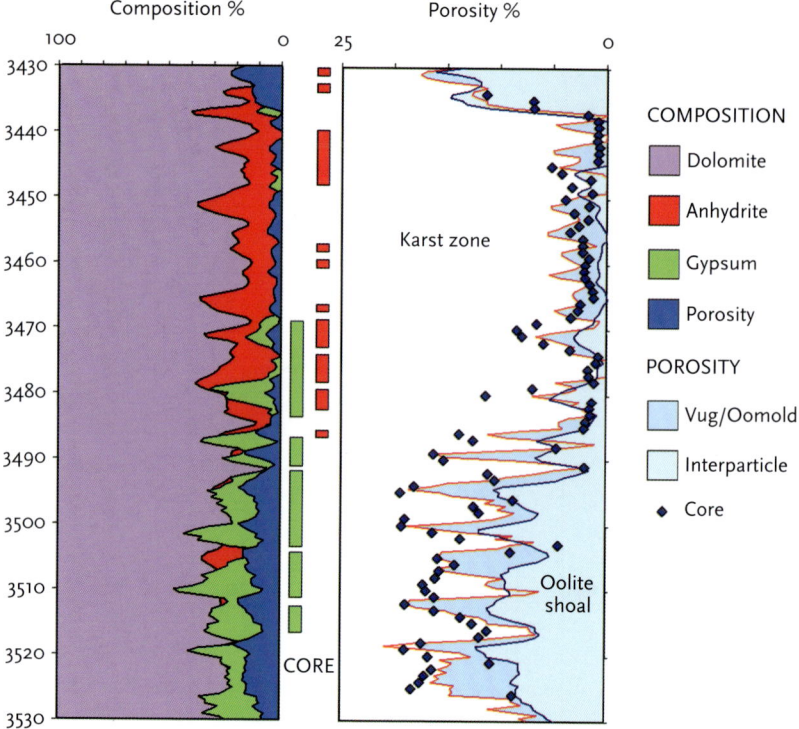

Figure 4.3: Compositional log analysis of a San Andres formation section of the Waddell Field, Texas in terms of dolomite, gypsum, and anhydrite, pore space partitioned between interparticle and vuggy/moldic porosity, and comparison with core measures of porosity and the presence of anhydrite and gypsum. From Nissen et al. (2008).

mathematical error, apparently anomalous zones are located outside the composition space defined by the mineral endmembers as vertices. Consequently, the generation of negative proportions is a perfectly natural consequence of the model and can contain useful feedback information. If the negative values are small, then this is usually caused by the stochastic nature of the input nuclear logs coupled with the borehole rugosity perturbations. If they are large, the possibility of washouts and gas effects should be examined before evaluating the possibility of another mineral that is not included in the composition model.

If these explanations are not sufficient, then negative proportions of the components have a role as a basic check on the validity of the model used for compositional analysis. As such, they are diagnostic errors with an information content to be used to guide the analysis to a better solution. The distinction between errors that are acceptable as minor, random, measurement noise and systematic deviations is best made by a comparison between the original logs and the logs predicted by the model solution. The predictions are given by:

$$\hat{L} = CV$$

If the inverse procedure has generated zone solutions with proportions that are negative or exceed unity, then the adjustment to rational proportions will result in log predictions that will deviate from the original logs. The deviations between measurements and predictions can then be examined to differentiate minor measurement errors from systematic perturbations that require intervention and correction. In the more sophisticated models to be reviewed, tool-response errors are actively incorporated within the solution algorithm, together with constraints that preclude irrational compositional proportions. However, if the solution results in compositional proportions that are all positive, then there will be an exact match between the logs and model predictions. This equivalence does not imply that the result is geologically correct; it simply means that the solution is rational and consistent with the choice of components and their properties. There may be other satisfactory solutions based on alternative mineral suites.

UNDERDETERMINED SYSTEMS

The basic compositional inversion procedure requires a precise match between the number of knowns and unknowns. This situation is called a "determined system." The alternative possibilities are that the number of logs is insufficient to provide a unique resolution of the proportions of the components (an underdetermined system), or that the number of logs exceeds the number of components (an overdetermined system). In reality, it is likely that most formations present underdetermined compositional problems, if all the constituents are counted and matched against the number of logs run in a typical borehole. As counterpoint, many of the minerals will be found in small quantities, and the overall composition dominated by a few components.

McCammon (1970) and Harris and McCammon (1971) considered alternative model procedures for the estimation of mineral compositions from logs in underdetermined cases. Although their algorithms have been superseded by optimization procedures, their approach is instructive concerning the role of information in log compositional analysis and the potentially competing criteria of mathematical optimality and geological reality. McCammon (1970) considered the underdetermined system in terms of classical information theory, which proposes that the least-biased solution is the one that maximizes the entropy function:

$$E = \sum p_i \log p_i$$

where p_i is the proportion of the ith component. This equation for entropy is closely approximated by that for proportional variance:

$$P = \left(\frac{n}{n-1}\right) \sum v_i (1-v_i) = \left(\frac{n}{n-1}\right)\left(1 - \sum v_i^2\right)$$

The maximum of the variance function, P, is close to the condition of maximum entropy, and the resulting optimal solution is easier to compute using the matrix algebra equation:

$$V = C^t \left(CC^t\right)^{-1} L$$

where V is a vector of unknown proportions, C is the matrix of component log properties, t signifies a matrix transpose, and L is the vector of the zone log responses (Doveton and Cable, 1979).

The compositional solution from the proportional variance algorithm is optimal from a classical statistical viewpoint: the average squared errors between the estimates and the real compositions should be the minimum possible.

This is a conservative philosophy that aims to be least wrong or risk averse, with a minimum error as penalty. However, mineral proportions are frequently distributed in a highly unequal manner. Therefore, the real rock composition will often be one of several extreme possibilities, rather than the less likely seemingly homogeneous composition that can result from a minimum-variance solution. The correct interpretation of a bland compositional solution is that it represents the average of a range of possibilities. As such, it is a good estimate of the average, but may be a very poor prediction of the particular: the composition of the zone in question. Such a result is a useful diagnostic that suggests that several extreme alternatives should be reviewed and that extra information is required. The information can take a variety of forms, such as explicit geological knowledge of the range of actual compositions or the use of additional constraints that preclude impossible solutions.

The interrelationships between component volumetric estimates and logs at differing degrees of indeterminancy can be illustrated by an example case study. The gamma-ray, photoelectric-factor, density, and neutron porosity logs of a Mississippian section (Figure 4.4) show variations in porosity in a succession of cherty carbonate lithologies that are the final result of episodes of deposition, diagenesis, and weathering. The section is capped by spiculitic chert underlain by dolomitic cherty limestone and cherty limestone, which becomes progressively less cherty with depth.

The minimum-variance compositional-analysis solution for the mineral and porosity variation of the Mississippian section are shown in Figure 4.5. In the extreme limiting case, no logs are available and so the solution partitions equal proportions between the dolomite, chert, calcite, and pore components. When one log is used, for example, the density log, information is introduced that is reflected in the compositional solution. Notice that the porosity estimate is reasonable (even if inaccurate) because of the major difference between the pore-fluid density and the mineral densities, as contrasted with the relatively smaller density differences between calcite, dolomite, and quartz, which causes poorly resolved variation in the mineral composition. Note that this would not be the case in a San Andres formation application because of the comparatively high-density contrast between anhydrite, dolomite, and gypsum, which, of course, was the original work with composition analysis

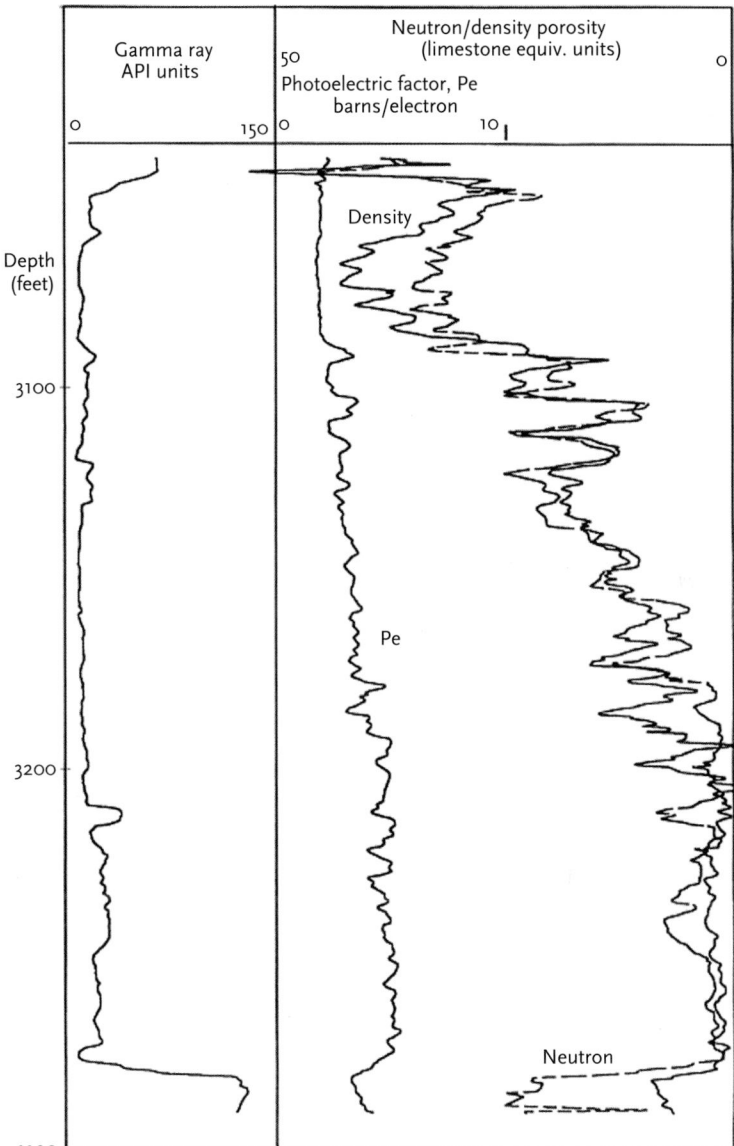

Figure 4.4: Gamma-ray, photoelectric-factor, density, and neutron porosity logs of a Mississippian section of cherty carbonate lithologies in southern Kansas.

that provided sound estimates of porosity. With the application of both density and neutron logs to the Mississippian section, there is an improvement in the mineral volumetric estimates, with better differentiation of dolomite from chert, but with some degree of ambiguity that is finally resolved by adding the photoelectric factor to create a fully determined solution system.

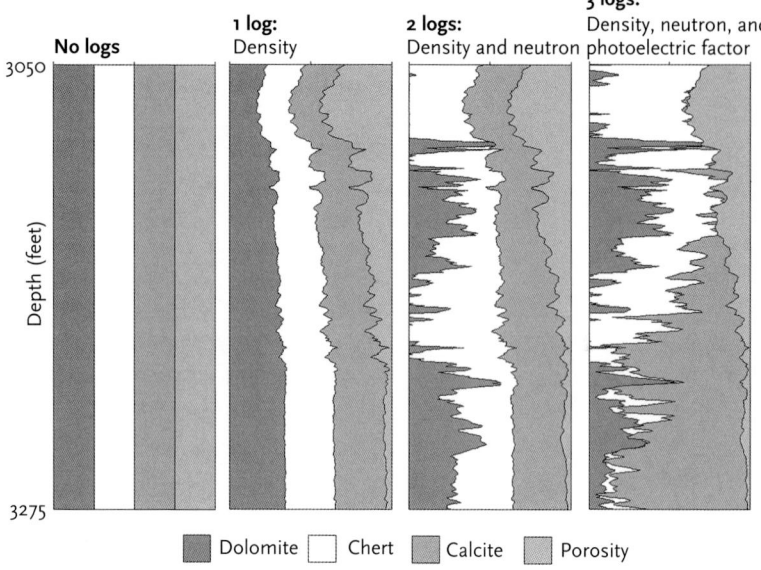

Figure 4.5: Minimum-variance compositional-analysis solutions for the mineral and porosity variation of the Mississippian section for underdetermined systems of no logs, one log (density), two logs (density plus neutron), compared with a determined system solution from density, neutron, and photoelectric-factor logs.

OVERDETERMINED SYSTEMS

Many rocks are dominated by a relatively small number of components, so that the number of logging-tool measurements may exceed the number of significant lithological components. The situation becomes overdetermined when the number of log-response equations is greater than the number of components. The appropriate solution is then the one that most accurately reproduces the original logs when the logs are calculated as predictions from the compositional solutions. Using conventional statistical theory, this solution is the one that minimizes the sums of squares of the deviations between the original logs and their predictions. The least-squares solution is given readily by the matrix algebra equation:

$$V = \left(C^t C\right)^{-1} C^t L$$

where the terms are the same as those in both the determined- and underdetermined-matrix algorithms given earlier. The matrix formulation requires an additional weighting function to allow for the fact that the logging measurements are recorded in radically different units. Without any weighting, the error minimization is predicated on equal units and results in a solution that preferentially honors logs with the highest range of data. The modified least-squares algorithm is then:

$$V = (C^t W C)^{-1} C^t W L$$

where W is a diagonal matrix that contains the elements of a weight vector (Harvey et al., 1990). The weights may be assigned based on physical first principles or by a standardization scheme, such as a transformation from the original measurement to a scale anchored to the mean and counted in units of standard deviation.

For any given zone, the sum-of-squares error is given by:

$$e = (L - \hat{L})^t (L - \hat{L})$$

where $\hat{L}$ is the vector of log responses associated with the least-squares solution. The error term can be plotted as a monitor log to highlight zones where there are striking inconsistencies between the model and the log responses. The overall performance of an algorithm may be judged from the standard error, computed from the summed zone errors as:

$$s_e = \sqrt{\frac{\Sigma e}{(n - m - 1)}}$$

where n is the number of observations and m is the number of logs.

OPTIMIZATION MODELS FOR COMPOSITIONAL SOLUTIONS

Current compositional-analysis procedures have moved beyond the simple inversion algorithms described above, so that constraints and tool-error functions have been incorporated as part of the solution process. The methodology was first developed by Mayer and Sibbit (1980), who applied modified steepest-descent strategies to hunt for an optimal solution that minimized the "incoherence" between the logs and their predicted values. For any given log, the incoherence function is given by:

$$I_A = \frac{(a - \hat{a})^2}{(\sigma_A^2 + \tau_A^2)}$$

where I_A is the incoherence for log A, a is the log response for the zone, $\hat{a}$ is the prediction of a, and σ_A^2 and τ_A^2 are the uncertainties associated with the log measurement and the response equation, respectively.

The uncertainty term for each log measurement is compounded from the sources of sensor error, data acquisition, and the dispersions associated with environmental corrections. The response-equation dispersion represents the uncertainties introduced by linear approximations, erroneous choices of component log responses, and hidden factors, such as the influence of textural parameters. It seems reasonable to suppose that these two types of uncertainty are independent, so that they can be summed as one total error term for each tool:

$$u_A^2 = \sigma_A^2 + \tau_A^2$$

The total log incoherence for any particular depth zone is the sum of the separate log incoherences:

$$I_t = I_A + I_B + I_C + \cdots$$

The form of the equation shows that the solution will tend to be most strongly influenced by the logs to which the most confidence can be attributed. Logs with large errors will have greater incoherences and will contribute more to the total incoherence term.

Constraints are also included and take the general form of:

$$g_i(v_i) \geq 0$$

where g_i is some function that constrains the value of the unknown proportion of the ith component. Rigid, mathematical constraints are those that preclude the occurrence of proportions that are negative or those that exceed unity. Geological and local constraints incorporate relations that conform to general geological principles or prior knowledge of local geology. These geological constraints are more generalized, so that appropriate uncertainties are assigned to them. The constraint dispersions generate additional incoherence terms to be considered. A combined-incoherence function is then the sum of the log and constraint incoherences:

$$I_t = \sum \frac{(a_i - \hat{a}_i)^2}{(\sigma_i^2 + \tau_i^2)} + \sum \frac{g_j(v_j)^2}{\tau_j^2}$$

Notice that if the system is fully determined, then the total incoherence will be zero, provided that no constraints are violated. This special situation is the limiting case of applications that are otherwise presumed to be overdetermined. In a routine application of the optimization algorithm, the number of logs would be expected to exceed the number of components. In part, this is feasible because the bulk of rock compositions tend to be dominated by relatively few components. In addition, the range of wireline measurements used today typically extends beyond the traditional porosity logs to resistivity, spectral, gamma-ray, and geochemical logs.

The optimization method of Mayer and Sibbit (1980) is an iterative search procedure. The system model of input logs and output components are first defined, then the incoherence values associated with each log type are entered, together with the constraints to be met. For each zone, an initial composition is estimated by an approximate method and used as the starting point for a sequence of intermediate solutions. At each step, the incoherence is calculated between the input log responses and those predicted from the solution. A gradient is also computed as the means to generate the next solution, using a steepest-descent technique. The process terminates when it is determined that convergence has been satisfied, at which time there is no appreciable difference between successive solutions. The final solution will be approximate, but the total incoherence between the logs and the compositional estimate will be the minimum possible. The combined display of real and theoretical logs is invaluable as a quality control mechanism to alert the user to problem zones that may be optimal, but are flatly wrong. The generality of the approach allows alternative and remedial attempts to be made without major difficulty.

In further refinements, Gysen et al. (1987) described an extension of the method to the simultaneous optimization of component proportions and response parameters. Moss and Harrison (1985) also reported a technique for solving the uncertainty multipliers that contain the total error associated with each tool. Although the errors cannot be solved for every depth zone, they can at least be estimated for selected intervals and assumed to be effectively constant between zones.

MULTIPLE-MODEL SOLUTIONS OF ROCK COMPOSITION

Clearly, it would be both impractical and unreasonable to incorporate all minerals that could possibly occur into a single and universal solution. Instead, mineral associations that typify common lithological successions are grouped as the basis for alternative compositional models. The fundamental subdivision differentiates between clastic, carbonate, and evaporite models. A specific model may be selected based on prior geological knowledge, or several alternative models may be evaluated and the choice of the final model made from a mix of statistical and geological judgment.

The mineral properties that are listed represent ideal values with differing degrees of variability, especially since field occurrences of minerals often differ from museum-quality specimens. So, for example, the density of dolomite can be quite variable because the mineral ranges from calcium magnesium carbonate towards ankerite with increasing amounts of iron substitution. The tabulated value of 2.88 gm/cc is high and more typical of an iron-rich dolomite from West Texas, but grain densities of about 2.85 gm/cc are more common in Paleozoic dolomites of the midcontinent, as shown by core grain-density measurements.

Phyllosilicate minerals pose a difficult problem because their composition is so variable. However, the clay-mineral properties listed provide a useful reference standard for estimating the hypothetical composition volumes in the absence of explicit information keyed to the formation that is being analyzed. The estimates can be considered as normative, as contrasted with modal predictions of clay-mineral proportions based on X-ray diffraction analyses from the core.

Optimal, minimum-error solutions are worthless if the component model is incorrectly specified. Meaningful results are best obtained by patient geological evaluation of a sequence of solutions where the results of each are used to improve the successive solution. Modern compositional analysis software utilizes the power of the error-minimization method, but allows user interaction so that alternative geological models can be compared.

Quirein et al. (1986) described the use of quadratic programming techniques and linearized response equations as an improvement over the penalty-constraint approach used by earlier methods. In addition, they incorporated a program to solve for poorly known log responses of a component subset, as an optimization procedure applied to specific depths that could be used for calibration. These calibration intervals are those in which both logs and compositions are known and are most typically those that have been cored. In addition, knowledge of composition can be utilized from other sources. Not all component log responses need to be estimated, since their properties are restricted to a limited range. However, a subset of mineral components has ambiguous and locally variable properties. The most notorious of these components are clay minerals, and these will be discussed more fully in the following section.

In common with earlier optimization methodologies, the system is assumed to be either determined or overdetermined. The use of multiple alternative models then allows a more realistic treatment of this assumption, in which common associations can be modeled in parallel and a final selection made between them at any depth. Wherever possible, each separate model is designed to be close to fully determined, in an attempt to find a good match and to sidestep problems associated with the estimates of log and equation dispersions (Marett and Kimminau, 1990). The appropriate logs for each model are clearly those that discriminate well between the separate components. If a poor choice of logs is made, then the model is ill conditioned. The model structure can be checked through the computation of the condition number of:

$$C^t DC$$

where C is the matrix of component log responses and D is a matrix of uncertainty values. The condition number is higher for ill-conditioned models and gives a measure of the sensitivity of proportion estimates to small changes in component log responses (Quirein et al., 1986). The choice between alternative models for any zone can be made by the user, based on an assessment of the relative incoherence of the solutions and their feasibility as reasonable geological descriptions. Alternatively, the decision can be made on the basis of probability established either from comparison of alternative solutions or the use of a Bayesian prior probability.

While generally still applied to an overdetermined system, the multiple models are not far removed from determined matches of components and logs. Where a model becomes determined, the solution is that of a simple and fast matrix inversion with zero incoherence, provided that the nonnegative constraint is not violated. The analysis of the relative conditioning of the model system is a valuable mathematical contribution to the determination of which logs provide the maximum discrimination of model components that will lead to the most stable estimates of volumetric proportions.

ELUCIDATION OF CLAY MINERALS

Shales are composed typically of a mixture of clay minerals, quartz, carbonates, and iron minerals, as well as other accessory components. Clay minerals are markedly different from other rock-forming minerals in terms of both their complexity and their variability. Shales present special problems for log interpretation, and while many algorithms have been designed for their volumetric estimation, the meaning and limitations of their results should be understood.

The log most commonly used to estimate shale content is the gamma-ray log, which sums the natural radiation from potassium-40 and isotopes of the uranium and thorium series. The majority of shales are mixtures of clay minerals and a silt fraction that is typically composed of quartz, calcite, feldspar, iron oxides, and other materials. Most, but not all, of the radioactivity is associated with the clay-mineral content. Yaalon (1962) found an average silt content of 41 percent from his study of thousands of shales. The standard (but arbitrary) scale that is universally used for gamma-ray logs is set by the primary calibration test pit at the University of Houston, where a radioactive cement calibrator is assigned a value of 200 API units; the scale was conceived originally so that a typical midcontinent shale would register at about 100 API units (Ellis, 1987).

In cases where a spectral gamma-ray log has been recorded, the measured gamma rays are subdivided between the contributions from potassium (in percent or proportion), uranium, and thorium (the last two in parts per million). Expressed in these units, logs of subsurface shales can be compared with laboratory data for elemental concentrations in shales measured from a variety of geological studies.

A conventional gamma-ray log in API units can be approximately reconstructed from the elemental abundances by multiplying their estimates by eight for uranium (ppm), four for thorium (ppm), and sixteen for potassium (%), and then summing their contributions (Luthi, 2001). This relationship provides a useful method for predicting subsurface gamma-ray logging values of shale samples from outcrop and core, based on laboratory geochemical measurements. Analyses of the North American shale composite (NASC) reference standard (Gromet et al., 1984) reported values of Th 12.3 ppm, U 2.66 ppm, and K 3.2 percent, which converts to an equivalent standard gamma-ray (SGR) log reading of 121.7 API units. By way of comparison, the Marine Sciences group's black shale composite (BSC) described by Quinby-Hunt et al. (1989) is characterized by values of Th 11.6 ppm, U 15.2 ppm, and K 2.99 percent, which is equivalent to an SGR log reading of 215.84 API units. As might be expected, the potassium and thorium contents of this composite are not markedly dissimilar from the NASC standard. However, the high variability of the content of uranium among the black shales results in a broader range for this standard, so that a maximum equivalent of about 3,700 API units would be expected for the black shale with the most elevated uranium content that was reported by Quinby-Hunt et al. (1989).

In more detailed work, the older and broader methods of shale evaluation have been expanded to the quantitative assessment of clay-mineral species. Clay minerals show differing degrees of variability, but are generally subdivided between four major types: illite, smectite, kaolinite, and chlorite. Clay-mineral typing is based on several log criteria, which must be considered carefully and collectively. Ellis (1987, p.460–461) noted that the four principal clay-mineral types could be combined into two types, based on their hydroxyl content. Kaolinite and chlorite have eight hydroxyls, as contrasted with four for smectite and illite. The neutron log is sensitive to this difference, which can be used as one diagnostic guide, through comparison of the neutron and density porosities when they are both scaled with respect to a quartz matrix. The photoelectric factor is also a useful clay discriminator because of its control by the aggregate atomic number. Ellis (1987, p.451–454) pointed out that iron-free aluminosilicate clays would have photoelectric absorption characteristics that are virtually the same as for quartz. Therefore, variations in the photoelectric factor within shales are primarily a reflection of iron content. Overall, there is a tendency for a progressive increase in iron from low values in kaolinite, through smectite and illite, to high values for iron-bearing chlorite. Distinctions between clay minerals can also be made on the basis of spectral gamma-ray logs, particularly in the differentiation of relatively potassium-rich illites, from low-potassium kaolinite and chlorite.

The quantitative estimation of clay-mineral abundances from the neutron, density, photoelectric-factor, and spectral gamma-ray measurements is fraught with difficulty. Wide compositional changes within clay-mineral groups pose special problems. Useful quantitative models are not easy to define and are frequently ambiguous in their interpretation. The most realistic approach would be to coordinate log measurements with laboratory analyses of core samples. The core values may be idealized as a calibration standard in the development of a statistical prediction model for clay minerals from logs. Even this strategy must be considered thoughtfully and honestly. The most widely used laboratory method for estimating quantities of clay minerals is that of X-ray diffraction. Even with careful sample preparation procedures, the error of clay-mineral estimates from X-ray diffraction can be routinely expected to be 50 percent or more of the reported value (Eslinger and Pevear, 1988, p.A-24). Nevertheless, an important result is that at least the appropriate mineral subset can be identified with some confidence. This ensures that the correct components will be selected for compositional analysis from logs. Reconciliation of the log estimates with X-ray diffraction analyses should then be made within a model that attributes appropriate error magnitudes to both data sources.

A case study of this type of application was made in a well that penetrated a Lower Cretaceous clastic succession with a wide range of clay-mineral types. The Lower Cretaceous Dakota formation in Kansas is the record of a complex of deltaic deposits, with fluvial sandstones, alluvial-plain clays, and estuarine and paralic units that underlie extensive marine shales. The logs and core summary from a hydrology observation well (Figure 4.6) show a fairly typical succession of interbedded sandstones and shales formed in a variety of depositional environments. Clay mineralogy within the shales is highly variable, so that there is relative enrichment in kaolinite within the thick paleosols, which contrasts with the more illitic marine shales. Smectite is common throughout the succession, and bentonites occur at some horizons as records of volcanic ash falls. Sample analyses of mineral composition from X-ray diffraction were used to calibrate the logs in the well (Hoth and Doveton, 1999). The results could then be used to interpolate composition in this well between sample points and to provide a composition predictor in neighboring wells that were logged, but either not cored or not sampled. The mineral composition of the Dakota formation based on X-ray diffraction analysis is shown in Figure 4.7, together with sedimentary environment interpretations from core observations.

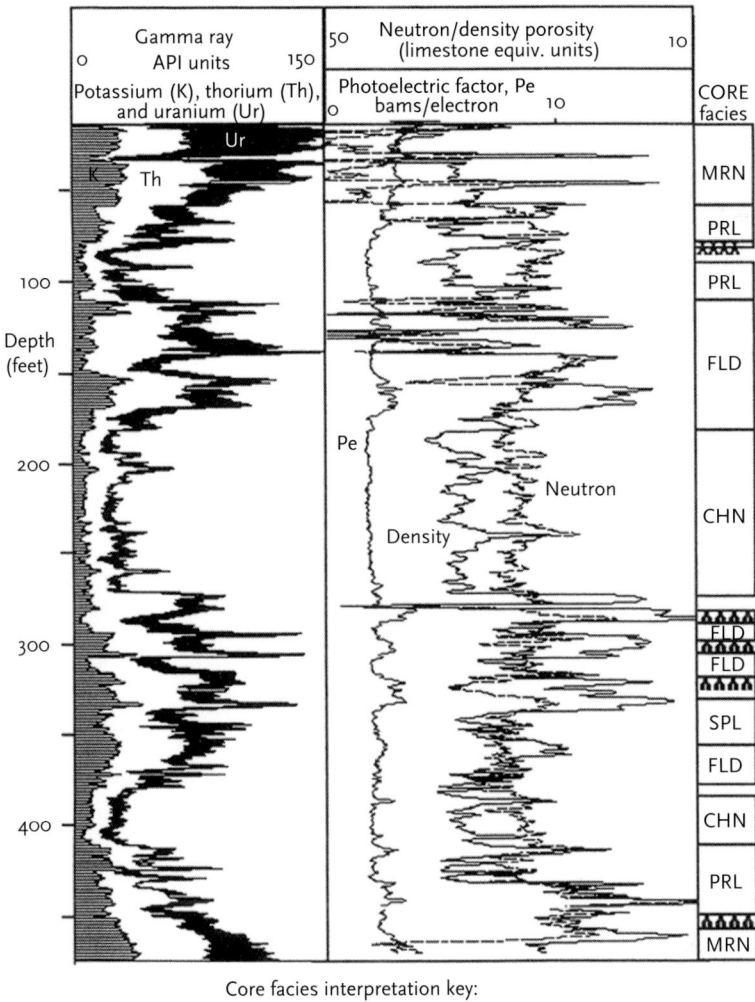

Figure 4.6: Spectral gamma-ray, photoelectric-factor, density, and neutron porosity logs of the Dakota formation (Lower Cretaceous), together with sedimentary environment interpretations of core in a Kansas hydrologic observation well.

Potassium, uranium, thorium, density, neutron porosity, and photoelectric-factor logs were used as predictors to estimate proportions of organic carbon, illite, smectite, kaolinite, iron minerals, and quartz. The model is fully determined, but the mineral properties have different degrees of uncertainty associated with them; for example, quartz coefficients can be set with some confidence, while clay-mineral properties are more speculative. The initial values of the coefficient matrix, drawn from various sources, are shown in Table 4.1. The basic inversion algorithm described can be adapted to include an iterative procedure in which the log properties of the

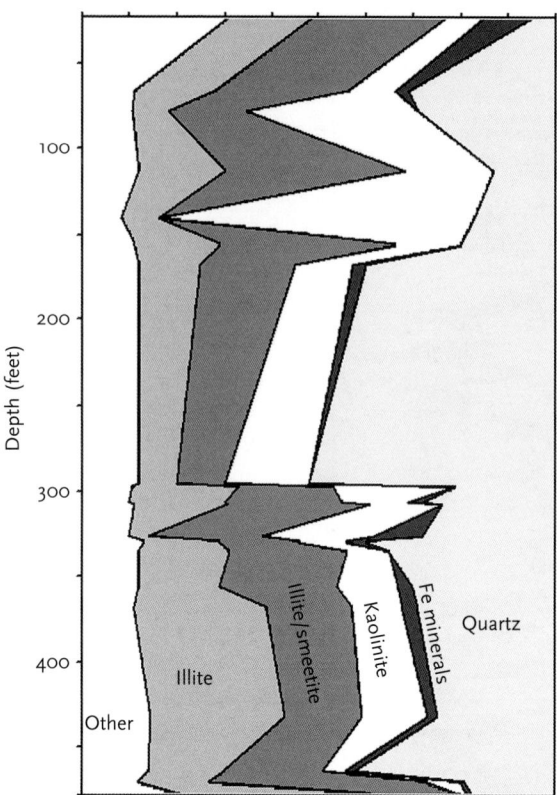

Figure 4.7: Modal mineral model: X-ray diffraction analyses from core samples of the Dakota formation.

component assemblage are successively modified within physically reasonable constraints to produce an optimal least-squares match with core measurements. The iterative search can be made by a Newton-Raphson method, which is coded as the Solver option in the Excel software package. The best-fit least-squares solution was matched with the modified coefficient matrix, as shown in Table 4.1. The overall performance of this matrix as a predictor can be gauged by creating a crossplot of the log predictions of mineralogy with their X-ray diffraction (XRD) measurements (Figure 4.8). Discrepancies are inevitable, not only because of logging errors, but because of residual depth misregistration, differences in sample size, statistical errors of XRD analyses, and other factors. However, attention should be focused on the general power (or lack thereof) of prediction and the distinction between minerals that are estimated crisply and those that defy estimation.

Following the optimal least-squares calibration of the logs with XRD volumetric estimations to establish the mineral-component log properties, the entire section was processed using the standard inversion procedure. The compositional profile (Figure 4.9) shows patterns of variation that supply useful supplementary information to the features seen in the core. An illite-smectite marine shale at the base

Table 4.1. INITIAL AND MODIFIED COEFFICIENT MATRICES OF URANIUM (*URAN*), PHOTOELECTRIC-FACTOR (*PEF*), POTASSIUM (*POTA*), THORIUM (*THOR*), NEUTRON POROSITY (*NPHI*), AND DENSITY (*RHOB*) VALUES OF ORGANIC CARBON (*ORG*), IRON MINERALS (*FE*), ILLITE (*I*), ILLITE-SMECTITE (*I-S*), KAOLINITE (*KA*), QUARTZ. (*Q*), AND POROSITY (*PHI*)

	org	Fe	I	I-S	Ka	Q	phi
Initial Coefficient Matrix:							
URAN	130	0	4.5	3.5	5.5	0	0
PEF	0.20	13	4.5	3	1.9	1.8	0
POTA	0	0	5.5	4	0	0	0
THOR	0	0	18	18	15	9	0
NPHI	50	13	20	45	35	-2	100
RHOB	1.20	4.90	2.66	2.61	2.59	2.65	1
UNITY	1	1	1	1	1	1	1
Modified Coefficient Matrix:							
URAN	130	0	8.3	2.8	9.7	0	0
PEF	0.50	20.99	6.76	1.80	2.88	1.80	0.50
POTA	0	0	6.25	3.80	0	0	0
THOR	0	0	37.47	12.52	43.82	1	0
NPHI	45.0	15.0	20	55.3	30.0	-5.0	100
RHOB	1.20	4.90	2.66	2.61	2.59	2.65	1
UNITY	1	1	1	1	1	1	1

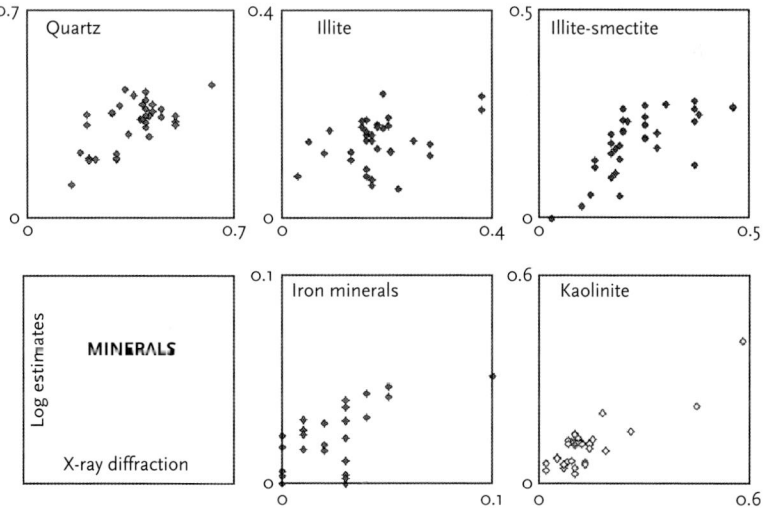

Figure 4.8: Crossplots of mineral analysis from X-ray diffraction (X) versus prediction from calibrated inverse solution prediction from logs (Y).

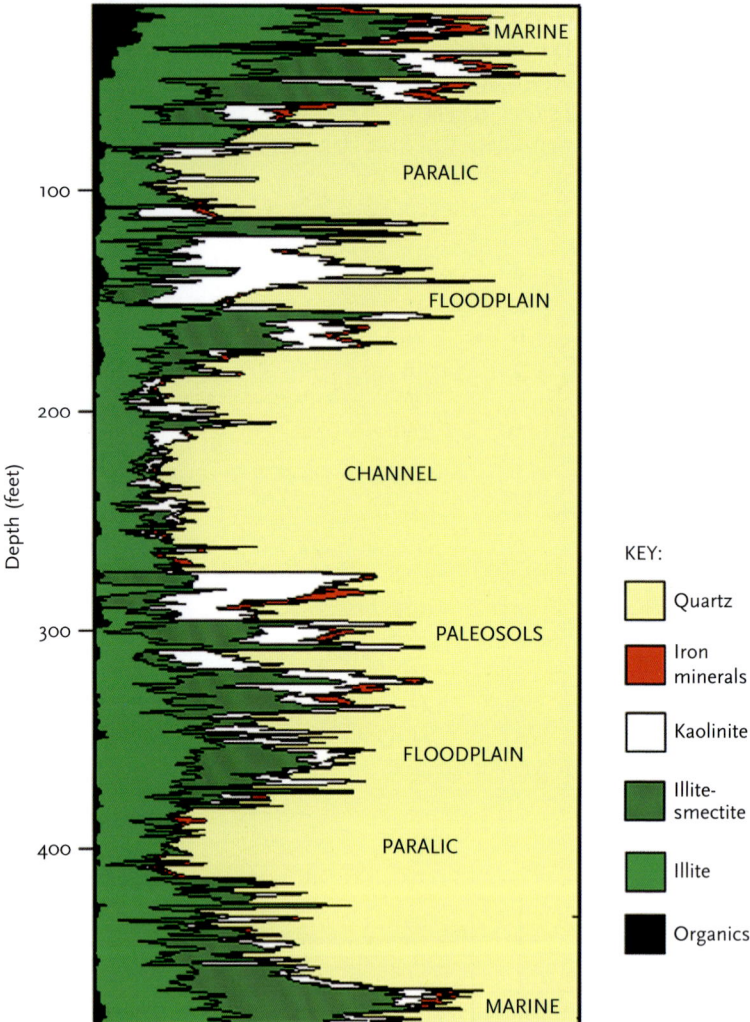

Figure 4.9: Composition log of Dakota formation section based on inversion of spectral gamma, photoelectric, density, and neutron porosity logs calibrated to mineral estimates from X-ray diffraction of core samples. Sedimentary environments were assigned from visual core interpretation. From Hoth and Doveton (1999).

is succeeded by paralic and channel sandstones, which are overlain by a stacked paleosol sequence of hematite-stained paleosols, with a pronounced upward trend of increased kaolinite/decreased illite that probably reflects soil-leaching processes. The paleosol sequence is cut unconformably by a large channel sandstone, which is overlain by flood-plain facies with variable clay composition. Finally, the sequence is terminated by a paralic sandstone and a major transgressive marine shale in which organic carbon contents are elevated through preservation under reducing conditions on the sea-floor. The data from this well provides both calibration and prediction, but the calibrated inversion matrix can also be applied to Cretaceous sections in other Kansas wells that were logged but not cored, in order to track regional changes in mineral assemblages in both time and space.

COMPOSITIONAL ANALYSIS FROM GEOCHEMICAL LOGS

Geochemical logging tools measure induced gamma-ray spectra that are created when a formation is bombarded by high-energy neutrons from a pulsed electronic source. A matrix-inversion spectral-fit algorithm then separates the spectrum into individual elemental sources. The major rock composition elements of silicon, calcium, magnesium, iron, sulfur, titanium, and carbon are estimated together with the rare earth element, gadolinium. In addition, potassium, thorium, and uranium can be estimated from the natural gamma rays emitted by formations and measured by the spectral gamma-ray log. As a consequence of the direct relationship between elemental data and mineral compositions, more realistic mineral transforms have been developed that are a major improvement on models based on mineral properties. However, a distinction must be made between normative minerals that are computed from transforms of elemental data and modal minerals that are observed visually or by petrographic laboratory methods, such as X-ray diffraction or infrared spectroscopy. Clearly, the fundamental goal of an effective transform is to provide a close match between normative mineral solutions and modal mineral suites.

The calculation of "normative" minerals from oxide analyses has been a standard procedure in igneous petrology since the Cross-Iddings-Pirsson-Washington (CIPW) norm was introduced by Cross et al. (1902). These normative minerals are contrasted with modal compositions that are commonly measured by point-counting of minerals in thin sections of rock. The normative concept has also been extended to sedimentary rocks in attempts to compute realistic mineral assemblages. Krumbein and Pettijohn (1938, p.490–492) explained the molecular ratio method to calculate the probable mineral composition of a rock, based on chemical analyses of oxide percentages. As a first step, the minerals to be resolved are identified from thin-section observation or other sources of information. The molecular ratios are then assigned in a stepwise fashion to the minerals. The process consists of a logical order of steps that first accommodates unique associations between oxides and certain minerals, and then allocates the remainder to other components. Imbrie and Poldervaart (1959) described a commonly used method of sedimentary normative analysis and then compared the results with modal estimates from mineralogy. From a detailed study of the Permian Florena shale, they concluded that estimates of the chert, calcite, dolomite, and clay had errors of less than 5 percent. However, there was little agreement between the clay-mineral proportions that were computed and those produced from X-ray diffraction analysis. Imbrie and Poldervaart (1959) were not surprised by this discrepancy, but attributed it to the known high variability of the composition of clay-minerals, which is due to isomorphous substitution.

Essentially the same problems are tackled in the computation of sedimentary normative minerals when these are based on elements measured by geochemical logs (Herron, 1986). However, many of the older normative methods predated computers. The classical norm calculation is subtractive, deterministic, and rigidly leveraged. As discussed by Harvey et al. (1990), the method can be useful when certain elements can be assigned totally to single individual minerals. These assignments can then be made by following an ordered protocol of analysis to partition them between the mineral species. Otherwise, the use of simultaneous equations to link mineral compositions with elemental measures is a much more general and powerful method. The speed of modern software also allows real-time interaction between petrophysicist and machine, so that alternative models can be evaluated quickly and decisions can be made that blend mathematical optimality with geological credibility. Any analysis should be preceded by some notion of what constitutes a fit-for-purpose estimation. Less accuracy is needed if the intent is for a generalized semiquantitative description of variation, rather than more rigorous estimates for use in quantitative basin modeling or physical property predictions (Harvey et al., 1998).

A model that links minerals with elements can be set up as a fully determined system and solved by standard matrix inversion using methods described previously. Whenever the components are computed as positive proportions, then the compositional solution is rational and honors the analysis perfectly. However, in common with the normative model, any apparent precision read into the result is illusory because the determined system makes no allowance for analytical error. It is usually practical to model a rock with a set of minerals that are fewer in number than the elements available from geochemical logging. The system is then overdetermined and can be resolved by one or another of a variety of optimization techniques. The additional complexity in computation is offset by several distinct advantages. The overdetermination allows constraints and error functions to be incorporated, both for optimal solution control and for diagnostic evaluation of the sources of analytical error. The choice of an overdetermined system also provides better assurance of a stable solution in situations where the mineral response matrix becomes sparse or there are potential compositional colinearities that link some of the mineral subsets (Harvey et al., 1990).

Strictly speaking, there will almost always be more minerals than elements that can be used to solve for them, so the problem is always underdetermined. However, as Herron (1988) noted, the overwhelming majority of sedimentary rocks are composed of only ten minerals: quartz, four clays, three feldspars, and two carbonates. In practice, reasonable compositional solutions can be generated using relatively small subsets of minerals, provided that they have been identified correctly and that the compositions used are both fairly accurate and constant. Alternatively, the inversion procedure can be run as an unconstrained procedure in which components with negative proportions are eliminated from the model. Harvey et al. (1998) found this approach to be successful but cautioned that negative components should be eliminated one at a time, starting with the largest negative component, because of interactions between the components.

Mineral solutions may be calculated by two alternative strategies. In the first, the average chemical compositions of minerals drawn from a large database are used as endmember responses and resolved by standard matrix-inversion procedures. This result is normative and generic in the sense that it is based on a sample drawn from a universal mineral reference set and applied to a specific sequence where local mineral compositions may deviate from the global average. The result is hypothetical but has the particular advantage that comparisons can be made between a variety of locations and do not require expensive ancillary core measurements. New methods of classification may also be necessary, as discussed by Herron (1988) in his study of terrigenous sands and shales in terms both of core and geochemical log data.

In a second approach, the solution is calibrated to core data, where laboratory determinations of mineralogy and elemental geochemistry are analyzed by multiple regression techniques to determine local mineral compositions. This result is linked to petrography and so is philosophically closer to an estimated modal solution rather than the more hypothetical normative model. As mentioned earlier, realistic statistical calibration models should incorporate error terms from all sources of measurement. When geochemical logging was first introduced, several detailed studies were made to assess the strengths and limitations of borehole geochemistry through exhaustive comparisons with core elemental and mineralogical analyses. These included comparisons in the Conoco Research well, Ponca City, Oklahoma, by Hertzog et al. (1987); the discussion of the results from an Exxon research well that penetrated Upper Cretaceous siliciclastic rocks in Utah, by Wendlandt and Bhuyan (1990); and an assessment of data from three Shell wells in the Netherlands, Oman, and the United States, by van den Oord (1990).

There are several ways to assess modal mineralogy, so which constitutes the most accurate method to use as a standard for determining the real mineral composition? Harvey et al. (1998) addressed this problem when they compared core data from the spectral measurements of quantitative X-ray diffraction and infrared spectroscopy, as well as micrometric analysis from thin-section point counts. Overlapping peaks and poor resolution at low-resolution pose special problems for the spectral methods, while appropriate sample sizes must be observed to produce robust statistics in micrometric analysis. Also, the distinction between volume percentage and weight percentage must be observed when interrelating modal and normative compositions. Harvey et al. (1998) concluded that the results of their study did not favor one method over another, but pointed out that their comprehensive analysis demonstrated the difficulty of obtaining accurate modal estimates and even the notion of what constitutes the "real" mineral composition. This is certainly worth bearing in mind when making a judgment about the "accuracy" of a normative mineral solution from inversion of log responses. So, for example, mismatches in clay-mineral estimates by log inversion in the Dakota formation described earlier represents a failure to reproduce the results of quantitative X-ray diffraction, which are themselves only estimates of the true composition.

A major obstacle in the production of unique mineral transformations from element concentrations has been the problem of compositional colinearity. Ambiguities in the separate resolution of illite, mica, kaolinite, and K-feldspar by silicon, aluminum, and potassium can be understood when these minerals are plotted on a ternary diagram (Figure 4.10). Illite is located at a position that is intermediate between K-feldspar and kaolinite. If precisely colinear, then an

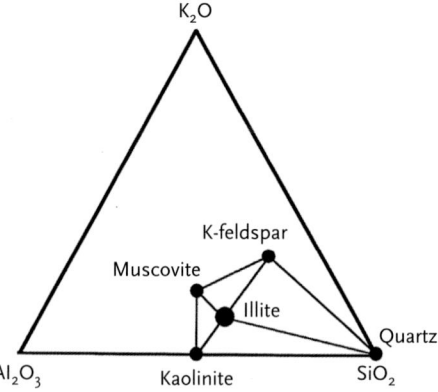

Figure 4.10: Example of a compositional colinearity problem. From Harvey et al. (1995), courtesy the Society of Petrophysicists and Well Log Analysts (SPWLA).

infinite range of solutions is possible, causing a matrix singularity and a breakdown of the inversion procedure. If average mineral compositions are used, a solution becomes possible, but it may be unstable (Harvey et al., 1995). Wendlandt and Bhuyan (1990) found that the use of silicon, potassium, and aluminum tended to result in overestimates of kaolinite; the use of iron to predict illite content caused underestimates of kaolinite. However, effective discrimination between illite and kaolinite contents became possible when dry density was applied as an extra constraint.

There are numerous potential applications of mineral transforms of geochemical logging data in addition to the immediate quantitative rendition of lithofacies. These include quantitative estimates of grain size, cation-exchange capacity, and permeability, using the minerals as surrogates for other petrophysical properties (Chapman et al., 1987). Accurate clay-mineral typing and geochemical clues as to diagenesis have immediate obvious consequences as tools to improve reservoir engineering practice. Selley (1992) considered that the "third age of log analysis" had arrived with the introduction of geochemical logs and that they could be useful discriminators of a variety of diagenetic effects of cementation and solution, especially when used in conjunction with other logs.

The analysis of gas-shale compositions presents special challenges, but the results have great potential economic significance. The distinction between different clay minerals, as well as the nonclay components of quartz, carbonate minerals, and pyrite, adds some complexity to the evaluation of kerogen content and the evaluation of adsorbed and free gas. Quirein (2010) noted that it was imperative to develop an orderly workflow, hopefully guided by core data, in order to avoid "a never-ending journey" and, instead take one of "finite duration." The most difficult aspect is the selection of an appropriate mineral model. An example of the results of a compositional analysis from geochemical and other logs of an organic-rich shale from a southern Kansas well is shown in Figure 4.11.

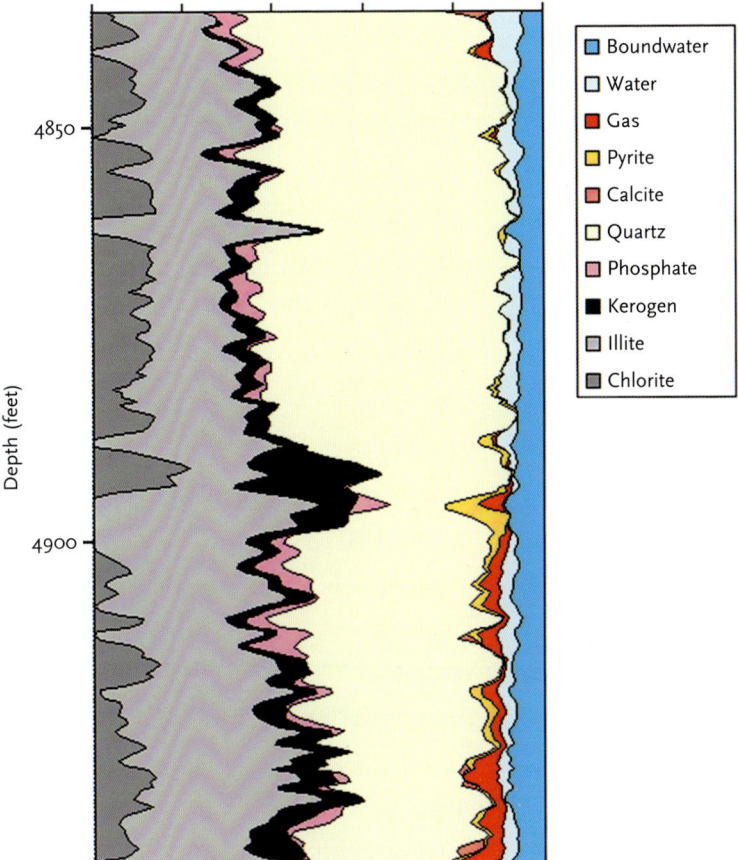

Figure 4.11: Compositional analysis of a Kinderhook/Chattanooga shale sequence in southern Kansas from Schlumberger Spectrolith inversion processing of geochemical logs.

INVERSION MAPPING OF COMPOSITIONS

The conventional result of a compositional analysis from simple inversion is equivalent to a transformation of log curves into a compositional profile graphed as a function of depth. The input vectors of log responses represent zones that are digitally sampled by depth. However, instead of vectors sampled vertically along a depth axis, vectors of log responses can be input from geographic locations across a stratigraphic unit and the compositional results ten interpolated laterally to produce a lithofacies map.

Bornemann and Doveton (1983) described a case study of the application of this mapping paradigm to the lithofacies of the Middle Ordovician Viola limestone in south-central Kansas. Density, neutron, and sonic logs were used to estimate the proportions of calcite, dolomite, and chert, and the pore volume. The logs were first normalized, based on the results of a trend surface analysis applied to a calibration unit (Doveton and Bornemann, 1981). Average normalized log values of the Viola were then interpolated between well controls, using a standard automated contouring

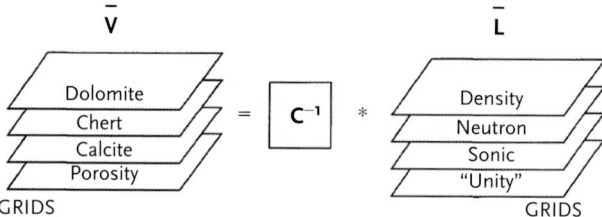

Figure 4.12: Composition inversion of density, neutron, sonic logs, and unity areal grids into four map grids of mineral components and porosity applied as a grid-to-grid operation in a computer contouring software package.

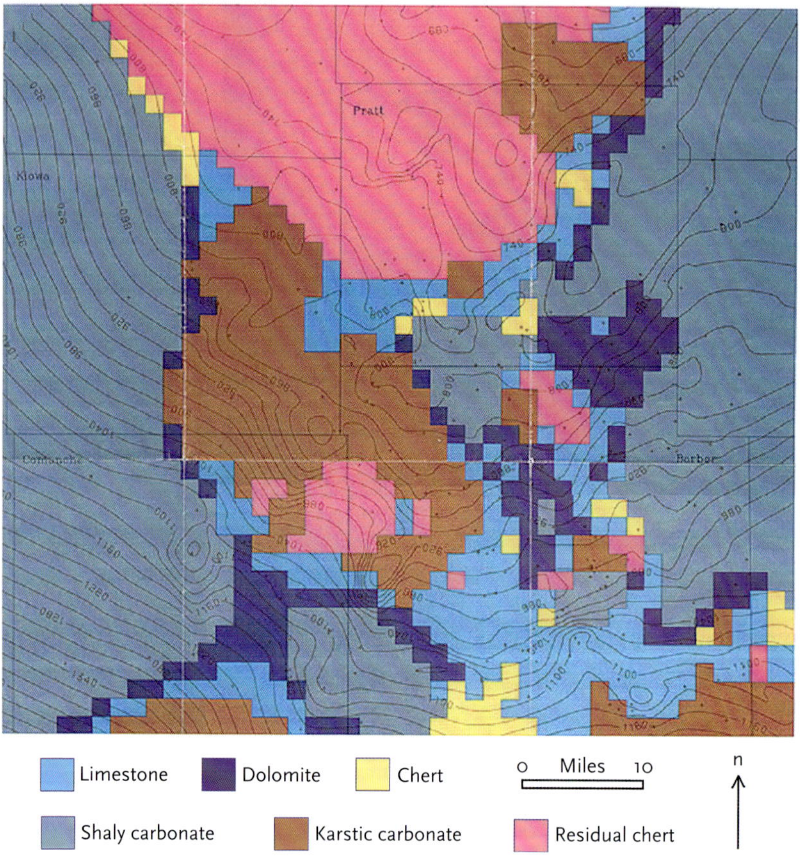

Figure 4.13: Lithofacies map of the Viola limestone in south-central Kansas computed from the inversion of grids of average neutron, density, and sonic transit time.

package. The result of this step was the generation of three grids of average log values. A particular advantage of this approach is that all three logs are not required in any of the wells, because the inverse transformation is applied to the gridded values from interpolation, rather than the values at individual well locations.

The matrix algebra inversion procedure was then applied to the three grids supplemented by a grid of unit values. Each cell log-response vector was processed by the grid-to-grid operation in the production of four solution grids of the compositional proportions (Figure 4.12). The grid values were combined in a single map that is a compositional expression of the lithofacies (Figure 4.13). Each cell was assigned a symbol according to whether the dominant component was calcite, dolomite, or chert. The map was carefully validated, using standard lithological information available from drill cuttings and core records. Three additional lithofacies were identified as a result of negative proportional solutions at a number of the cell locations. Negative dolomite and negative quartz were found to have an excellent match with areas of residual chert and karstically weathered sections. The association is caused by the insensitivity of the sonic measurement to larger pores in these facies. Negative calcite solutions reflected significant occurrences of shale as an additional component in shaly carbonate facies.

REFERENCES

Alger, R.P., Raymer, L.L., Hoyle, W.R., and Tixier, M.P., 1963, Formation density log applications in liquid-filled holes: Transactions of the American Institute of Mining, Metallurgical, and Petroleum Engineers, v. 228, no. 1, pp. 321–332.

Bornemann, E., and Doveton, J.H., 1983, Lithofacies mapping of Viola limestone in south central Kansas based on wireline logs: American Association of Petroleum Geologists Bulletin, v. 67, no. 4, pp. 609–623.

Cannon, D.E., and Horkowitz, J.P.,1997, Complex reservoir evaluation in open and cased wells: Transactions of the Society of Professional Well Log Analysts, 38th Annual Logging Symposium, Paper DD, 14 p.

Chapman, S., Colson, J.L., Flaum, C., et al., 1987, The emergence of geochemical well logging: The Technical Review, v. 35, no. 2, pp. 27–35.

Cross, W., Iddings, J.P., Pirsson, L.V., and Washington, H.S., 1902, A quantitative chemico-mineralogical classification and nomenclature of igneous rocks: Journal of Geology, v. 10, no. 6, pp. 555–690.

Doveton, J.H., and Bornemann, E., 1981, Log normalization by trend surface analysis: Log Analyst, v. 22, no. 4, pp. 3–8.

Doveton, J.H., and Cable, H.W., 1979, Fast matrix methods for the lithological interpretation of geophysical logs, in D. Gill and D.F. Merriam, eds., Geomathematical and petrophysical studies in sedimentology: Computers in Geology Series, Pergamon Press, Oxford, pp. 101–116.

Ellis, D.V., 1987, *Well logging for earth scientists*: Elsevier Science Publishing, New York, 532 p.

Eslinger, E., and Pevear, D., 1988, Clay minerals for petroleum geologists and engineers: Socity of Economic Paleontologists and Mineralogists, Short Course No. 2, 410 p.

Gromet, L.P., Dymek, R.F., Haskin, L.A., and Korotev, R.L., 1984. The "North American shale composite": Its compilation, major and trace element characteristics: Geochimica et Cosmochimica Acta, v. 48, no. 12, pp. 2469–2482.

Gysen, M., Mayer, C., and Hashmy, K.H., 1987, A new approach to log analysis involving simultaneous optimization of unknowns and zoned parameters: Transactions of the Canadian Well Logging Society, 11th Formation Evaluation Symposium, Paper B, 20 p.

Harris, M.H., and McCammon, R.B., 1971, A computer-oriented generalized porosity-lithology interpretation of neutron, density, and sonic logs: Journal of Petroleum Technology, v. 23, pp. 239–248.

Harvey, P.K., Brewer, T.S., and Lovell, M.A., 1998, The estimation of modal mineralogy: a problem of accuracy in core-log calibration, *in* P.K. Harvey and M.A. Lovell, eds., Core-log integration: Special Publication 136, Geological Society, London, pp. 25–38.

Harvey, P.K., Bristow, J.F., and Lovell, M.A., 1990, Mineral transforms and downhole geochemical measurements: Scientific Drilling, v. 1, no. 4, pp. 163–176.

Harvey, P.K., Lofts, J.C., and Lovell, M.A., 1995, The characterization of reservoir rocks using nuclear logging tools: Evaluation of mineral transform techniques in the laboratory and log environments: Log Analyst, v. 36, no. 2, pp. 16–28.

Hedberg, H.S., and May, D.H., 1990, Gypsum determination in shallow West Texas carbonates: Society of Petroleum Engineers, SPE 20106-MS, 8 p.

Herron, M.M., 1986, Mineralogy from geochemical well logging: Clays and Clay Minerals, v. 34, no. 2, pp. 204–213.

Herron, M.M., 1988, Geochemical classification of terrigenous sands and shales from core or log data: Journal of Sedimentary Petrology, v. 58, no. 5, pp. 820–829.

Hertzog, R., Colson, L., Seeman, B., O'Brien, M., Scott, H., McKeon, D., Grau, J., Ellis, D., Schweitzer, J., and Herron, M., 1987, Geochemical logging with spectrometry tools, Society of Petroleum Engineers, SPE 16792-MS; Later published in 1989, SPE Formation Evaluation, v. 4, no. 2, pp. 153–162.

Hoth, P., and Doveton, J.H., 1999, Clay mineral estimation from nuclear petrophysical logs using a calibrated numerical inversion procedure: Geophysical Research Abstracts, v. 1, no. 1, pp. 179.

Imbrie, J., and Poldervaart, A., 1959, Mineral compositions calculated from chemical analyses of sedimentary rocks: Journal of Sedimentary Petrology, v. 29, no. 4, pp. 588–595.

Krumbein, W.C., and Pettijohn, F.J., 1938, Manual of sedimentary petrography: Appleton-Century-Crofts, New York, 549 p.

Lucia, F. J., 1999, *Carbonate reservoir characterization*: Springer, Berlin, 226 p.

Luthi, S.M., 2001, *Geological well logs: Their use in reservoir modeling*: Springer-Verlag, Berlin, 373 p.

Marett, G., and Kimminau, S., 1990, Logs, charts, and computers—the history of log interpretation modelling: Log Analyst, v. 31, no. 6, pp. 335–354.

Mayer, C., and Sibbit, A., 1980, GLOBAL: A new approach to computer-processed log interpretation: Society of Petroleum Engineers, SPE 9341-MS, 12 p.

McCammon, R.B., 1970, Component estimation under uncertainty: In: D.F. Merriam, ed., Geostatistics, a colloquium: Plenum, New York, pp. 45–61.

Moss, B., and Harrison, R., 1985, Statistically valid log analysis method improves reservoir description: Society of Petroleum Engineers, Offshore Europe Conference, Aberdeen, SPE 1398-MS, 32 p.

Nissen, S.E., Doveton, J.H., and Watney, W.L., 2008, Petrophysical and geophysical characterization of karst in a Permian San Andres reservoir, Waddell field, West Texas (abs): American Assocociation of Petroleum Geologists Convention Abstracts.

Quinby-Hunt, M.S., Wilde, P., Orth, C.J., and Berry, W.B.N., 1989, Elemental geochemistry of black shales-statistical comparison of low-calcic shales with other shales, *in* R.I. Grauch and J.S. Leventhal, eds., Metalliferous black shales and related ore deposits: US Geological Survey Circular 1037, pp. 8–15.

Quirein, J., Kimminau, S., Lavigne, J., Singer, J., and Wendel, F., 1986, A coherent framework for developing and applying multiple formation evaluation models: Transactions of the Society of Professional Well Log Analysts, 27th Annual Logging Symposium, Paper DD, 16 p.

Quirein, J., Witkowsky, J., Truax, J., Galford, J., Spain, D., and Odumosu, T., 2010, Integrating core data and wireline geochemical data for formation evaluation and characterization of shale gas reservoirs: Society of Petroleum Engineers, SPE134559-MS, 18 p.

Savre, W.C., 1963, Determination of a more accurate porosity and mineral composition in complex lithologies with the use of the sonic, neutron, and density surveys: Journal of Petroleum Technology, v. 15, no. 9, pp. 945–959.

Selley, R.C., 1992, The third age of wireline log analysis—application to reservoir diagenesis, *in* A. Hurst, C.M. Griffiths, and P.F. Worthington, eds., Geological applications of wireline logs II: The Geological Society, London, Special Publication No. 65, pp. 377–387.

van den Oord, R.J., 1990, Experience with geochemical logging: Transactions of the Society of Professional Well Log Analysts, 31st Annual Logging Symposium, Paper T, 25 p.

Wendlandt, R.F., and Bhuyan, K., 1990, Estimation of mineralogy and lithology from geochemical log measurements: American Association of Petroleum Geologists Bulletin, v. 74, no. 6, pp. 837–856.

Yaalon, D.H., 1962, Mineral composition of average shale: Clay Minerals Bulletin, v. 5. no. 27, pp. 31–36.

CHAPTER 5

Petrophysical Rocks: Electrofacies and Lithofacies

FACIES AND ELECTROFACIES

Many years ago, the classification of sedimentary rocks was largely descriptive and relied primarily on petrographic methods for composition and granulometry for particle size. The compositional aspect broadly matches the goals of the previous chapter in estimating mineral content from petrophysical logs. With the development of sedimentology, sedimentary rocks were now considered in terms of the depositional environment in which they originated. Uniformitarianism, the doctrine that the present is the key to the past, linked the formation of sediments in the modern day to their ancient lithified equivalents.

Classification was now structured in terms of genesis and formalized in the concept of "facies." A widely quoted definition of facies was given by Reading (1978) who stated, "A facies should ideally be a distinctive rock that forms under certain conditions of sedimentation reflecting a particular process or environment." This concept identifies facies as process products which, when lithified in the subsurface, form genetic units that can be correlated with well control to establish the geological architecture of a field. The matching of facies with modern depositional analogs means that dimensional measures, such as shape and lateral extent, can be used to condition reasonable geomodels, particularly when well control is sparse or nonuniform. Most wells are logged rather than cored, so that the identification of facies in cores usually provides only a modicum of information to characterize the architecture of an entire field. Consequently, many studies have been made to predict lithofacies from log measurements in order to augment core observations in the development of a satisfactory geomodel that describes the structure of genetic layers across a field.

The term "electrofacies" was introduced by Serra and Abbott (1980) as a way to characterize collective associations of log responses that are linked with geological attributes. They defined electrofacies to be "the set of log responses which characterizes a bed and permits it to be distinguished from the others." Electrofacies are clearly determined by geology, because log responses are measurements of the physical properties of rocks. The intent of electrofacies identification is generally to match them with lithofacies identified in the core or an outcrop. However, correspondences are frequently blurred and care should be taken to select petrophysical measurements for the electrofacies model that are most likely to differentiate the target set of lithofacies.

An important philosophical distinction between electrofacies and "classical" geological facies is that electrofacies are primarily observational in origin, and facies are traditionally rooted in genesis. Electrofacies can be distinctive empirical log-response associations, but whether they reflect depositional environment, diagenetic overprints, or other processes requires careful consideration on a case-by-case basis.

DUNHAM TEXTURES AND ELECTROFACIES

The relationship between lithofacies and petrophysical properties that could be used to construct a matching set of electrofacies can be illustrated by a simple example. The Dunham classification of limestones (Figure 5.1) has been used for many years in the textural description of core and outcrop samples and can be related easily to analogs of modern carbonate environments. The basic range from grainstone to mudstone reflects a decrease in depositional energy and a corresponding increase in mud content. Consequently, the first common petrophysical measurement to be considered would be the gamma-ray log because of its sensitivity to potassium and thorium contents associated with clay minerals. Lucia (1999) was able to distinguish mud-dominated packstones, wackestones, and mudstones from grainstones and grain-dominated packstones in the San Andres and Grayburg formations by using a gamma-ray cut-off of 30 API units, with an 80 percent success rate when matched with core descriptions. He cautioned that the computed gamma-ray (potassium and thorium sources) should be used rather than total gamma rays, since uranium content reflects diagenetic effects.

Grain density and porosity are also useful diagnostic limestone fabric indicators, as shown in Figure 5.2, where average values for Lower Permian Chase Group

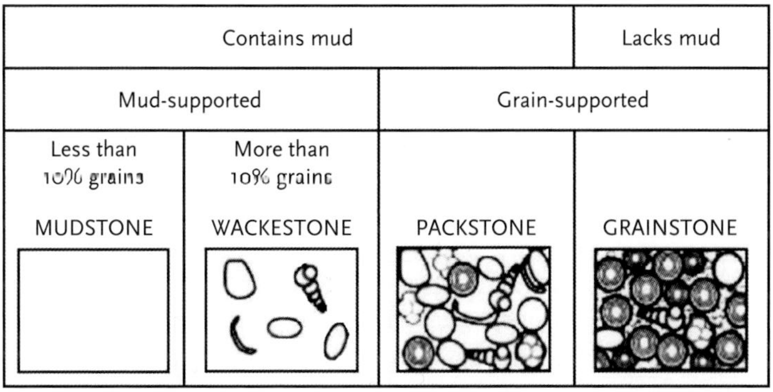

Contains mud			Lacks mud
Mud-supported		Grain-supported	
Less than 10% grains	More than 10% grains		
MUDSTONE	WACKESTONE	PACKSTONE	GRAINSTONE

Figure 5.1: Basic Dunham textural classification of limestones.

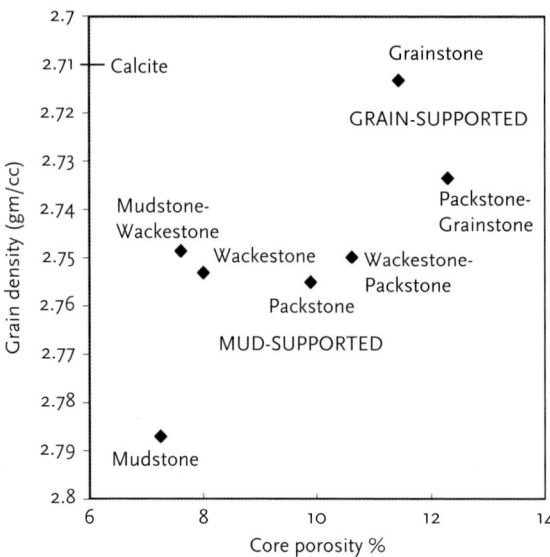

Figure 5.2: Average core grain densities and porosities for limestones from the Lower Permian Chase Group subdivided between Dunham textural classes.

limestones are plotted for Dunham textural types described in the core. Notice that the high-energy, mud-free grainstones have an average grain density close to the density of calcite, but they are contrasted with higher grain densities for packstones and wackestones, with another sharp increase within mudstones. Differentiation within the packstone-to-wackestone trend appears to be linked mainly with a progressive decline in porosity. Since both of these properties are measured by logs, they could be combined with gamma-ray measurements to predict limestone rock fabric, utilizing the core observations for specific assignments and associated probabilities.

PETROPHYSICAL RECOGNITION OF LITHOFACIES

So far in this discussion, we have considered facies variation within a single lithology, but most analyses will include multiple lithologies in logged sequences. Earlier methods of lithology differentiation from logs used crossplots as a pattern-recognition tool, based on density, neutron porosity, and sonic transit-time measurements. The three log responses were reduced to two by elimination of the porosity components and computation of matrix properties, either by an M-N plot (Burke et al., 1969) or an MID plot (Clavier and Rust, 1976). A major drawback of this log combination in both methods was the close correlation between density and transit time for many common minerals, leading to ambiguous mineral resolution because of this tendency to colinearity. Also, the relative insensitivity of the sonic measurement to larger pores, as contrasted with either density or neutron logs, would need to be addressed in vuggy or moldic carbonates.

The introduction of the photoelectric factor log marked a major advance in mineral identification because of its direct relationship with the aggregate atomic number and led to the development of the RHOmaa-Umaa plot. The RHOmaa-Umaa crossplot utilizes the photoelectric index, neutron porosity, and bulk-density curves for matrix mineral evaluation. As with the M-N and MID plots, three logs are condensed to two variables through the partition and elimination of the pore-fluid component, so that matrix properties are computed as estimates of the apparent grain density (RHOmaa) and matrix volumetric photoelectric absorption (Umaa).

As a first step, the photoelectric factor, PeF, recorded in barns per electron, must be converted to a volumetric measure, U, measured in barns per cc. This conversion is made by multiplying by the electron density, ρ_e:

$$U = PeF \cdot \rho_e = \frac{PeF(\rho_b + 0.1883)}{1.07}$$

which is more commonly approximated by multiplying by the bulk density:

$$U = PeF \cdot \rho_b$$

The bulk density, ρ_b, and the volumetric photoelectric absorption, U, are properties of both the matrix and the pore fluid. The elimination of the contribution of the pore fluid to these quantities will yield estimates of the apparent density (RHOmaa) and photoelectric absorption (Umaa) of the matrix. In order to do these estimations, the true volumetric porosity, Φ_t, must first be interpolated between lithology lines on a neutron-density crossplot or be approximated by an average. RHOmaa and Umaa can then be calculated as follows. Because:

$$\rho_b = \Phi_t \rho_f + (1 - \Phi_t) RHOmaa$$

then:

$$RHOmaa = \frac{(\rho_b - \Phi_t \rho_f)}{(1 - \Phi_t)}$$

and because:

$$U = \Phi_t U_f + (1 - \Phi_t) Umaa$$

then:

$$Umaa = \frac{(U - \Phi_t U_f)}{(1 - \Phi_t)}$$

The density of the pore fluid, ρ_f, can be taken to be that of the mud filtrate, which will be about 1 gm/cc in a fresh water mud. The fluid photoelectric absorption, U_f, will also reflect the fluid character of the flushed zone, which for mud filtrate is approximately 0.5 barns/cc.

As an example of interrelating electrofacies and lithofacies on a *RHOmaa-Umaa* plot, a logged (Figure 5.3) and cored Chase Group sequence from the Mobil #1-2 Brown well is shown in Figure 5.4, subdivided between four plots according to the rock types identified in the core. Following the earlier discussion in this chapter, the cloud of limestone points can be interpreted as facies ranging from grainstones to mudstones, and the Chase Group core grain densities shown in Figure 4.2 can be applied to discriminate grain-supported from mud-supported limestones. An alternative interpretation that some of these points represent partially dolomitized limestones highlights the fundamental limitation of a crossplot, which is that it is constrained to two dimensions. The addition of a third dimension introduces more information so that, for example, the inclusion of a computed gamma-ray (CGR) log, with its sensitivity to clays, would help distinguish whether the cloud dispersion reflected potential limestone textural changes, dolomitization, or both. The interpretation of the processes, both depositional and diagenetic, that account for the data cloud morphologies of the other lithologies are also various and ambiguous.

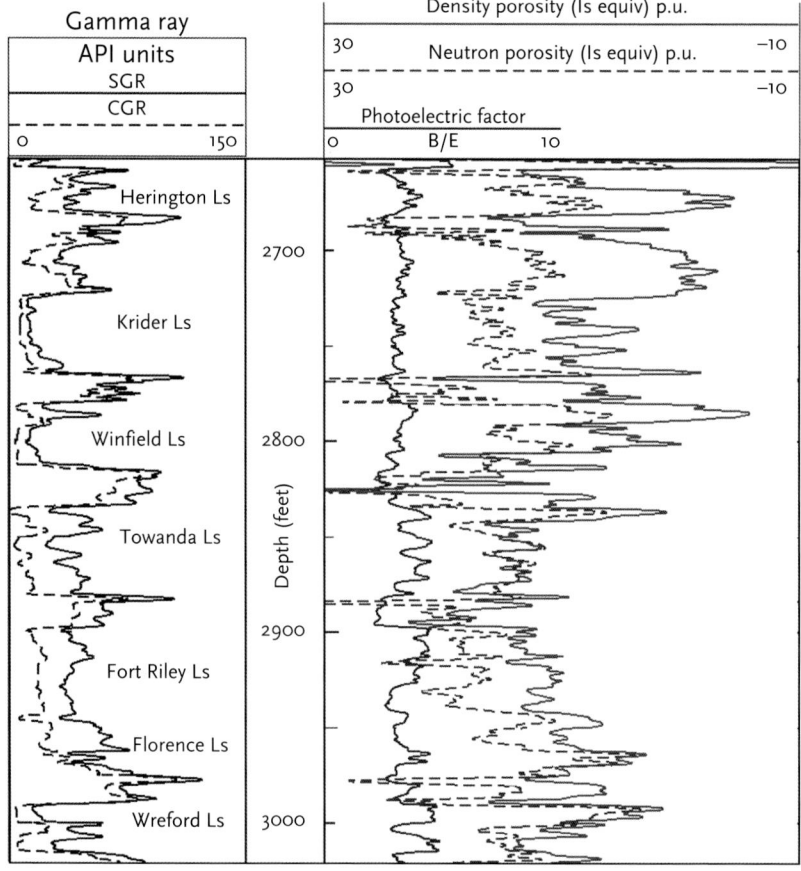

Figure 5.3: Gamma-ray, photoelectric factor, density, and neutron porosity logs of a Lower Permian Chase Group succession in the Mobil #1-2 Brown well in southwest Kansas.

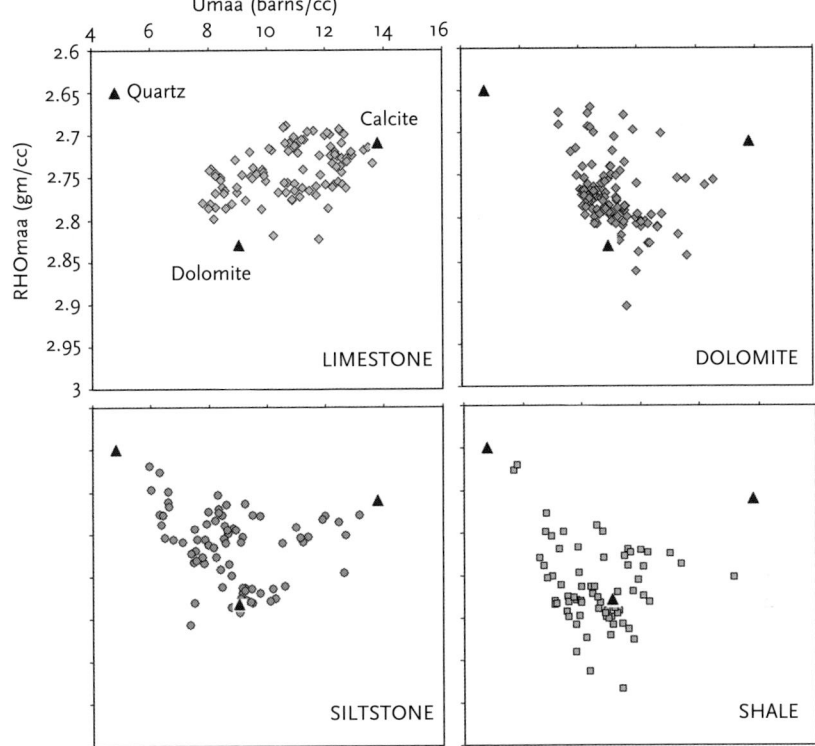

Figure 5.4: *RHOmaa—Umaa* crossplot of lithologies identified from core in a Lower Permian Chase Group succession in a well in southwest Kansas.

Finally and importantly, depth information is lost, although it could be implied through the use of an intelligent color scale. So, although the *RHOmaa-Umaa* plot is a valuable tool for pattern recognition, compositional analysis from multiple logs plotted with respect to depth is a necessary step forward to encompass all available and pertinent information.

With the addition of the computed gamma-ray log, the basic mineral composition of the sequence can be estimated by the standard inversion method described in the previous chapter (Figure 5.5). The solution is determined and applies *RHOmaa*, *Umaa*, and CGR values to the resolution of shale, quartz, dolomite, and calcite. On the composition profile, the silica component has been partitioned between quartz in the siltstone beds and chert within the carbonates, because the log responses make no distinction between them. Carbonate zones where no chert appears to be present are often marked by distinctive negative silica estimates that are caused by occurrences of anhydrite. Accordingly, the system could be expanded in a more detailed model, where anhydrite, particularly in small quantities, could be estimated by the addition of the sulfur curve from a geochemical log (Cannon and Horkowitz, 1997). Shale estimates are based on a rather generalized "shale" component, whose properties are selected from the log response of shale zones in the sequence, rather

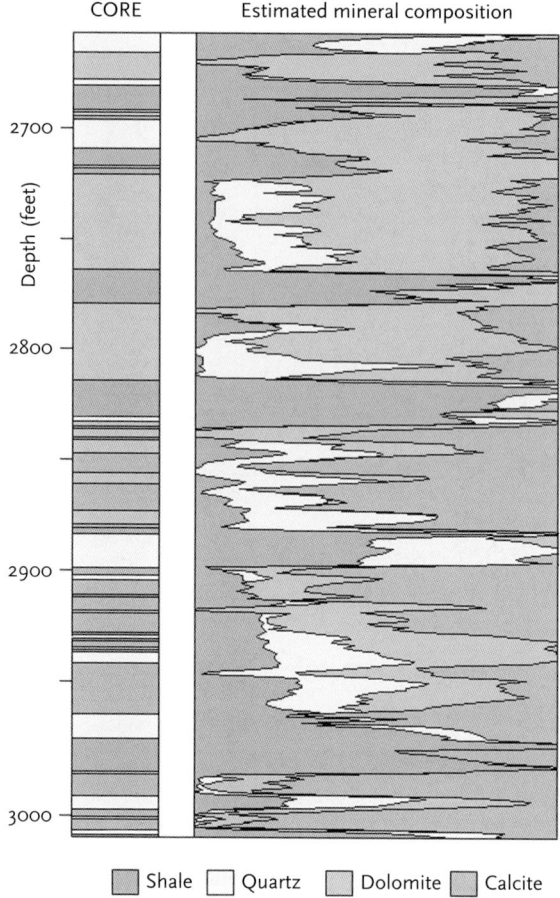

Figure 5.5: Core lithology classification matched with mineral composition analysis from inversion of *RHOmaa, Umaa,* and computed gamma-ray (CGR) values of a Lower Permian Chase Group succession in a well in southwest Kansas.

than distinctive mineral properties. A more comprehensive analysis would require quantitative X-ray diffraction analyses and geochemical logs in an expanded model of clay minerals, as discussed in the preceding chapter.

ZONATION BY CLUSTER ANALYSIS

Compositional analysis provides the basis for one method to establish rock-related electrofacies in which the sequence is zoned so that composition variability is minimized within the zone, but maximized between adjacent zones.

Given a set of compositional estimates, depth-constrained cluster analysis can be applied to segment the sequence into intervals that are as homogeneous as possible and as distinct as possible from each other in terms of their composition. Each of the logs employed is first standardized to zero mean and unit standard deviation before clustering in order to ensure that they all have approximately equal weight in the analysis. The clustering employs Ward's method, which, at each step of the process, joins the two subintervals that are most alike in a least-squares sense. The process applies the analysis-of-variance concept of classical statistics, in that it joins the two groups whose merger produces the least possible increase in the total within-groups sum of squares. The sum of squares for a single group, k, is given by:

$$W_k = \sum_{i=1}^{n_K} \| X_i - \bar{X}_K \|^2$$

where the squared distances are between the vector of standardized log values for data point i, x_i, and the vector mean x_k for group k. The within-groups sum of squares, W, is the sum of the W_k values over all groups. At each step of the clustering process, the number of groups is reduced by one, and the within-groups sum of squares increases. Depth-constrained cluster analysis only allows vertically adjacent subintervals to be joined, producing a sequence of zone memberships.

By examining a crossplot of the number of zones versus R-squared (the percentage of the variance within the zones divided by the total variance of the log values) as a "scree plot," the fundamental subdivisions that account for systematic components of the log variability can be assessed. The depths of these zones establish boundaries that identify stratal units. The application of depth-constrained cluster analysis for the zone subdivision of a single log was first described by Gill (1970), who later extended the method to the simultaneous segmentation of multiple well logs (Gill et al., 1993).

The scree plot for the Chase Group compositional-analysis profile is shown in Figure 5.6. Breaks in the slope reflect distinctive clustering levels, which can be seen more easily on the crossplot as the relative change in R-squared and indicate fundamental clustering at partitions of three, sixteen, and twenty-one zones. These

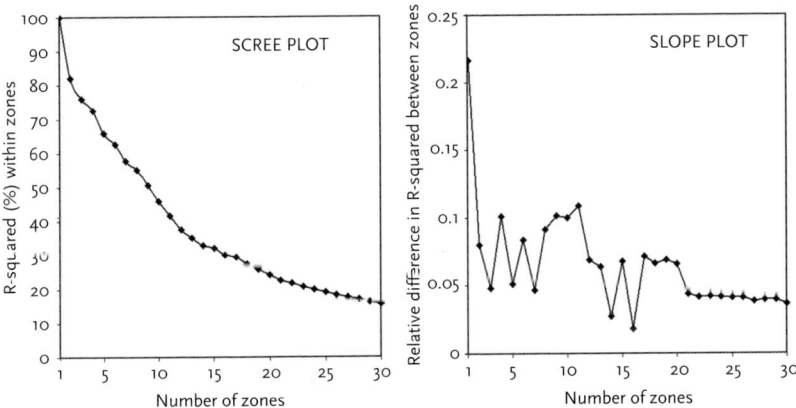

Figure 5.6: Scree and slope plots of variance partitioning by depth-constrained clustering of a composition analysis profile from logs of a Lower Permian Chase Group succession in a well in southwest Kansas.

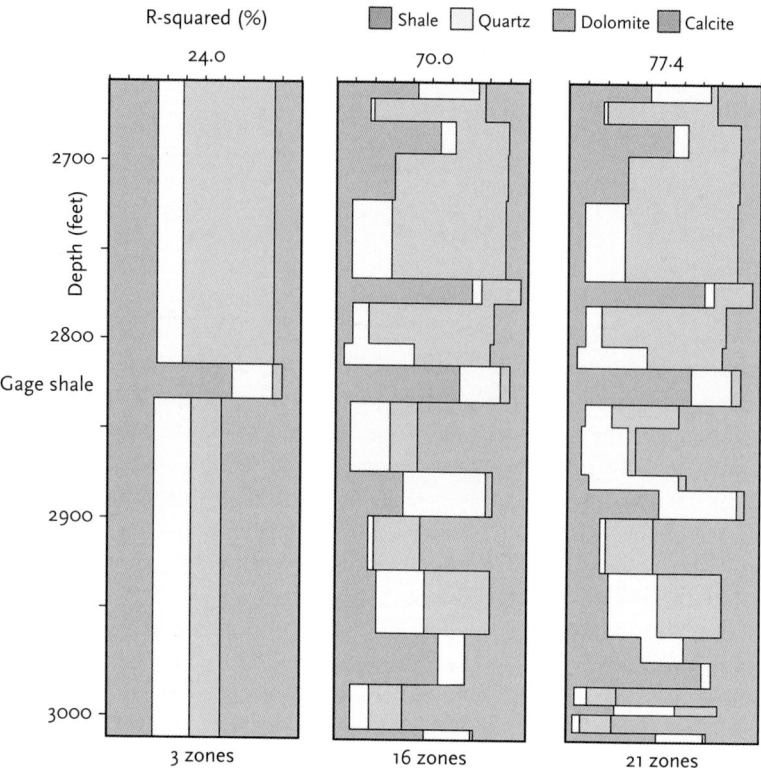

Figure 5.7: Depth-constrained clustering of three, sixteen, and twenty-one zones with their associated variance absorption of the total variability of a composition analysis profile from logs of a Lower Permian Chase Group succession in a well in southwest Kansas.

zonations are shown in Figure 5.7, in which the average composition is plotted for each zone together with the associated *R-squared* value, which expresses how much of the total compositional variability is accounted for by the zoned profile. The tripartite zonal division shows an upper Chase Group section, separated from a lower Chase Group by a shale zone, which is the Gage Shale. The zone compositions match the general lithological character of the Chase Group as described by Siemers and Ahr (1990) as: "the carbonate rocks below the Gage Shale are composed mostly of limestone; those above the Gage are dolostones." In a comparison of this subdivision with the gamma-ray logs in Figure 5.3, it is interesting to note the differentiation of relatively high uranium in the lower carbonates from lower uranium in the upper-zone carbonates, as shown by the separation of the standard gamma-ray (SGR) logs from the computed gamma-ray (CGR) logs. The uranium content was specifically excluded from the compositional analysis to screen out obvious diagenetic features, but the diagenetic overprint may show concordance with the gross compositional facies. Luczaj (1998) concluded that uranium mineralization in the Chase Group was independent of lithology, but uranium in the upper part was contained in pervasive dolomite cements that were a product of regional brine reflux from overlying Late

Permian evaporates, and contrasted with uranium concentrated in the lower section that was less well understood and may partly be related to fluid precipitation in fractures such as occurs in the Austin chalk (Fertl et al., 1980).

The finer subdivisions at clusters of 16 and 21 zones pick up stratal divisions that match with Chase Group formations and formation subdivisions. When depth-constrained clustering is applied to logs at this scale, then the coarser zones will match conventional stratal units of groups, formations, and members, so that the outcome is one that mirrors conventional subsurface stratigraphy based on logs. However, the application of statistics formalizes the boundary picks in a repeatable process, and the inversion of the logs to composition provides a more systematic base than subtle features of raw log curves that are sometimes chosen in more nuanced stratigraphic picks. At a finer scale, and with a judicious selection of log and log-transform inputs, the depth-constrained clustering can isolate distinctive "electrobeds" that represent occurrences of separate electrofacies. Electrofacies identification for lithofacies interpretation are described in the next section, where electrofacies are considered as complex log associations rather than as "petrophysical rocks" built from log-derived mineral compositions.

THEORETICAL, EMPIRICAL, AND INTERPRETIVE ELECTROFACIES METHODS

In the earliest applications of electrofacies, glyphs of "spider webs" and "ladders" were used as alternative ways to condense the multilog signature of an electrofacies into two-dimensional graphic forms (Serra and Abbott, 1980). However, glyphs are restricted to simple, visual comparisons, and their use reflected the constraints of computer technology at the time that they were introduced. In more systematic analyses, electrofacies must be mapped in the dimensional framework set by the log measurements, of which there are often many so that the nuances of potential electrofacies can be captured. A simple model of an electrofacies can be approximated by a cloud of points that are concentrated in the center and diffuse in density outwards. A multivariate normal distribution provides a convenient means to model such a distribution efficiently. It also provides the basis for a probability model that allows statistical classifications to be made from the electrofacies database.

For many electrofacies, there may be no genetic reason for the data points to be normally distributed about their mean. However, they commonly appear to be normal, probably as a result of compounded random measurement errors and independent systematic deviations. In other cases, obviously asymmetrical shapes can be normalized effectively through scale transformations, such as a logarithmic conversion. Alternatively, extended clouds may be partitioned into smaller clusters if they are separated by relatively diffuse regions. If approximately normal, then the expected density of points at any coordinate location can be specified completely by the statistics of the multivariate mean and the matrix of variances and covariances between the logs. The vector of mean values gives the location of the center of the cloud; the variance-covariance matrix gives its relative degree of dispersion and orientation within the multivariate log space.

The multivariate normal cloud is hyperellipsoidal in shape, but has no discrete surface because the normal distribution is continuous in all directions. However, the shape of the distribution ensures that the majority of points are confined to within a few standard deviations of the cloud centroid. Many representations of the normal ellipsoid set the 95 percent probability contour as an outer boundary to the cloud. Beyond this surface, any point has a rapidly decreasing likelihood of being drawn from the ellipsoid population. The situation can be visualized fairly easily in two dimensions (Figure 5.8), and the geometrical concepts are equally applicable to higher dimensions.

As an example of the development of an early electrofacies database, Delfiner et al. (1987) compiled log-response parameters for thirty sandstones, twenty-five shales, thirty limestones, twenty-five dolomites, twenty-five evaporites, three coals, ten igneous rocks, and four miscellaneous rocks. Local databases can be designed to include unusual lithologies and to fine-tune electrofacies parameters to specific rock types. So, for example, Stowe and Hock (1988) developed a Zechstein database for applying classification procedures to the Permian gas-bearing formations in northern Germany. Their Zechstein reference set consisted of electrofacies for forty-eight carbonates and twenty-four evaporites (Figure 5.9). Delfiner et al. (1987) described three different methodologies for the design of an electrofacies reference database: theoretical, empirical, and interpretive approaches. In practice, the database is

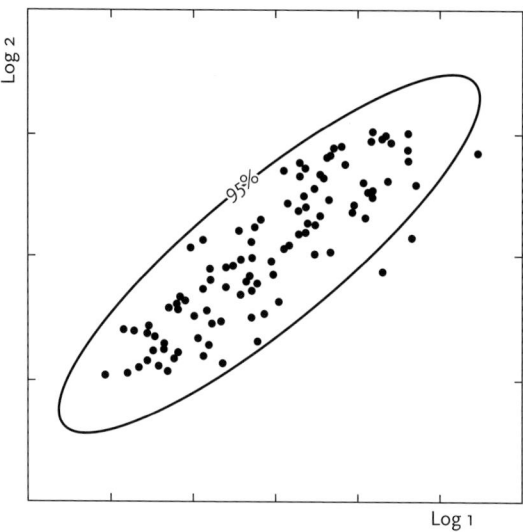

Figure 5.8: Representation of hypothetical electrofacies by 95 percent density contour of a bivariate normal ellipse plotted with reference to two logs set as orthogonal axes. From Doveton (1994), © 1994 American Association of Petroleum Geologists (AAPG), reprinted by permission of the AAPG, whose permission is required for further use.

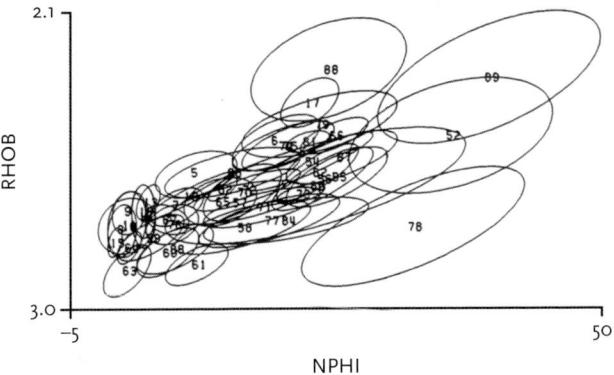

Figure 5.9: Density-neutron crossplot of Zechstein carbonate electrofacies. From Stowe and Hock (1988), courtesy the Society of Petrophysicists and Well Log Analysts.

built using all three procedures as a mix of predictions from tool-response equations and interpretations from crossplots integrated with core observations.

The theoretical approach to electrofacies definition is that of forward modeling, which was reviewed in detail by Quirein et al. (1986). While the determined-system solution of compositional proportions is obtained by inversion of the log-response equations, the forward model is the reverse of this process. The log responses of an electrofacies are predicted by multiplying its component log properties by the proportions with which they occur in an equivalent lithofacies:

$$CV = L$$

where C is the matrix of component properties, V is the vector of component proportions within the lithofacies, and L is the vector of electrofacies log responses. The lithofacies compositions can come from a variety of sources, such as generalized estimates from sedimentary geology textbooks (e.g., Pettijohn et al., 1972) and be convolved with mineral logging parameters (such as those listed in Edmundson and Raymer, 1979). Alternatively, electrofacies can be created from local lithofacies determined by core analyses or based on other sources of geological information. These customized electrofacies accommodate local variability, and the added precision inspires more confidence in the results of subsequent classifications. In all cases, the analysis ranges, porosity distributions, accessory-mineral influences, and data-acquisition errors are incorporated in the computation of descriptive normal ellipsoids. The process that underlies the theoretical type of electrofacies can be seen to be deductive and model driven.

The empirical approach relates lithofacies observed in core to the set of log responses measured over the cored interval. The log-response statistics of means, variances, and covariances then describe hyperdimensional ellipsoids for each electrofacies and are tagged to a specific lithofacies. This process of creation can be viewed as inductive, but supervised. The major drawback to this approach is economic, because of the costs incurred by coring long sections and matching observed responses logged by a full suite of tools.

The interpretive approach to electrofacies generation is based on the examination of log crossplots for clusters that can be identified geologically or validated by cores or cuttings. On each cluster, an ellipse is circumscribed that is set by the range on each log and the log-pair correlation coefficient for the cluster. Histograms and Z-plots are also used for the location of ellipse boundaries (Delfiner et al., 1987). However, in practice, the edge of an ellipse is usually chosen to be an informal estimate of the 95 percent probability density contour. Crossplots for all possible pairs of logs are examined to determine the parameters of the multivariate ellipsoid.

Serra and Abbott (1980) stressed that an important prior step to the generation of log crossplots was the segmentation or blocking of the original logs. As a result, data sampled from curve features that were transitions between electrobeds would be eliminated, while electrobed measurements would be retained. This procedure trims much of the diffusion from the data clouds and improves the discrimination of the electrofacies. The interpretive approach can be highly labor intensive if the electrofacies are built by examining all possible crossplots. For example, using five logs, Stowe and Hock (1988) analyzed some 4,500 crossplots in the construction of a Zechstein carbonates database of seventy-two electrofacies. The number of crossplots is partly contingent on the number of logs and can be calculated as the number of combinations of n logs, taken a pair at a time:

$$N = \frac{n!}{(n-2)!2!}$$

For five logs there are ten possible crossplots, which represent ten alternative and orthogonal views of the hyperdimensional data clouds. If the number of logs is expanded to eleven, then the total set of crossplots expands dramatically to fifty-five. This problem was commented on briefly by Serra et al. (1985), who bluntly called it "the curse of dimensionality." On one hand, a sufficient number of logs is required to distinguish between electrofacies with minimal ambiguity. On the other hand, the multidimensional space created by a large number of logs becomes difficult to handle by traditional methods.

PRINCIPAL COMPONENT ANALYSIS (PCA) OF ELECTROFACIES

Wolff and Pelissier-Combescure (1982) described a strategy to circumvent this dilemma through the use of principal component analysis (PCA) to condense the high dimensionality introduced by multiple logs. Although the locations of data points in multilog space collectively delineate clouds with the same number of dimensions as logs, they can often be mapped effectively in a much reduced dimensionality. This is because intercorrelations between the log variables cause the clouds to be extended along certain trends. Principal component analysis computes an ordered set of orthogonal axes that absorb the variation in a systematic manner. As implemented in the method described by Wolff and Pelissier-Combescure (1982), the clustering is run in two phases. In the first phase, the clustering is used to isolate local modes as zonal representatives of the digital data. In the second phase, the local modes are agglomerated into clusters that are equated with electrofacies. Decisions concerning potential cluster subdivision or fusion are monitored by referral to geological information from the core or from geological experience. The method was implemented in the 1980's as the Schlumberger computer-processed log product of *Faciolog*, and an example is shown in Figure 5.10 for a section of the Chase Group.

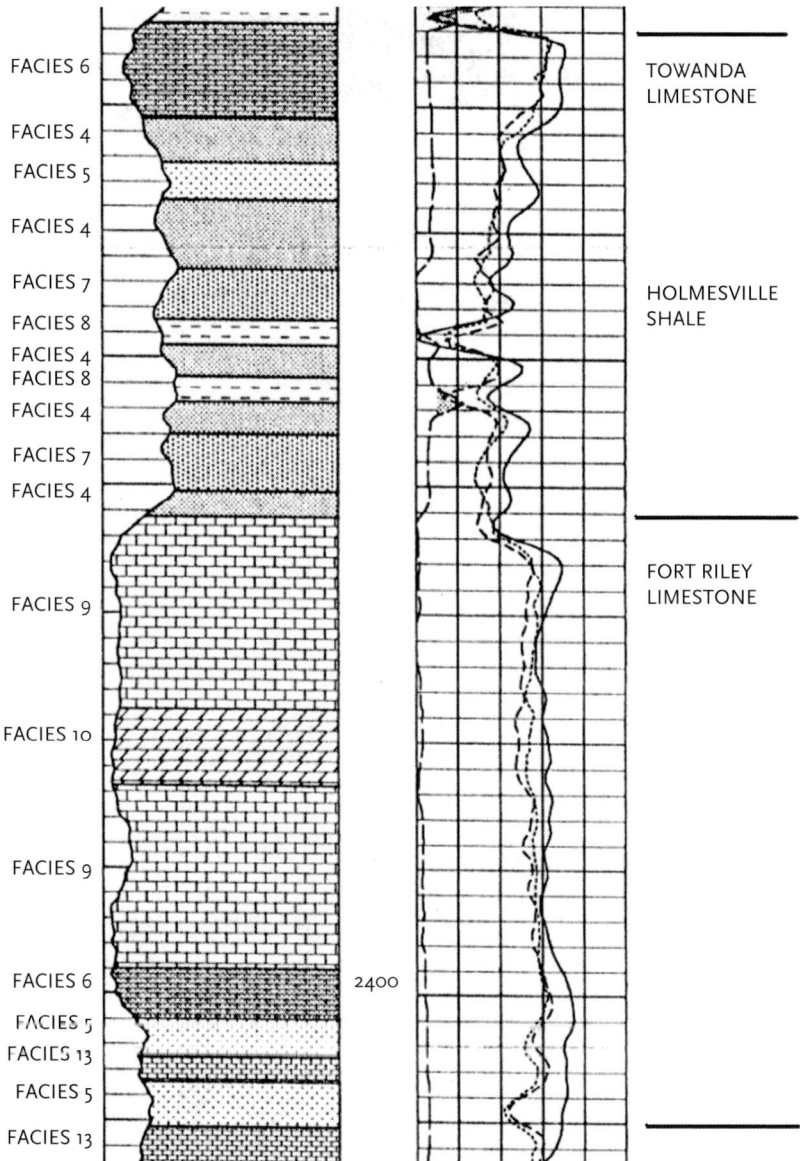

Figure 5.10: *Faciolog* (1983) presentation of facies predictions from spectral gamma-ray, photoelectric factor, density, sonic, and neutron porosity logs of a Chase Group (Lower Permian) section in Amoco #1 Hayward, southwest Kansas.

Computation of the principal components is simply a geometric rotation in multidimensional space that locks onto the orthogonal axes of relative elongation in a cloud of data points. If m log responses from a sequence of zones are plotted as points in a space with mutually orthogonal axes, they form a cloud in m-dimensional space. The raw data cloud is modeled by a single hyperellipsoid, whose center is at the multivariate mean value, and inflation is characterized by the variances and covariances of the measurement variables. Principal components are the eigenvectors of this cloud, computed to locate the major axes in order of importance based on their associated eigenvalues, and they reflect systematic relationships between the logs. These axes provide a new framework of reference that is aligned with the natural axes of the cloud ("eigen" is German for "intrinsic"), rather than the original log-measurement axes. The orientations of the principal components are computed from either the covariance or correlation matrix of the zone-log data. The correlation matrix is the more common choice, because most logs are recorded in radically different units. In order to avoid artificial and undue weighting by any of the logs, the original data should be standardized to dimensionless units by subtracting the mean and dividing by the standard deviation. The covariance matrix of standardized data is the correlation matrix. A simple depiction of the basic geometrical concept of this procedure is shown in Figure 5.11.

A compression of dimensionality is made possible because the components are based on the intercorrelations between the variables. The total variance of the original set of m variables is the sum of their separate variances. This quantity is absorbed by m possible principal components. In practice, many measurement variables show a significant degree of intercorrelation, and principal component analysis

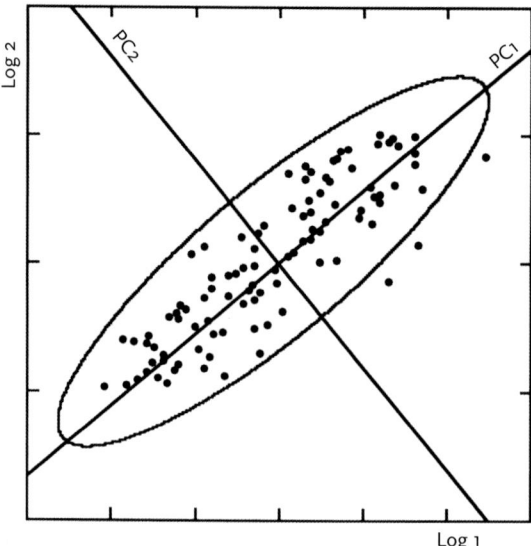

Figure 5.11: Principal-component axes of a hypothetical electrofacies data cloud of points (Figure 5.8) reference to two logs set as orthogonal axes. From Doveton (1994), © 1994 American Association of Petroleum Geologists (AAPG), reprinted by permission of the AAPG, whose permission is required for further use.

highlights the amount of information redundancy within the logs. If the majority of the variability is picked up by p principal components, then the dimensionality has been shrunk from m to p. The collapse reflects the dimensionality of the information content of the variables as a replacement for the original reference framework. It is not uncommon for the first two principal components to account for most of the variability, thus a multilog dataset can be mapped on a crossplot with little loss of information.

The derivation of the principal components follows from a property of matrix algebra that a symmetric, nonsingular matrix, S, can be converted into a diagonal matrix, L, by multiplying by an orthonormal matrix, U, through the following equation:

$$U^T S U = L$$

where T signifies the transpose of a matrix. If S is the covariance matrix, then the conversion to a diagonal matrix is the geometrical equivalent of a rotation of the original axes to new descriptive axes. A diagonal matrix has zeroes in the off-diagonal elements, which means that the new axes are independent of one another. The values of the diagonal elements register the eigenvalues of the principal components that express their variances. The sum of these eigenvalues is then the same as the sum of the variances of the original variables. The relationship gives an immediate measure as to how much variability is assigned to each principal component. The numbers are particularly easy to follow when the correlation matrix is selected. The variance of each variable is then unity, and the total variability equals m (the number of variables). Each eigenvalue divided by m is the proportion of a principal component's share of the total variability.

The fact that U is an orthonormal matrix leads to the useful result that the inverse of U is the same as the transpose of U. This means that both the transformation from the measurement space to the principal-component space and the reverse mapping are variations of the same operation. The matrix U contains the loadings that relate the eigenvectors to the original variables. The location of any point within the data cloud can be related to the principal-component axes by the transformation:

$$Z = U^T X$$

where X is a vector of the zone log responses and Z is a vector of the principal-component scores. This means that the score of the ith zone on the pth principal component is given more simply by:

$$z_{pi} = u_{1p} x_{1i} + u_{2p} x_{2i} + \cdots + u_{mp} x_{mi}$$

where the u coefficients are loadings from the pth principal component. The original variables can be recovered from the principal-component scores through the inverse of this procedure:

$$X = UZ$$

So, the gth log response of the ith zone can be computed by the equation:

$$x_{gi} = u_{g1}z_{1i} + u_{g2}z_{2i} + \cdots + u_{mp}z_{mi}$$

The loadings of the U matrix summarize the relationships between the log variables and the principal components. Consequently, they can often be "read" for their geological or petrophysical meaning. In the eigenvector analysis of the correlation structure of log relationships, the principal components will often reveal these properties implicitly. Common sense must be used in such interpretations. The computation of principal components is simply a geometrical operation that relocates the reference axes to the apparent axes of elongation of the data cloud. The preceding explanation only covers the bare bones of the mathematics of principal component analysis. The ideas and further ramifications are best understood by consideration of a case-study example.

Principal component analysis is a standard option on almost all statistical software packages, so that pioneering work on applications to electrofacies analysis can be made easily on a personal computer and tailor-made to local geology, as shown in the following case-study. Here we apply the interpretive approach to the isolation of distinctive clusters as separate electrofacies in multivariate log space, followed by the assignment of these electrofacies to matching lithofacies. The Chase Group shows a distinctive cyclic depositional pattern that has been extensively studied and described because it is the host for the Hugoton field, which is the largest onshore gas field in North America. A characteristic Chase Group cyclothem was described by Caldwell (1991) as being initiated by a transgressive gray siltstone and sandstone, followed by transgressive mudstones, wackestones, and packstones, which then coarsen upwards into regressive packstones and grainstones, before terminating with regressive reddish-brown mudstones, siltstones, and sandstones, with desiccation features. This basic model provides a useful genetic framework in the interpretation of electrofacies and their sedimentological implications.

The target interval of the case study is the Towanda Limestone, Holmesville Shale, and Fort Riley Limestone (Chase Group) subsection of the Chase Group in the Amoco #3HI Montgomery well (Figure 5.12). This matches a lower interval of the Chase Group in the Mobil #1-2 Brown well (Figure 5.3) and coincides with the section in the Amoco #1 Hayward well, used as an example of a *Faciolog* output (Figure 5.10). The basic elements of the cyclothem model described by Caldwell (1991) can be deduced from the log curves of Figure 5.12, but the purpose of the principal component analysis is to dig deeper, in the recognition of textural classes within the carbonates and sedimentological variability in the clastics. With these goals, the selection of logs and computed logs is a crucial initial step, so that emphasis is placed on variables that are most likely to reflect depositional controls, as contrasted with those that are likely to be linked with diagenetic overprints. With this in mind, the log-response variables of apparent matrix density (*RHOmaa*), matrix volumetric photoelectric absorption (*Umaa*), total porosity (*PHIt*), thorium (*THOR*), and potassium (*POTA*) were extracted from the spectral gamma-ray, photoelectric factor, density, and neutron porosity curves as suitable measures for electrofacies discrimination.

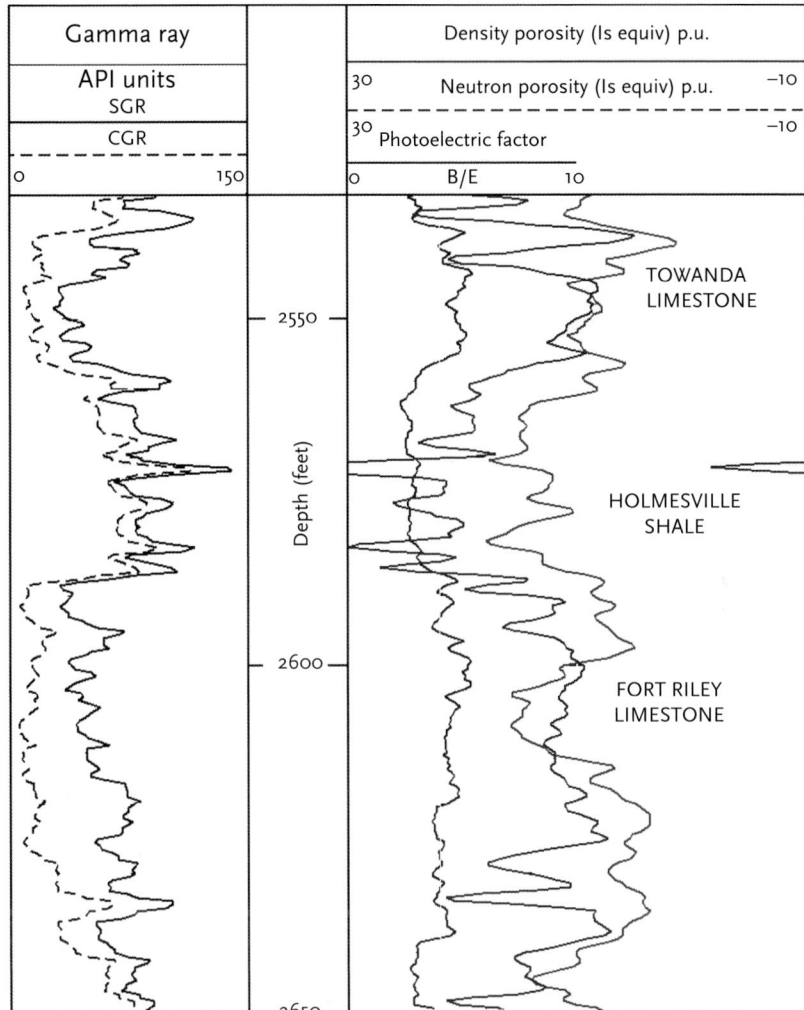

Figure 5.12: Gamma-ray, photoelectric factor, density, and neutron porosity logs of the Towanda limestone, Holmesville shale, and Fort Riley limestone (Chase Group) in the Amoco #3HI Montgomery well in southwest Kansas.

A correlation-coefficient matrix was computed as the first step in principal component analysis (Table 5.1). The relatively high values (both positive and negative) are clear indications of information redundancy and that much of the variability in five measurement dimensions can probably be absorbed by a few principal components. A reading of the correlation interrelationships shows a thorium-potassium-porosity association related to clastic zones and contrasted with negative correlations against matrix volumetric photoelectric absorption, which reflects carbonate occurrence. The results from a principal component analysis provide a much expanded interpretation from this preliminary assessment of pairwise variable interrelationships. The

Table 5.1. CORRELATION-COEFFICIENT MATRIX OF APPARENT MATRIX DENSITY (*RHOMaa*), MATRIX VOLUMETRIC PHOTOELECTRIC ABSORPTION (*UMaa*), TOTAL POROSITY (*PHIt*), THORIUM (*THOR*), AND POTASSIUM (*POTA*) FOR THE CHASE GROUP SECTION IN THE AMOCO #3HI MONTGOMERY WELL

	Rhomaa	Umaa	Phit	Thor	Pota
RHOMAA	1.00	-0.51	0.53	0.67	0.68
UMAA	-0.51	1.00	-0.66	-0.75	-0.76
PHIT	0.53	-0.66	1.00	0.82	0.83
THOR	0.67	-0.75	0.82	1.00	0.89
POTA	0.68	-0.76	0.83	0.89	1.00

Table 5.2. EIGENVALUES, VARIANCES, AND LOADINGS OF PRINCIPAL COMPONENTS REFERENCED TO APPARENT MATRIX DENSITY (*RHOMaa*), MATRIX VOLUMETRIC PHOTOELECTRIC ABSORPTION (*UMaa*), TOTAL POROSITY (*PHIt*), THORIUM (*THOR*), AND POTASSIUM (*POTA*) FOR THE CHASE GROUP SECTION IN THE AMOCO #3HI MONTGOMERY WELL

	Principal Components				
	1	2	3	4	5
RHOMAA	0.387	-0.879	0.101	-0.255	0.050
UMAA	-0.426	-0.354	-0.793	0.250	-0.048
PHIT	0.448	0.311	-0.575	-0.605	0.072
THOR	0.482	0.061	-0.137	0.592	0.628
POTA	0.485	0.043	-0.110	0.394	-0.772
Eigenvalue	3.86	0.53	0.35	0.15	0.10
Variance %	77.2	10.5	7.1	3.1	2.1
Cumulative variance %	77.2	87.8	94.9	97.9	100.0

eigenvalues and eigenvector loadings (Table 5.2) show that the first two principal components account for 88 percent of the total variability, so that when used as axes of a crossplot (Figure 5.13), almost all of the input log information is preserved on this two-dimensional condensation. The plot shows two distinct electrofacies clouds, which are separated primarily by the first principal component. The eigenvector loadings show the cloud with positive scores to be clastics (higher thorium, potassium, and porosity), and the cloud with negative scores to be carbonates (higher matrix volumetric photoelectric absorption).

It should be noted that the principal components of a correlation matrix are computed as the eigenvectors of a single multivariate normal ellipsoid centered at the origin. However, in this style of application, the primary purpose is to distinguish between several electrofacies' data clouds. Consequently, the first few principal components will tend to be located to absorb the maximum variability between the data

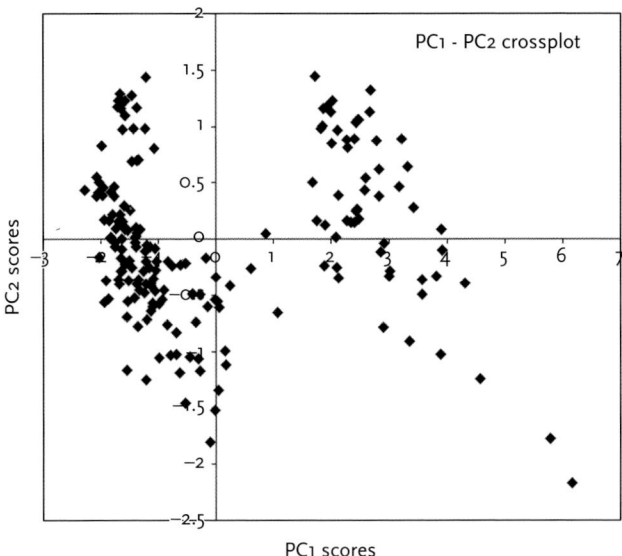

Figure 5.13: Crossplot of scores of first two principal components computed from log variables of the Towanda limestone, Holmesville shale, and Fort Riley limestone (Chase Group) in the Amoco #3HI Montgomery well in southwest Kansas.

clouds before accounting for the variability within the clouds. It would be tempting to believe that the principal components with the largest eigenvalues account for the largest discrimination between data clouds, but this is not always the case, as demonstrated both mathematically and by simulation (Chang, 1983). It is therefore important to couple the assessment of the eigenvalues with the potential petrophysical meaning of the eigenvector loadings as a guide to electrofacies differentiation. In a more detailed analysis, principal components can then be applied to each electrofacies. This extension was discussed by Brandsegg et al. (2010), who suggested that, since principal component analysis of long logged sections mixed variability between and within lithologies, subsets of fairly uniform lithologies should be analyzed separately, in a procedure they termed "structured principal component analysis." As a result, small-scale variability could be captured that was masked by unstructured PCA. Essentially, structured PCA honors the descriptive model of a multivariate normal ellipsoid. In the Chase Group example, the two electrofacies could be modeled separately by principal component analysis. However, as can be seen on the principal component (PC) scores crossplot (Figure 5.13), the first principal axes would be subparallel to the second principal component of this analysis. The loadings of the second principal component therefore reflect the major variability within both electrofacies, keyed most strongly to changes in matrix density.

A depth plot of PC score logs for the first two principal components (Figure 5.14) shows the fundamental partition between clastics and carbonates on the first component, while the second component score log is keyed mainly to increases in matrix density within each facies. Depth-constrained clustering was applied to the

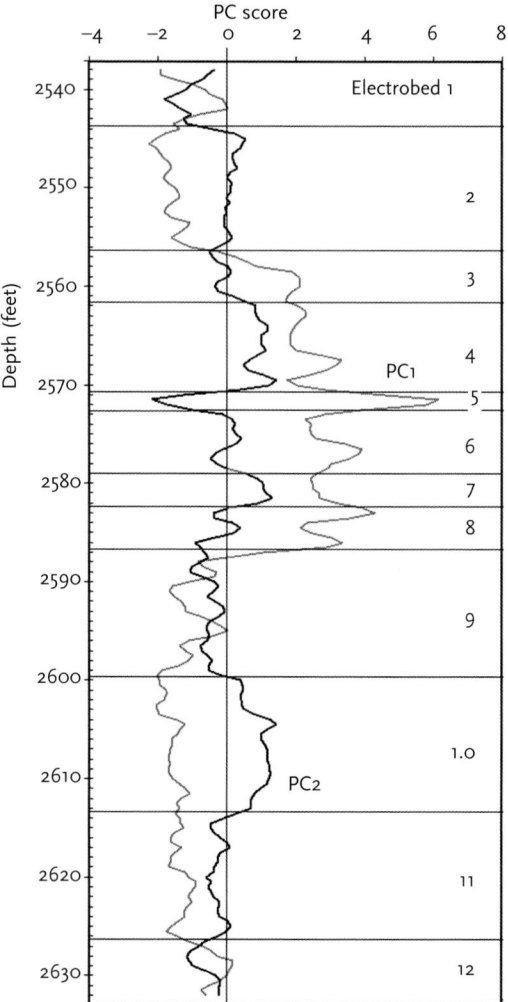

Figure 5.14: PC score logs of first two principal components subdivided into electrobeds by depth-constrained clustering of the Towanda limestone, Holmesville shale, and Fort Riley limestone (Chase Group) section in the Amoco #3HI Montgomery well in southwest Kansas

principal-component score logs and twelve distinctive electrobeds selected for further interpretation. Notice that the isolation of electrobeds was performed after the principal component analysis in this case study, rather than segmentation of the original logs as the initial step. This is because the depth-constrained clustering will differentiate systematic zones with the suppression of transitional curve features. However, essentially the same result would occur if depth-constrained clustering was applied to the input log variables prior to PCA, although a sufficiently fine zonation would be needed to generate viable correlation coefficients.

By plotting the electrobed principal-component scores on a crossplot (Figure 5.15), the electrobeds can be aggregated into electrofacies subdivisions that pick up

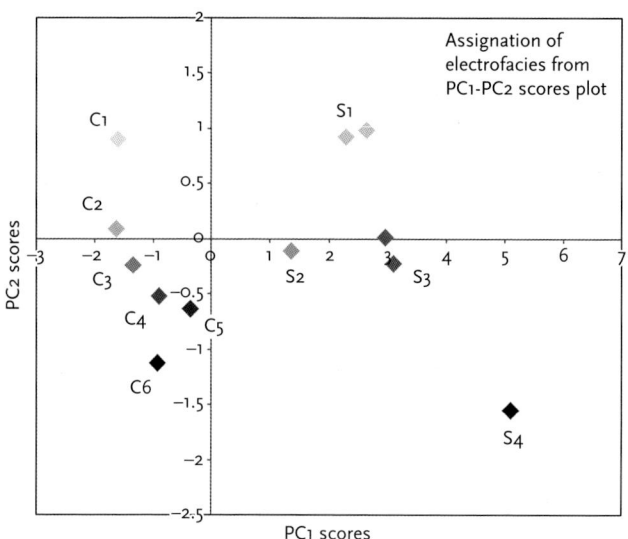

Figure 5.15: Crossplot of scores of first two principal components with electrobed locations and electrosubfacies labeled for clastic and carbonate electrofacies of the Towanda Limestone, Holmesville shale, and Fort Riley limestone (Chase Group) in the Amoco #3HI Montgomery well in southwest Kansas.

Table 5.3. MEAN VALUES OF LOG RESPONSES IN CARBONATE AND CLASTIC ELECTROSUBFACIES MATCHED WITH ELECTROBED ZONES OF THE TOWANDA LIMESTONE, HOLMESVILLE SHALE, AND FORT RILEY LIMESTONE (CHASE GROUP) IN THE AMOCO #3HI WELL IN SOUTHWEST KANSAS.

Average Log Response

Facies	Zone	Rhomaa	Umaa	Phit	Thor	Pota	Interpretation
C1	10	2.69	13.1	12.5	1.7	0.22	GST
C2	2	2.72	13.6	9.5	2.0	0.38	PKST-GST
C3	11	2.75	12.0	8.1	1.9	0.25	WKST-PKST
C4	9	2.77	12.6	10.5	2.3	0.43	WKST-MDST
C5	12	2.79	11.5	10.3	2.6	0.72	MDST-WKST
C6	1	2.80	13.0	9.0	2.4	0.42	AN-LS
S1	4&7	2.77	7.8	17.8	7.0	1.78	CSLTST
S2	3	2.80	8.9	13.0	5.2	1.42	SLTST
S3	6&8	2.83	8.9	18.4	7.9	2.15	SHSLTST
S4	5	2.94	9.8	24.1	11.0	2.66	SLTY-SH

textural, mineralogical, and pore-volume changes. The next step is transcription, that is, to assign lithofacies equivalents to each of the clusters. Essentially, we are looking for the lithofacies synonym of the electrofacies. Generalized interpretations are generally easy to make, but assignations that are locked into local core observations and measurements are clearly preferable. In Table 5.3, the mean values of the carbonate and clastic subfacies are shown and ranked with respect to apparent matrix density, the strongest loading on the second principal component that is keyed to variation within the clastic and carbonate electrofacies. Grain-density and porosity measurements from a large Chase Group core database (Figure 5.2) were used for interpretative lithofacies transcription. The carbonate subfacies were assigned textural equivalents within the grainstone to mudstone range and a discrimination of an anhydritic limestone. Increases in apparent matrix density within the clastic electrofacies was matched by increases in potassium and thorium contents, reflecting finer grain sizes and increasing clay contents in the coarse siltstone to silty shale range.

The final interpreted electrofacies log from principal component analysis and depth-constrained clustering (Figure 5.16) is drawn to conform loosely with graphic

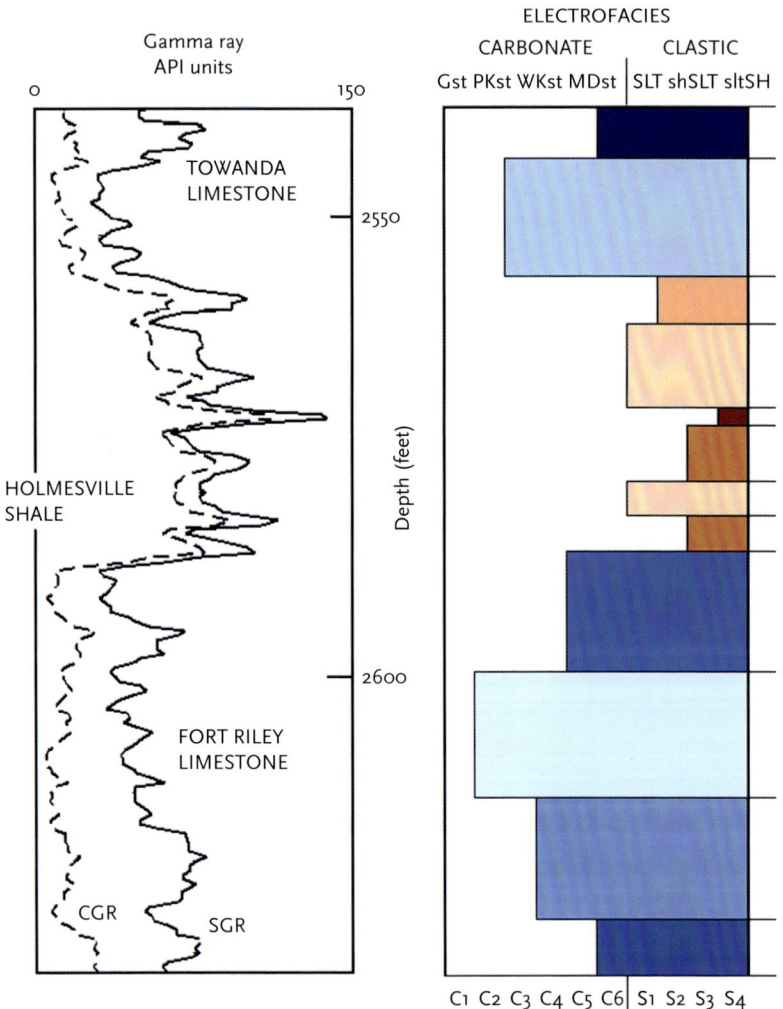

Figure 5.16: Interpreted electrofacies log from PCA and depth-constrained clustering of the Towanda limestone, Holmesville shale, and Fort Riley limestone (Chase Group) in the Amoco #3HI Montgomery well in southwest Kansas.

profiles commonly used to record lithologic observations made from core. By way of comparison, a core description and sequence stratigraphic interpretation of the core from this well (Winters, 2007) serves both as a validation test and useful insight on visual assessments and petrophysical measurements (Figure 5.17). The basic Chase Group cyclothem of Caldwell (1991) described earlier has been studied in more detail by Dubois et al. (2006), who integrated detailed petrophysical analysis and core studies as contributory elements to the development of a regional geomodel. Used in that

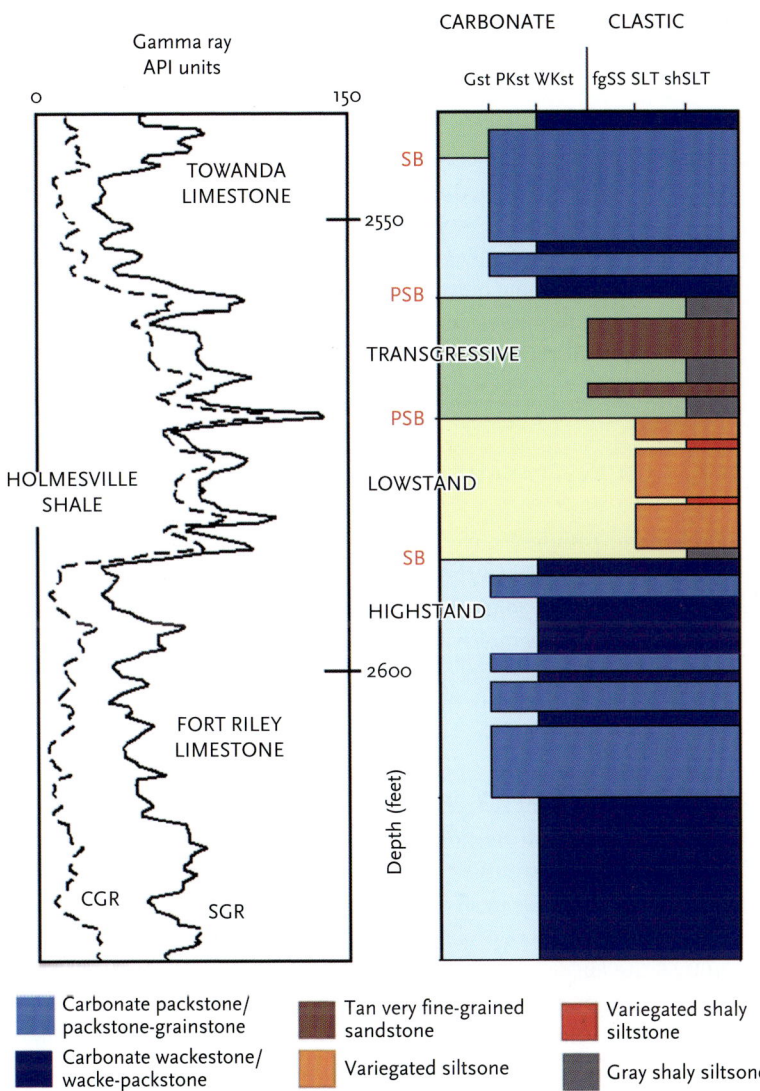

Figure 5.17: Core description and interpretation of the Towanda limestone, Holmesville shale, and Fort Riley limestone (Chase Group) in the Amoco #3HI Montgomery well in southwest Kansas. After Winters (2007).

sense, comparison between electrofacies and core description is not a competitive process to judge which is most "correct." Instead, the appropriate philosophy is one of a collaborative integration, in which features observed visually are augmented by petrophysical measurements, and vice versa. The final synthesis provides a learning experience about genetic rock properties and how they relate to petrophysics, but, importantly, they provide a more sophisticated prediction model to apply in wells that are logged but not cored.

Clearly, this case study represents a labor-intensive effort that would be unrealistic to apply to the evaluation of a large field. However, most systematic field studies identify a small number of key wells characterized by extensive core and full suites of modern logs that are used as the touchstones for field characterization. It is in these wells that studies of this kind are particularly useful. The unsupervised approach allows the recognition of distinctive electrofacies whose lithofacies equivalents have not yet been recognized in the core; the core validation tests whether observed lithofacies are recognized by electrofacies or whether analysis should be repeated with the incorporation of new log variables. In the process, a keener understanding of the relationships between petrophysics and the geology of the field are gained, with an improvement in predictive power and modeling on turnkey processing from generalized electrofacies databases. Once a satisfactory electrofacies model has been developed and validated from the key wells, prediction methods utilizing probability can then be applied to the bulk of the field wells in an efficient manner, as described in the next section.

CLASSIFICATION BY A PARAMETRIC ELECTROFACIES DATABASE

If a comprehensive parametric electrofacies database is established that links all available core observations of lithofacies to their associated logs, then it can be used as a means to classify zones, based solely on their log responses. The problem becomes that of allocating a multivariate coordinate location to one or another of a number of electrofacies hyperellipsoids. Because these hyperellipsoids are located at their multivariate log means, and their shape and orientation are specified by their multivariate covariances, the database is parametric, that is, defined by the parameters of means and covariances. The effective boundaries of the hyperellipsoids are commonly set by 95 percent density contours. This convention is a simple way to illustrate what are really diffuse clouds of points and allows their graphical display on log crossplots for inspection by petrophysicists (e.g., Figure 5.9). The diffused density of the hyperellipsoids in all directions means that the allocation of a zone to any electrofacies is a matter of probability.

The multivariate normal distribution associated with each electrofacies is infinite in all directions, although realistic probability contours are more localized around the electrofacies centroid. Consequently, all zone log-response coordinates have a finite probability of belonging to any of the electrofacies. The probability of observing a set of log responses, L, given an electrofacies i, is symbolized as:

$$P(L|F_i)$$

and is given by the normal probability distribution structure of the ith electrofacies hyperellipsoid. But the actual problem is the reverse: What is the probability of being a product of electrofacies i, given a set of log responses, L? This posterior probability, $P(F_i/L)$, is given by Bayes' theorem as:

$$P(F_i / L) = \frac{p_i \cdot P(L / F_i)}{\sum p_j \cdot P(L / F_j)}$$

where p_i is the prior probability of electrofacies i.

The prior probabilities are determined by geological experience and are an important means for excluding lithologies that do not occur in the analytical sequence, or for weighting electrofacies that occur particularly commonly. The prior probabilities can be based on frequency of occurrence observed in core and outcrop studies or expressed as likelihood from judgments formed from observations of rocks or modern depositional analogs. With no prior information, equal prior probabilities can be assigned to all reasonable electrofacies. Posterior probabilities of zero value are an immediate means to extract a subset of relevant electrofacies from a large database and thus to speed computations. The classification step follows from the computation of the posterior probabilities, with electrofacies assignment dictated by the maximum probability value. The probability figure gives the degree of confidence associated with the classification decision. In cases where the zone falls outside the 95 percent limits of all electrofacies ellipsoids, the zone is normally considered to be "unidentified."

SUPERVISED ELECTROFACIES ANALYSIS METHODS

An electrofacies analysis that is given no prior information concerning group membership of individual observations is an "unsupervised" approach, where trends and clusters that are perceived in petrophysical measurements are subsequently assigned geological meaning. This inductive philosophy "allows the data to speak for themselves" and operates from the "bottom-up," because the analysis originates with the observational data rather than being driven top-down by a deductive model. Advantages include the element of surprise, in that new associations may be revealed that represent useful information to augment the current geological model. A common disadvantage in complex lithologies is that multiple and competing interpretations, both real and unrealistic, may create a seeming wilderness of mirrors with the risk of bewilderment rather than meaningful geological insight.

In a "supervised" method, different categories or groups are specified before the analysis; the goal of the method is to find the best function to distinguish the categories, based on the characteristics of the data. Subsequently, the function can be used to classify unknown observations on the basis of likelihood of membership in one or another of the groups. In this case, lithofacies are identified in the core and linked with log data from a cored well as a "training set," from which a multivariate statistical method can "learn" the relationships between logs and core lithofacies.

A satisfactory classification function can then be applied to the analysis of other wells that have been logged but not cored. The success of the classification function must first be assessed in the training well, by matching lithofacies occurrence with log prediction. Particular attention should be paid to differentiating lithofacies with a high prediction success rate from those that are difficult to discriminate. The transcription between lithofacies and electrofacies is not necessarily one of exact correspondence because they are based on separate criteria of visible descriptions and invisible properties. Because the match of prediction to observation within the training well is recursive, an independent verification of predictive success should be made in a "validation well" that supplies both core observations and log measurements. If the validation phase is satisfactory, then the classification method can be applied to logged but uncored wells for predictions that are tagged with their associated probabilities of correct assignment.

ELECTROFACIES CLASSIFICATION BY DISCRIMINANT FUNCTION ANALYSIS (DFA)

The most commonly used supervised parametric method for probabilistic classification of unknown observations between groups is discriminant function analysis (DFA). Discriminant analysis covers a wide range of techniques aimed at the classification of unknown samples to one of several possible groups or classes. Classical discriminant function analysis has the tighter focus of attempting to develop a linear equation that best differentiates between two different classes. Fisher (1936) first derived the linear discriminant function as a statistical method to separate two populations by a weighted linear function of their measurement variables. The method is a supervised technique that requires a training data set for which assignments to the two populations are already known. The data consist of multivariate values for every individual in both population samples. If the two groups were plotted in multidimensional space, they would appear as two clouds of data points with either a distinctive separation or some degree of overlap. An axis is located on which the distance between each cloud is maximized, while the dispersion within each cloud is simultaneously minimized. This axis defines the linear discriminant function and is calculated from the multivariate means, variances, and covariances of the two groups. The data points of the two groups may be projected onto this axis as locations on a single line. This operation results in the collapse of the many variable dimensions of the recorded data into a single, composite variable that best discriminates between the two groups.

The discriminant function is then the equation of the axis that cuts obliquely across the crossplot at an angle determined by the relative contribution of the variables to the discrimination. In the computation of the function, the two clouds are modeled by multivariate normal distributions whose probability density contours map out hyperellipsoids. This is the same representation used in the electrofacies data banks described previously.

The equation of the discriminant function, Z, is:

$$Z = \lambda_1 X_1 + \lambda_2 X_2 + \cdots + \lambda_m X_m$$

where X_1 to X_m are the m variables used for discrimination and λ_1 to λ_m are the weighting coefficients to be applied to each of the m variables. The optimum discriminant function is set at the location that maximizes the value of the distance between the cloud centroids divided by the dispersion of the two clouds. This condition occurs when the following equation is satisfied:

$$S\Lambda = D$$

where S is a matrix of pooled variances and covariances from the two groups, Λ is a vector of the unknown λ coefficients to be solved, and D is a vector of the differences between the means of the two groups with respect to the m variables.

Multiple discriminant analysis (or canonical discriminant analysis) is a multi-group extension of linear discriminant analysis for two groups. Instead of computing a single discriminant function, a solution is made for multiple functions that are orthogonal to each other. The first function locates the axis where the distances between the group centroids is maximized and the discriminant scores about the group means are minimized. The second function is orthogonal to the first and represents the second most powerful discriminator axis. The remaining functions account for successively lower discriminations. A schematic representation of the multiple discriminant process is shown in Figure 5.18. When observations are classified with respect to one or another of the groups, the assignment can be made in terms of probability, using a Bayesian estimation from the discriminant scores in conjunction

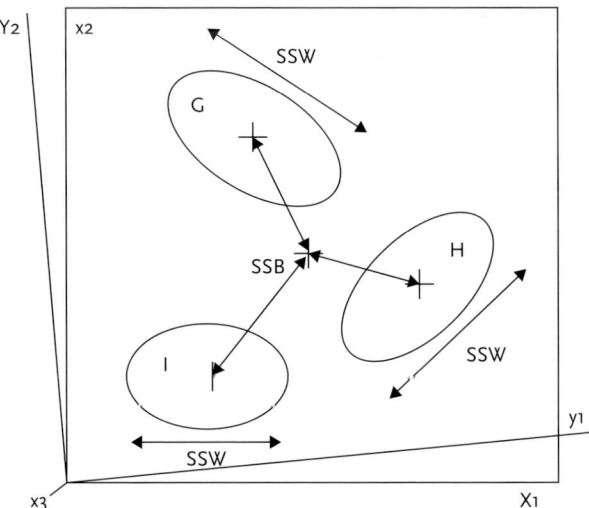

Figure 5.18: Schematic picture of multiple discriminant analysis applied to the differentiation of groups G, H, and I, in terms of variables $X1$, $X2$, $X3$, with location of new axes $Y1$ and $Y2$, from analysis of within-groups sums of squares (SSW) and between-groups sums of squares (SSB).

with a prior probability. The prior probability is conventionally chosen to be either an equal probability for all groups or the probabilities that correspond to the proportion of the observations in the total sample that are supplied by each group.

Logs (Figure 5.3) and core lithology assignments (Figure 5.4) of a Chase Group section from the Mobil #1-2 Brown well can be used as a demonstration of the application of multiple discriminant analysis, There are four lithologies (shale, siltstone, dolomite, and limestone) in this training well. By using the log measurements of *RHOmaa* (apparent grain density), *Umaa* (apparent matrix bulk photoelectric absorption), *PHIT* (average of the neutron and density porosities), *THOR* and *POTA* (thorium and potassium, respectively, from the spectral gamma-ray log) that were applied in the principal component electrofacies analysis, instructive comparisons can be drawn between the two approaches. A tally matrix of lithology predictions in the training well are shown in Table 5.4, based on maximum probability estimated from scores computed from multiple discriminant analysis. The statistics suggest that shale and dolomite are most easily discriminated and limestone predictions are weaker than might be expected, but limestone misclassifications are influenced by chert content and some dolomitization. The variety of cements, accessory minerals and compositional variability of the siltstones account for their broad spread of predictive assignments.

Results of the application of the training discriminants to logs in the validation well of Amoco #3HI Montgomery are shown as a probability profile and lithology assignment column in Figure 5.19. The probability profile can be considered from two perspectives. First, a zone may be thought of in probability terms as to the likelihood of any zone being one or another of the lithologies. Alternatively, the probabilities associated with a zone may be viewed as proportions of the endmembers. So, for example, a zone with equally high probability of being either a siltstone or a shale, may simply be a silty shale (or a shaly siltstone). In this sense, the probabilities for each zone characterize lithofacies that, if valid, are a truer rendition of the continuous variability of lithofacies expressed by the log measurements than the categorical assignment to one or another of the lithologies. The observations from the core interpretation of this section of the prediction well shown in Figure 5.17

Table 5.4. TALLY MATRIX OF LITHOFACIES PREDICTIONS IN THE CHASE GROUP TRAINING WELL MOBIL #1-2 BROWN BASED ON MULTIPLE DISCRIMINANT ANALYSIS. SH = SHALE; SLT = SILTSTONE; DOL = DOLOMITE; LS = LIMESTONE; (BROKEN) = MISSING CORE

		PREDICTED			
		SH	SLT	DOL	LS
OBSERVED	SH	60	10	0	1
	SLT	7	42	19	10
	DOL	1	8	94	12
	LS	0	20	11	66
	(BROKEN)	0	0	2	1

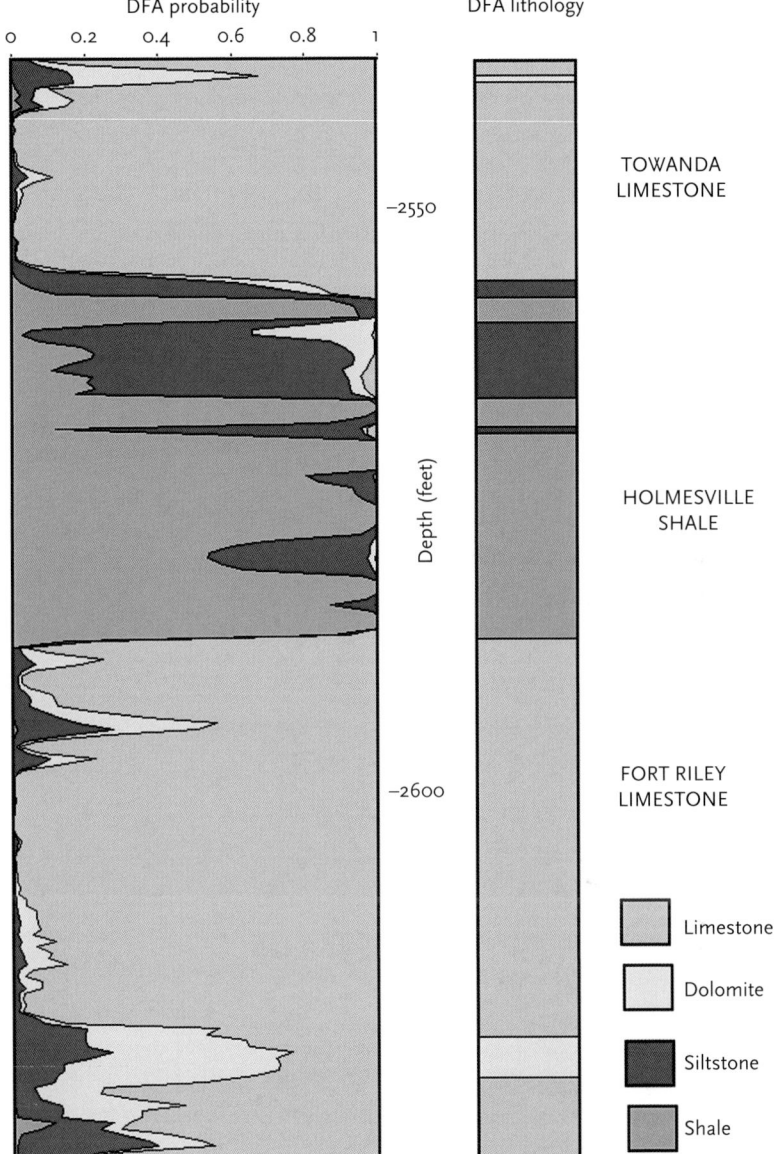

Figure 5.19: Multiple discriminant analysis prediction of lithofacies as probability (left) and assignment (right) in the Amoco #3HI Montgomery prediction well.

show good concordance with the discriminant predictions. Of particular interest is the apparent subdivision of the Holmesville Shale by the discriminant predictions into an upper part dominated by siltstones, from a lower part which is classified primarily as shale. This character matches core observations of variegated siltstones and shaly siltstones interpreted as lowstand deposits, succeeded by very fine-grained sandstone and siltstone of transgressive clastics.

NONPARAMETRIC DISCRIMINANT ANALYSIS

Both the supervised and unsupervised methods described up to this point have been parametric. In other words, the discrimination has been made in terms of parameters that summarize the distributions, rather than the original data clouds themselves. The parameters of each electrofacies are the multivariate means and the variances and covariances that describe the location and dispersion of a multivariate normal hyperellipsoid. Consequently, the normal distribution is a useful model, because it is specified by a few parameters and has a well-known probability density structure. However, it is not uncommon for electrofacies to take a variety of cluster shapes that are represented poorly by symmetrical ellipsoids. In an alternative strategy, their discrimination can be based on the actual distribution of their component data points rather than their summary parameters. This can be achieved by building a multivariate histogram into a computer database. The range of each log variable is partitioned into divisions. In multidimensional space, with logs as orthogonal axes, the divisions subdivide the space into a lattice of cells (Figure 5.20). The log responses for each zone from a training set are coordinates of a single data point that can be allocated to one of these cells. The frequencies of the points for each electrofacies are totaled for all cells to create the database histogram.

The database should be designed efficiently to avoid a huge and unnecessary allocation of computer memory. For example, the relatively coarse partition of each of six logs into ten subdivisions results in a framework of a million cells. If the training set has a thousand zones, then at least 999,000 cells will be empty. Many of these empty cells can also be discounted because they represent unreasonable

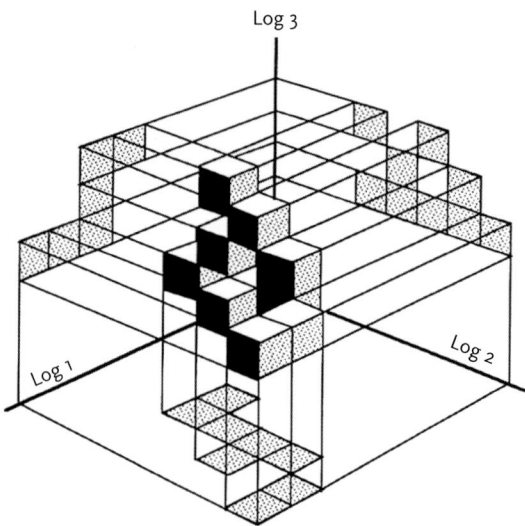

Figure 5.20: Representation of electrofacies in three-dimensional log space as a multivariate histogram of counts in a lattice work of cells. Adapted from Serra and Abbott (1980), Used with permission © 1980 Society of Petroleum Engineers.

combinations of log responses. In practice, efficient histogram databases store only information for those cells that are not empty. The database can then be used for classification of an unknown zone by inspection of the cell that matches the zone's log coordinates. If the cell is not empty, then the relative frequencies of each of the possible electrofacies can be used in the Bayesian formula described earlier to compute their respective posterior probabilities. The *a priori* probabilities can be taken from the training set as the proportion of zones that belong to a given electrofacies, or equal prior probabilities can be assigned to all electrofacies. If the unknown zone responses coincide with an empty cell, then a decision must be made by examining the neighboring cells. Tetzlaff et al. (1989) interpolated (or extrapolated) frequencies from neighboring cells that were weighted by their inverse squared distance. The sum of these interpolated values is a projected estimate of the electrofacies densities for the empty cell. If all the neighboring cells are empty, then the category of the zone remains unknown.

Zones can be misclassified by the two approaches of multiple discriminants and multivariate histograms, and this is often the result of different reasons. The parametric method generalizes by basing the probability density at any multivariate location on the normal statistics of the entire cloud. Misclassifications can occur for outlying points or spatial zones where the overlap of group clouds is poorly represented by multivariate normal ellipsoids. By contrast, the nonparametric method is highly specific, with its prediction based on the contents of an individual cell. When using training sets of a typical size, the cell frequencies may be small and poor estimates of their theoretical population values. Ideally, a perfect discrimination method would incorporate the strengths of both methods by sufficiently generalizing the data-point distributions to make predictions robust, but would retain localized features that reflect systematic, but smaller-scale effects.

The memory requirements of the multivariate histogram can be reduced considerably by the use of a "shingled block lattice" (SBL) structure of cell allocation, and the SBL method has some additional useful features. The structure of the SBL algorithm was inspired by the cerebellum model articulation controller (CMAC) model, which was introduced by Albus (1975) as a control mechanism for robots, and is modeled on the structure and function of the cerebellum. Although it would be considered today to be a type of neural network, the CMAC predates neural networks by years and is still virtually unknown outside robotics. Its greatly superior speed makes the CMAC preferable to conventional neural networks for the real-time demands of robot controllers. A geological application to mapping surfaces by using a CMAC was described by Hagens and Doveton (1991). According to Burgin (1992), a CMAC is most closely comparable to a feed-forward neural network that is trained by back propagation, but it almost always outperforms the neural network.

In this electrofacies application, the feedback learning process of the CMAC is eliminated, but the design features of a shingled block-lattice structure for coding inputs is retained. The *SBL* structure for coding inputs is a good choice for electrofacies distinctions because it maintains the advantages of the multivariate histogram approach, but its structure simultaneously results in huge savings in memory allocation and causes data-point densities to be generalized over localized neighborhoods.

These features combine to make the approach a simple and powerful method for electrofacies characterization and prediction. The CMAC design is retained, but the SBL components are used to sum frequencies rather than to adjust weights to match some localized target value.

The SBL structure is shown for two simple models in Figure 5.21, one with two inputs, the other with three. The concepts can be generalized to larger numbers of inputs in lattices of higher dimensions. The two-input lattice is a two-dimensional

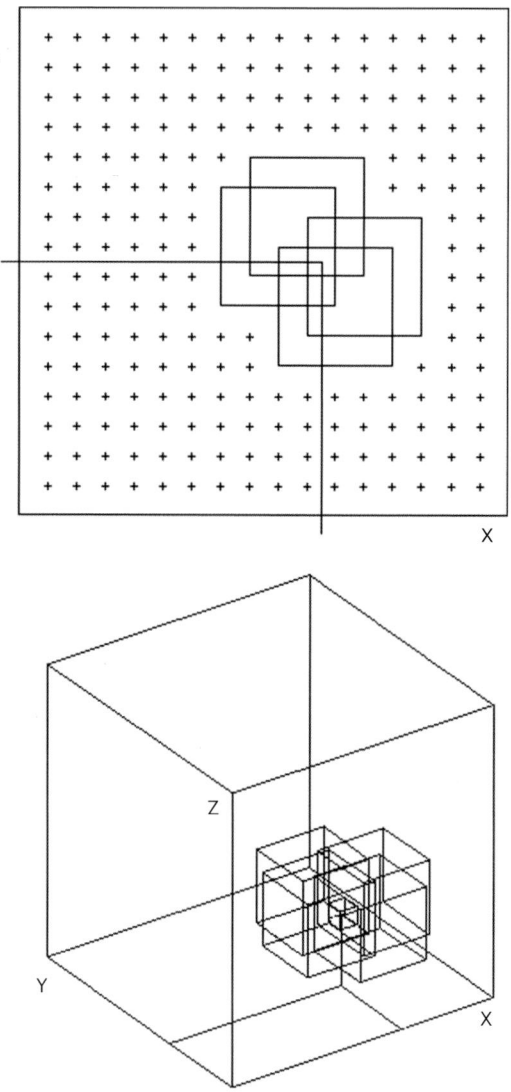

Figure 5.21: Basic structure of an SBL model with two inputs (above) and three inputs (below). In each case, log responses from a single depth zone are matched with a grid cell coded as the overlap of a unique set of blocks.

grid whose smallest element is a grid cell. However, rather than reference each cell with a specific location address, the cell is identified by the common overlap of four unique blocks. Notice that the arrangement of all possible blocks forms an intricate shingled tessellation whose overlapping structure causes each input vector of surface information to be generalized over a local neighborhood of blocks. The set of blocks is unique to each grid cell, but blocks are shared in common with adjacent grid cells. In the simple example, the surface values of 289 grid cells are encoded by the weights of one hundred blocks. This memory design differs markedly from conventional storage in both style and size requirements. Each input vector is tagged with several address locations rather than a single location. The overlapping of blocks allows generalization, rather than the "rote" memory of conventional storage. When used to encode data frequencies, the SBL operates in the same manner as an averaged shifted histogram (ASH), as described by Scott (1992).

For a two-dimensional square grid of c elements on each axis, a conventional memory of c^2 locations is required. This contrasts with rm^2 locations in the cerebellum model, where r is the width of the block measured in cells, and m is the number of blocks per layer on each axis. This allocation results in storage savings that increase dramatically for models with higher numbers of inputs. A three-input model can be shown conceptually as a three-dimensional construct in which a single grid cell is identified by the overlap of a unique set of surrounding blocks. Higher-dimensional models will take the form of nested hypercubes. For a prediction system with n inputs and c divisions per input, the memory size for a multivariate histogram would be c^n grid cells. At even a modest number of inputs, huge savings in memory would be made by the SBL model because only c^n/r^{n-1} blocks would be needed, where r is the cell activation range of the blocks. Added to this, the SBL has better predictive capabilities than the equivalent multivariate histogram.

As a demonstration case study, the SBL method of multivariate histograms was applied to the logs and core of the Chase Group sections in training and validation wells that were used earlier for the demonstration of multiple discriminant analysis. The same log inputs and lithology assignments were used so that useful comparisons could be made between the parametric and nonparametric approaches. The most fundamental decision concerns the number of layers to be used in the multivariate histogram. At one extreme, a single layer would be represented by one hyperdimensional cell, which would include all the log data points. Predictions of the probability of lithology assignment for any log input set would be the same and would be equal to a vector of the proportions of the lithologies observed in the core. At the other extreme, a high number of layers would ultimately partition the multivariate log space into such a fine mesh that each observation would be assigned a unique cell. In this case, the prediction in the training well would be perfect. However, in the validation well, predictions would be erratic and often incorrect. The cause of this behavior is that the model has "overfitted" the data (a frequent problem for naïve neural-network practitioners, as discussed later). In overfitting, the generalized qualities of predictive value in other wells are lost in increasingly detailed localized characterizations. As such, the multivariate histogram can reproduce training-set assignments as a rote procedure rather than a generalized prediction. Any measure of the real success of the application must be judged from comparison between predictions and core in a validation well and used as a basis for comparing results generated from SBL models with differing numbers of layers.

In the Chase Group application, a thirty-layer *SBL* model was chosen as the best predictor following a review of predictions in the Amoco #3HI Montgomery validation well, using alternative numbers of layers. A tally matrix of lithology predictions in the Mobil #1-2 Brown training well is shown in Table 5.5, based on maximum probability estimated from frequencies counted from the SBL multivariate histogram The tallies show improvement on the equivalent predictions by multivariate discriminant analysis (Table 5.4), which is to be expected because the multivariate histogram has greater flexibility in conforming to the shape of the electrofacies data cloud, rather than a hyperellipsoid. However, the lithofacies probability profile produced by the SBL multivariate histogram (Figure 5.22) does not show dramatic differences in overall predictive assignments compared with that of the multivariate discriminant analysis (Figure 5.19). Consequently, it confirms that multivariate discriminant analysis generally provides a robust and serviceable predictor of category.

The multivariate histogram approach refines the internal variability of composition within the lithologies, which can be useful in lithofacies interpretation. Core measurements of grain density and porosity are also plotted on Figure 5.22 for comparison, together with general lithology assignments observed and interpreted from the core. The core measurements can also be interpreted with respect to limestone textural changes, using the average values reported from Chase Group cores, that are shown in Figure 5.2. So, for example, the shoaling-upward limestone sequence in the Fort Riley limestone shows a clear upward trend in increasing porosity, which matches expectations of changes from a mud-supported to a grain-supported texture. The marine sandstones in the Holmesville Shale and the Fort Riley Limestone are matched by grain densities of quartz. The regressive lowstand shales and siltstones of the lower part of the Holmesville Shale are contrasted with the coarser-grained clastics of the succeeding transgressive deposits.

Table 5.5. TALLY MATRIX OF LITHOFACIES PREDICTIONS IN THE CHASE GROUP TRAINING WELL MOBIL #1-2 BROWN BASED ON NONPARAMETRIC *SBL* 30-LAYER MULTIVARIATE HISTOGRAM. *SH* = SHALE; *SLT* = SILTSTONE; *DOL* = DOLOMITE; *LS* = LIMESTONE; *(BROKEN)* = MISSING CORE

		PREDICTED			
		SH	SLT	DOL	LS
OBSERVED	SH	71	0	0	0
	SLT	4	58	8	8
	DOL	0	4	108	3
	LS	0	2	14	81
	(BROKEN)			2	1

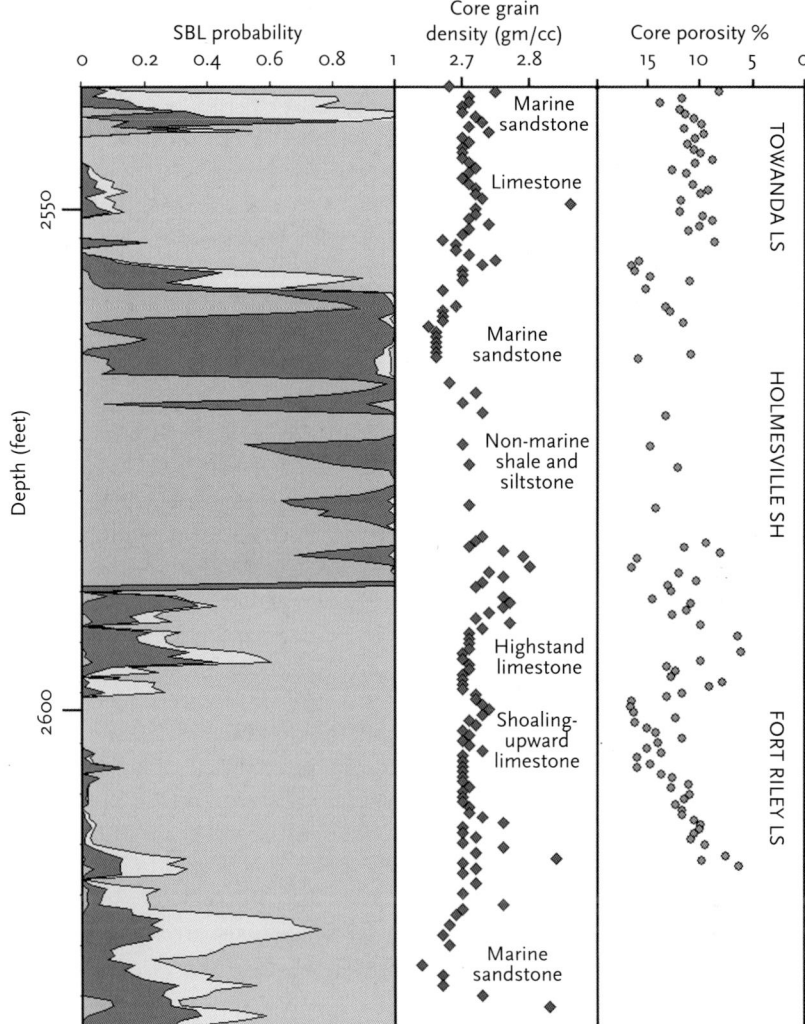

Figure 5.22: Nonparametric multivariate histogram prediction of lithofacies as probability (left) matched with core observations of grain density, porosity, and rocktype (right) in the Amoco #3HI Montgomery prediction well.

NEURAL-NETWORK PREDICTION OF LITHOFACIES FROM LOGS

The reservoir management of large oil and gas fields relies increasingly on the development of static geomodels as the precursor to dynamic models and simulations. Data for geomodels is drawn from existing wells, most of which have been logged but not cored because of the much higher costs that would be incurred in data acquisition. As a result, the effective prediction of lithofacies from logs has an important economic impact, rather than one of purely academic interest. The giant and mature Hugoton gas field provides an excellent example of where a geomodel was developed to provide a comprehensive lithologic and petrophysical realization of a complex gas system (Dubois et al., 2006). This goal required the creation of a fine-scale cellular petrophysical construct to model the thin layers of the Chase Group and the differentially depleted reservoir units within the field. Core-based calibration of neural-network predictions of lithofacies using log data were coupled with geologically constraining variables in order to provide accurate lithofacies models whose resolution ranged from well to field scales. Because of the high degree of variability within the key properties of porosity, permeability, and fluid saturation, eleven lithofacies were selected for prediction. The model represented more than 10,000 square miles and consisted of 169 layers, with cell dimensions of 660 x 660 feet, resulting in 110 million cells. A project of this magnitude required the flexibility of a neural-network application to accommodate the prediction of different lithofacies, coupled with extensive cross validation to ensure the closest match with known geometries and constraints.

Much of the theory that underlies neural networks was originally inspired by research by neuroscientists on the structure and function of the brain. The brain consists of a massively interconnected system of neurons that performs sensory processing, controls motor functions, and recognizes patterns. The synapses are the communication medium between neurons and are strengthened or weakened by electrochemical processes. It is theorized that we learn, store memories, and modify our behavior because of changes in the strengths of synapses within our brains. This phenomenon provides the core concept for artificial neural networks, which have been applied in a wide variety of disciplines, with probably the earliest application to petrophysics by Baldwin et al. (1989). Other early examples of petrophysical neural-network studies include Derek et al. (1990) and Rogers et al. (1992).

The structure of a neural network is conventionally drawn as a hierarchy of layers in which nodes (representing neurons) are connected by arcs (Figure 5.23). The arithmetic value of any node is equal to the sum of the values of the preceding nodes, each multiplied by the weight of the connecting arc:

$$y_i = \sum w_{ij} x_j$$

where y_i is the value of the ith node, x_j is the value of the jth node in the preceding layer, and w_{ij} is the weight associated with the arc that connects the two nodes. The output of a node is governed by an activation function of the summed input and a threshold that determines the initiation of output. The firing state of a node is either unity or zero, and it is determined by a sigmoidal function that models the transfer from input to output signals (see Figure 5.23). The general equation of the sigmoidal function takes the form:

$$P = 1/\left(1 + e^{-a/t}\right)$$

where P is the probability of the node firing, t is a constant that determines the steepness of the function, and a is the activation of the node. This feature attempts

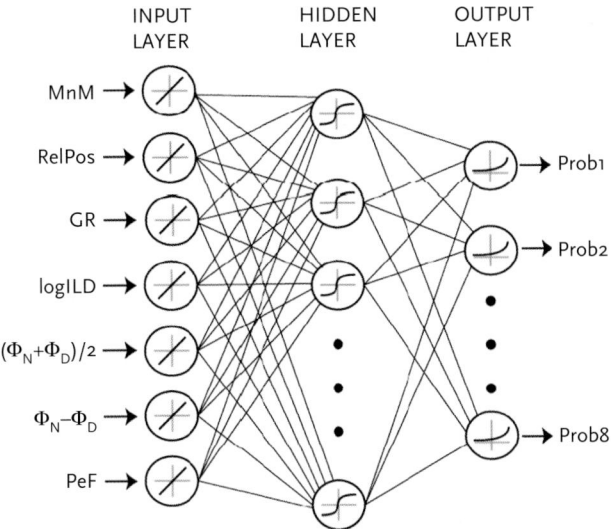

Figure 5.23: Schematic representation of single hidden-layer neural network used to estimate probabilities of Chase Group lithofacies from inputs of log parameters and geologic constraining variables.

to imitate the behavior of real neurons, which often tend to be either active or inactive. The steepness of the activation function will determine whether most of the input is transferred onward through the nodes or whether output is only initiated by stronger inputs.

A basic model consists of three layers of nodes. The first is an input layer that receives input data. The middle or hidden layer draws stimulation from the input nodes and transmits onwards to the final output layer, which is the result of the system. More complex neural networks incorporate multiple hidden layers. However, more hidden layers increase the likelihood that the network will simply memorize the input associations, rather than learn to generalize for useful predictions beyond the training set. By the same token, the number of nodes within a single hidden layer should be restricted in order to encourage generalization rather than memorization. In training the network, a set of patterns is presented repeatedly and the weights of the arcs are modified so that the output makes a better match with a desired result. Training is usually accomplished by the back-propagation of errors through the network, which distributes the difference between the desired result and the actual output as small incremental adjustments in the interconnection weights. The process is gradual and iterative; weights gradually converge to an equilibrium setting, at which point the network is trained.

For the prediction of lithofacies from logs in the Hugoton field, a standard single-hidden- layer neural network was applied to two computed geological variables of a depositional environment indicator (MnM) and a stratigraphic cycle relative position ($RelPos$), as supplements to the following log values: gamma-ray (GR), logarithmically-scaled deep-induction resistivity ($logILD$), averaged neutron and density porosity ($\Phi_n/2 + \Phi_d/2$), neutron and density porosity difference ($\Phi_n-\Phi_d$), and photoelectric factor (*PeF*). A diagrammatic representation of the neural-network operation and its inputs are shown in Figure 5.23. The eleven lithofacies used for prediction consisted of continental sandstone, continental coarse siltstone, continental fine siltstone, marine sandstone, marine siltstone, mudstone, wackestone, packstone-grainstone, bafflestone, very fine crystalline dolomite, and fine to medium crystalline dolomite. An example of lithofacies prediction in the Chase Group of a Hugoton well is shown in Figure 5.24 and compared with actual lithofacies assignments from the core.

Verification of performance by a validation set or sets is particularly important when using neural networks. Validation provides a more realistic test of effective prediction power than the statistics of learning generated by the training set. There are times when a network can be made so complex that it reproduces outputs from

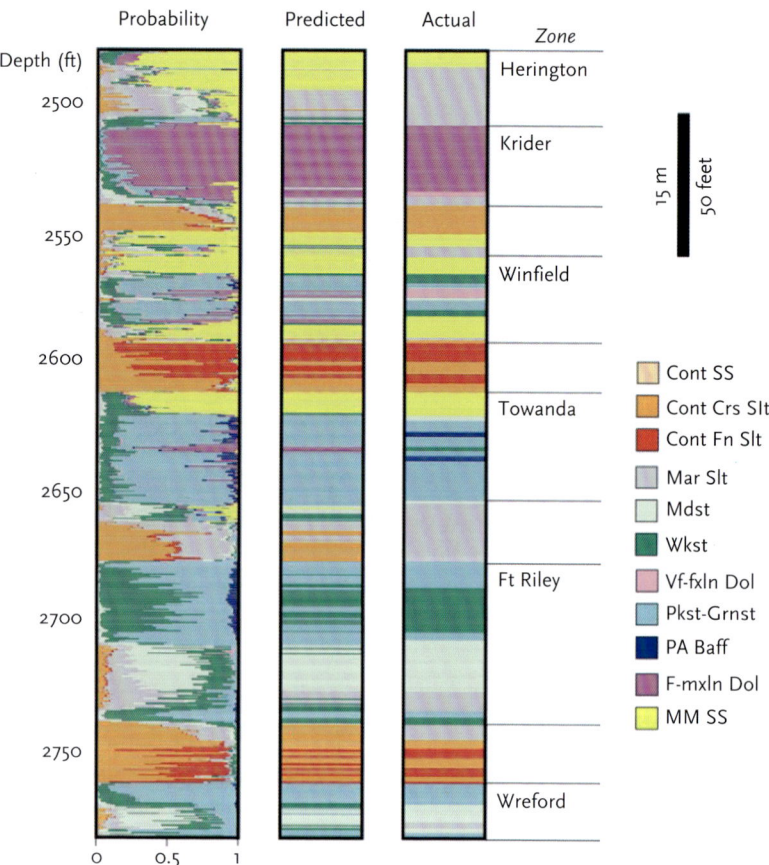

Figure 5.24: Example of lithofacies prediction in the Chase Group of a Hugoton field well by a neural network. From Dubois et al. (2006), ©2006 American Association of Petroleum Geologists (AAPG), reprinted by permission of the AAPG, whose permission is required for further use.

the training sets almost perfectly. However, this same network may perform worse than conventional statistical methods when it generates predictions from new observations. The purpose of the training is to acquire the ability to generalize from observations rather than to reproduce them by rote. Generalization will absorb the systematic trends that link observations and screen out the localized and random error components. If these components are absorbed from the training input, the weight configuration is distorted and may show erratic behavior in predictions outside the training set. Neural-network specialists are well aware of this problem, which is generally termed "overfitting," and they have suggested various strategies to minimize the effects (see, e.g., Smith, 1993).

In the development of the Hugoton geomodel, the training of the neural network was monitored by intensive and repetitive cross validation among the training set of cored and logged wells. The goal of obtaining a perfect predictive model is clearly unrealistic and even unnecessary, because the overriding purpose of the geomodel is to capture the petrophysical variability in porosity, permeability, and fluid saturations. Consequently, the cost of predicting the wrong lithofacies is lower if the predicted lithofacies has petrophysical properties that are similar to the actual lithofacies, than if their petrophysical properties differ substantially. This concept was incorporated within the cross validation procedure as a misallocation cost matrix. A fence-diagram representation of the final geomodel is shown in Figure 5.25, where continental siltstones (red and orange) separate the marine carbonate half cycles. Both continental (yellow-orange) and marine (yellow) sandstones predominate in the landward direction to the northwest, while marine carbonates (cooler colors) dominate in the basinward direction to the southeast.

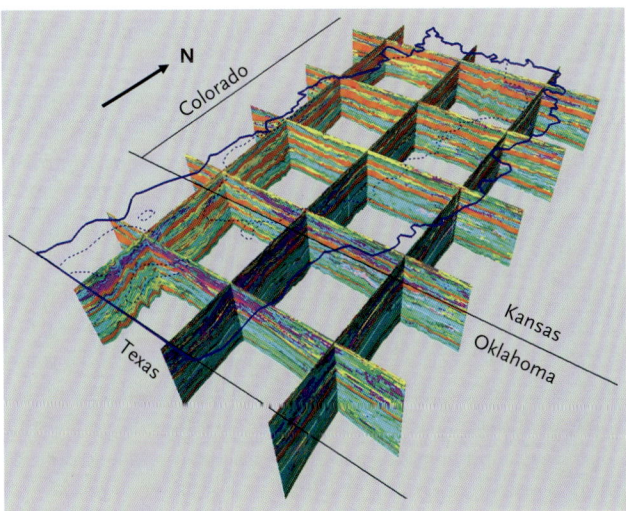

Figure 5.25: Stratigraphic cross-section fence diagrams of predicted lithofacies within the Hugoton Chase Group geomodel. From Dubois et al. (2006), © 2006 American Association of Petroleum Geologists (AAPG), reprinted by permission of the AAPG, whose permission is required for further use.

As a final note, most neural networks operate as supervised procedures, that is, they are trained on known material before being applied to problem data. Unsupervised learning is obviously a much more difficult process to emulate in a realistic and convincing manner. However, some useful progress has been made on self-organized networks, following the pioneering work of Kohonen (1984).

BEYOND PRODUCT FACIES TO THE PETROPHYSICAL PREDICTION OF PROCESS FACIES

In the applications described in this chapter, both supervised and unsupervised methods have been used to predict lithological categories on the basis of petrophysical logs. When the categories are lithologies, then the assignment is one of petrophysical rock typing. When the categories of lithology are subdivided into separable lithofacies, then the classification is keyed to textural and compositional characteristics. Because both lithofacies and their petrophysical measurements vary continuously, the probabilistic predictions of lithological categories are intrinsic indications of the variability of the lithofacies. The properties associated with each lithofacies are chosen selectively by the geologist as those that are most likely to be linked with the genetic process of either deposition or subsequent diagenetic overprint. If petrophysical measurements can be mapped successfully to separate lithofacies, then the geologist can then interpret process for a logged section where no core is available. The identification of lithofacies in outcrop, core, or by petrophysical prediction is what we mean by the classification of product facies, which are subsequently interpreted in terms of process facies. So, for example, a coarse-grained sandstone lithofacies generally reflects a high-energy depositional environment, but this could be assigned either to a channel deposit or to a beach sandstone. In the absence of definitive additional criteria, the determination is generally made by context, that is, the nature of the vertically adjacent lithofacies.

Walther (1894) was one of the first to realize that facies that succeed one another conformably must have been deposited in adjacent environments. This concept, known as Walther's law, was summarized by Selley (1976): "a conformable vertical sequence of facies was generated by a lateral sequence of environments." Because empirical electrofacies are the petrophysical expression of lithofacies, they are subject to the same constraints. It follows that, in a conformable sequence, certain electrofacies will never be expected to succeed others. Up to this point, we have considered the classification of any unknown zone in isolation. However, Walther's law implies that the nature of the electrofacies of the immediately preceding zone is an important piece of information. This can be incorporated easily as the prior probability term of a Bayesian equation, based on the assigned electrofacies probability for the preceding zone.

Posterior probabilities based on sequence position are easily generated by a Markov-chain analysis. Based on data from control wells, a tally matrix of observed vertical transitions between electrofacies types can be converted to a matrix of transition probabilities. Each transition probability is then a prior probability as to the electrofacies identity of a zone, given the identity of the immediately preceding zone. The transition matrix automatically quantifies Walther's law for insertion into the Bayesian classifier. It should be noted that Walther's law holds exactly only for conformable sequences. However, the exceptions that are caused by unconformities and other discontinuities are automatically accommodated within the transition statistics of the observational sequences. Transition probabilities of an electrofacies to itself capture the relative thickness of the electrofacies. Bed thicknesses predicted by this transition probability will follow a geometric distribution (Krumbein and Dacey, 1969). Although sedimentary bed thicknesses are commonly considered to be approximately lognormally distributed (see, e.g., Pettijohn, 1957; Potter and Siever, 1955), they are reasonably represented by the geometric distribution for the purposes of electrofacies prediction. Finally, the common tendency for loosely repetitive motifs of lithofacies (or cycles) is expressed by the structure of the transition probability. In the event of a randomly ordered succession of electrofacies, the transition probabilities are equal to the relative proportion of the occurrence of each of the electrofacies. These estimates would be appropriate prior probabilities for a Bayesian model that disregards vertical contiguities and is weighted by overall electrofacies abundances.

For a case-study application of this concept, which was described by Bohling et al. (1997), we revisit the hydrology observation well described in Chapter 4 for clay-mineral compositional estimation from logs calibrated to X-ray diffraction analyses. The logged section of the Lower Cretaceous clastic succession was cored and the depositional environments were interpreted using standard visual observations of the core (Figure 5.26). The core represents the record of a complex of deltaic deposits, with fluvial sandstones and alluvial-plain clays, and estuarine and paralic units. The core has been studied by several geologists independently, so that there are three different environmental interpretations. Although they are in essential agreement, there are local differences of opinion, which commonly occur between geologists. Consequently, sedimentary process interpretation of units in a core should be considered as a "ground interpretation," rather than the ground truth that would be expected from core descriptions of lithology and texture, which should be repeatable between competent geologists. Rather than presenting a drawback to a petrophysical application, this provides a new role for the integrative method to capture the log properties of the environment; to capture the consistencies, ambiguities, and contradictions of the audit; and to alert the geologist to zones of anomalous petrophysical character that were passed over in the initial visual assessment. The feedback loop between core examination and supervised log calibration coordinates visual observations with invisible petrophysical properties.

Six sedimentary facies were identified from core descriptions of the well: marine shale, paralic deposit, floodplain deposit, channel sandstone, splay sandstone, and paleosol. The "floodplain" and "paralic" terms were applied as broad "generic" facies, that is, mixtures of facies from a generalized environmental setting, as contrasted with the other "specific" facies, which have diagnostic features of distinct environments. This classification strategy cut down on the more speculative identification of sedimentary environments in cored sections whose features (or lack of them) led to

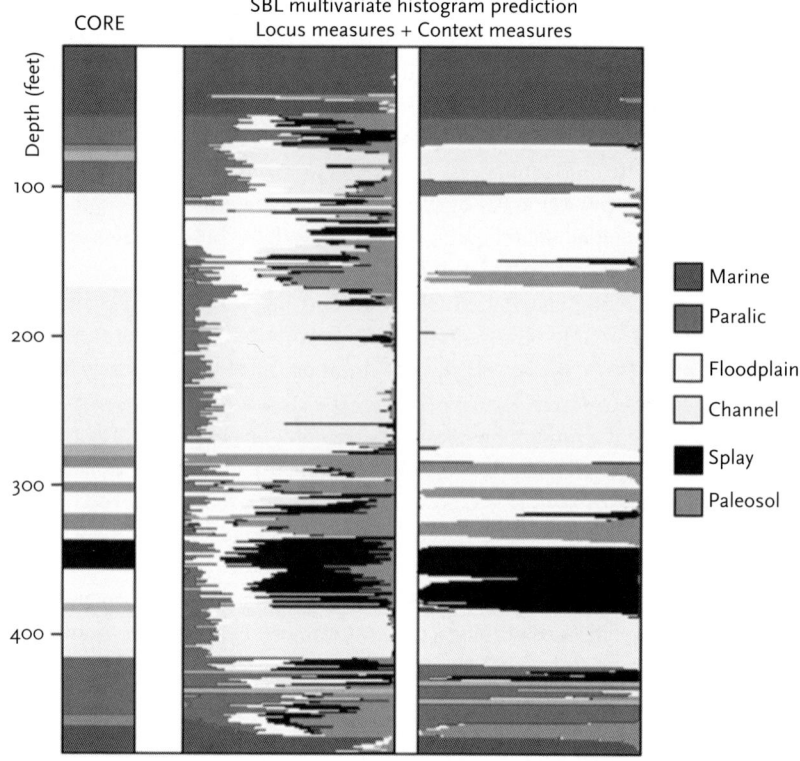

Figure 5.26: Interpreted depositional facies from core observation (left) matched with SBL multivariate histogram prediction based on log locus measures (center) then modified by the incorporation of transition probability context measures (right) of a Lower Cretaceous deltaic sequence. Courtesy Society for Sedimentary Research.

arbitrary or ambiguous calls. At the same time, the strategy provided an additional opportunity for a petrophysical classification method to either subdivide generic facies or to be as conservative as the geologist. So, for example, would the petrophysical classification identify all floodplain units as "floodplain," or would it be bolder, such as by identifying some of the sandstones as either channel or splay deposits? These considerations underscore the role of this approach as an integrative learning tool.

A SBL multivariate histogram was used to produce probability density estimates for each facies for all logged depths, including those with missing cores. These density estimates are combined using Bayes' theorem to produce probabilities of membership in each facies. These initial results are shown in Figure 5.27, where the probability profile is matched against the facies assignments from the core. In a second pass, transition probabilities generated from the succession of cored facies were substituted as prior probabilities for the classification procedure. By doing this, the locus measures that were based purely on their position in multivariate log space were augmented by context measures that reflected the nature of the underlying facies. At the most basic level, transitions that are never observed to occur are excluded by a prior probability of zero. In this sense, the transition probability matrix applies Walther's law.

The results from the addition of transition probabilities into the priors are shown in Figure 5.27 for further comparison and interpretation. The immediate difference that can be seen is that the predictions are much crisper than those restricted to locus measures, where all facies are considered to be possible. The log responses of paralic and channel sandstones are so similar that the SBL multivariate histogram has difficulties in distinguishing them until the context measure of transition probability is used as a modifier. A paleosol is now predicted in the lower part of the upper floodplain unit that was not observed originally, although comparison with the log-predicted clay mineral profile (Figure 4.9) suggests that this core section should be revisited for further evaluation. These and other observations illustrate the enhanced collaborative aspects of processed logs that are a significant advance on traditional core-log evaluations based on raw curves.

The interpretation of deposits from the ephemeral facies mosaic of a delta plain will always contain some ambiguity, regardless of whether judgments are based on core, logs, or both. A similar situation can be encountered in the interpretation of satellite imagery, where context is an important component of identification. Haslett (1985) showed how substantial improvements could be made in maximum-likelihood classifications of pixel elements in *LANDSAT* imagery by incorporating Markovian transition probabilities for the states of adjacent pixels. By this means, the vector of responses at each pixel location is not processed in isolation but is considered in the context of broader geographic elements and lateral relationships.

In a related core-log study, Bohling et al. (1996) related lithodensity-neutron and spectral gamma-ray measurements to the sequence stratigraphic framework interpreted from a 330-meter Pennsylvanian core from southwestern Kansas. Middle and Upper Pennsylvanian depositional sequences in the upper Marmaton, Lansing, and Kansas City groups have been defined in cores and outcrops across the Kansas shelf. The depositional sequences are generally thinner than sixty feet and are carbonate-dominated successions representing marine inundation and retreat. These depositional sequences fall below the resolution of conventional seismic methods, but are well within the domain of information provided by wireline logs. Log values of the sequence stratigraphic depositional facies, including paleosols, flooding units, condensed sections and late high-stand deep-water, peritidal and subtidal carbonate units were enumerated within a SBL multivariate histogram for analysis and prediction.

Transition probabilities based on sequence position were generated through Markov-chain analysis of the sequence facies observed in the long core (Table 5.6). The transition matrix was evaluated in terms of the sequence stratigraphic model, with expectations that the values reflected processes of regional scale, rather than the localized character of lateral facies of a delta complex. The depositional sequences developed on the Kansas shelf in the neighborhood of the training well are typified

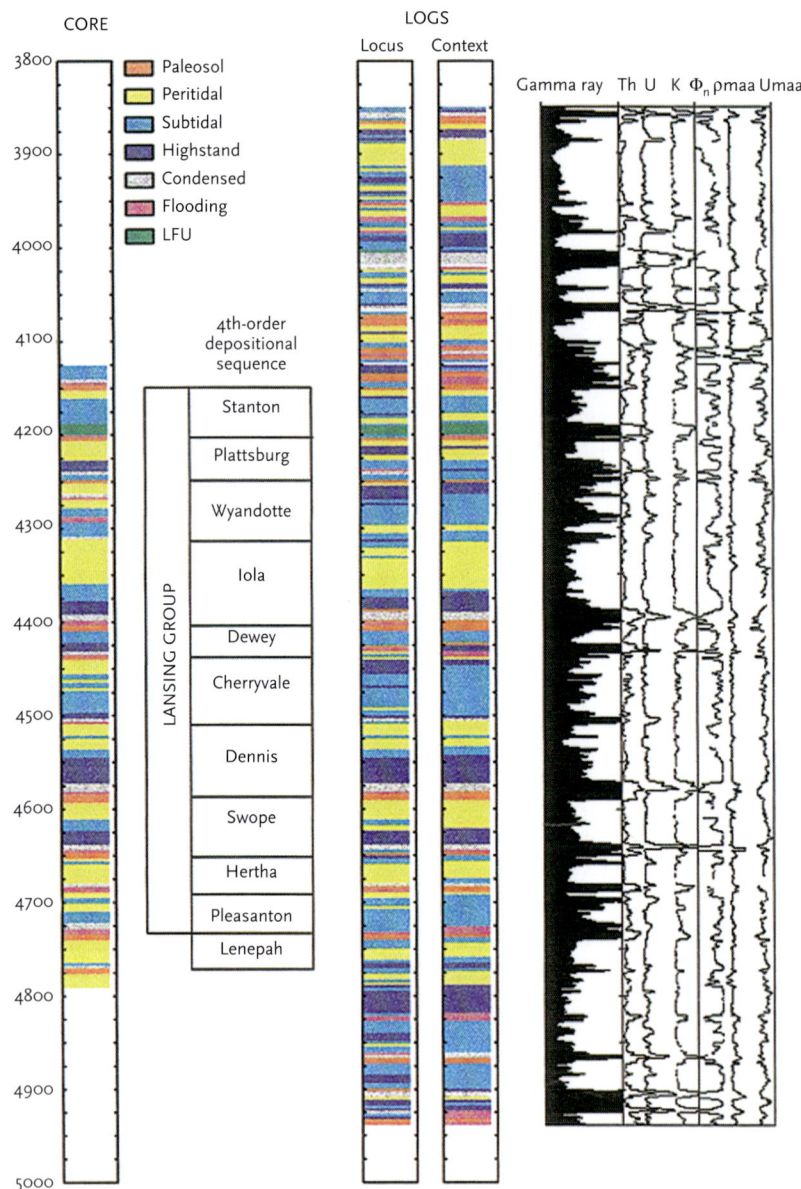

Figure 5.27: Sequence facies predictions using an SBL multivariate histogram applied to six log variables in a Pennsylvanian carbonate shelf sequence from Kansas. Sequence units observed in the core are compared with "locus" predictions (log responses alone) and "context" predictions (incorporating transition probabilities). From Bohling et al. (1996). Courtesy International Association for Mathematical Geosciences.

Table 5.6. TRANSITION PROBABILITY MATRIX OF PENNSYLVANIAN STRATIGRAPHIC UNITS OBSERVED IN CORE FROM A WELL IN SOUTHWEST KANSAS. FROM BOHLING ET AL. (1996), COURTESY SOCIETY FOR SEDIMENTARY RESEARCH

		FU	LFU	CD	HS	ST	PT	PLS
Flooding Unit (FU)	FU		0.08	0.92				
Lowstand FU (LFU)	LFU			1.00				
Condensed (CD)	CD				0.40	0.47	0.13	
Highstand (HS)	HS					0.71	0.29	
Subtidal (ST)	ST	0.07		0.07			0.79	0.07
Peritidal (PT)	PT	0.24				0.29		0.47
Paleosol (PLS)	PLS	0.89		0.11				

by thin flooding units and condensed sections interbedded with thick, regressive carbonates. The paleosols are generally well developed, but the subaerial exposure that terminates the sequence may be expressed only as an exposure surface and/or meteoric diagenetic overprinting of the underlying carbonate. The carbonate units are similar lithologically, but the highstand, subtidal, and peritidal facies are usually visually distinctive in core samples. However, it was anticipated that they may show only minor chemical and mineralogical differences and so be challenging for petrophysical differentiation.

Results of predictions by the SBL multivariate histogram are shown in Figure 5.27 and compared with core assignments. The paleosols in the sequence are mostly calcareous oxidized siltstones with some calcretes. Almost all of the paleosols observed in the core were correctly predicted using probabilities generated by the SBL multivariate histogram; this was an important facies to recognize, since these are principal horizons used to subdivide the depositional sequences. The incorporation of transition probabilities as prior probabilities eliminated most of the remaining misclassifications. Thicker flooding units and condensed sections were correctly predicted, but other, thinner units were often missed, partly because they were below the resolution of the well logs. The differentiation of highstand, subtidal, and peritidal carbonate facies proved to be more challenging, but the incorporation of transition probabilities improved the success of teh classifications. When examined in detail, many of the misclassifications were seen to be caused by diagenetic processes such as dolomitization and chert formation. With further training and evaluation, SBL multivariate histograms offer a fruitful integrative core-log strategy for the interpretation of stratigraphic sequences over large distances across the Pennsylvanian Kansas shelf, where cores are much less common than petrophysical logs.

REFERENCES

Albus, J.S., 1975, A new approach to manipulator control: The Cerebellar Model Articulation Controller (CMAC): Transactions of the American Society of Mechanical Engineers, v. 97, no. 3, September, pp. 220–227.

Baldwin, J. L., Otte, D.N., and Bateman, R.M., 1989, Computer emulation of human mental processes: application of neural network simulators to problems in well log interpretation: Society of Petroleum Engineers, SPE 19619-MS, pp. 481–493.

Bohling, G.C., Doveton, J.H., and Watney, W.L., 1996, Systematic identification of sequence stratigraphic units from wireline logs, in J.A. Pacht, R.E. Sheriff, and B.F. Perkins, eds., Stratigraphic analysis utilizing advanced geophysical, wireline, and borehole technology for petroleum exploration and production: Gulf Coast Section Society of Economic Paleontologists and Mineralogists 17th Annual Research Conference, pp. 29–37.

Bohling, G.C., Doveton, J.H., and Hoth, P., 1997, Probabilistic classification and prediction of facies types in a mid-continent Cretaceous deltaic-marine sequence from petrophysical log descriptors using a CMAC procedure, in V. Pawlowsky Glahn, ed., Proceedings of the 3rd Annual Conference of the International Association for Mathematical Geosciences, part 1, International Center for Numerical Methods in Engineering (CIMNE), Barcelona, pp. 242–247.

Brandsegg, K.B., Hammer, E., and Sinding-Larsen, R., 2010, A comparison of unstructured and structured principal component analyses and their interpretation: Natural Resources Research, v. 19, no. 1, pp. 45–62.

Burke, J.A., Campbell, R.L., Jr., and Schmidt, A.W., 1969, The lithoporosity crossplot: Transactions of the Society of Professional Well Log Analysts, 10th Annual Logging Symposium, Paper Y, 29 p.

Burgin, G., 1992, Using cerebellar arithmetic computers: AI Expert, v. 7, no. 6, June, pp. 32–41.

Caldwell, C.D., 1991, Cyclic deposition of the Lower Permian, Wolfcampian, Chase Group, Western Guymon-Hugoton field, Texas County, Oklahoma, in Midcontinent core workshop: Integrated studies of petroleum reservoirs in the midcontinent: Midcontinent American Association of Petroleum Geologists Section Meeting, Kansas Geological Survey, Wichita, Kansas, pp. 57–75.

Cannon, D.E., and Horkowitz, J.P.,1997, Complex reservoir evaluation in open and cased wells: Transactions of the Society of Professional Well Log Analysts, 38th Annual Logging Symposium, Paper DD, 14 p.

Chang, W.C., 1983, On using principal components before separating a mixture of two multivariate normal distributions: Journal of the Royal Statistical Society, Series C (Applied Statistics), v. 32, no. 3, pp. 267–275.

Clavier, C., and Rust, D.H., 1976, MID plot: A new lithology technique: Log Analyst, v. 17, no. 6, pp. 16–24.

Delfiner, P. C., Peyret, O., and Serra, O., 1987, Automatic determination of lithology from well logs: Society of Petroleum Engineers Formation Evaluation, v. 2, no. 3, pp. 303–310.

Derek, H., Johns, R., and Pasternack, E., 1990, Comparative study of back-propagation neural network and statistical pattern recognition techniques in identifying sandstone lithofacies, in Proceedings of the 1990 Conference on Artificial Intelligence in Petroleum Exploration and Production: Texas A&M University, College Station, Texas, p. 41–49.

Doveton, J.H., 1994, Geological log analysis using computer Methods: Computer Applications in Geology, No. 2, American Association of Petroleum Geologists, Tulsa, 169 p.

Dubois, M. K., Byrnes, A.P., Bohling, G.C., and Doveton, J.H., 2006, Multiscale geologic and petrophysical modeling of the giant Hugoton gas field (Permian), Kansas and Oklahoma, in P. M. Harris and L. J. Weber, eds., Giant hydrocarbon reservoirs of the world: From rocks to reservoir characterization and modeling: American Association of Petroleum Geologists Memoir 88, Tulsa, Oklahoma, pp. 307–353.

Edmundson, H., and Raymer, L.L., 1979, Radioactive parameters for common minerals: Transactions of the Society of Professional Well Log Analysts, 20th Annual Logging Symposium, Paper O, 25 p.

Fertl, W.H., Stapp, W.L., Vaello, D.B., and Vercellino, W.C., 1980, Spectral gamma-ray logging in the Texas Austin chalk trend: Journal of Petroleum Technology, v. 32, no. 3, pp. 481–488.

Fisher, R.A., 1936, The use of multiple measurements in taxonomic problems: Annals of Eugenics, v. 7, no. 2, pp. 179–188.

Gill, D., 1970, Application of a statistical zonation method to reservoir evaluation and digitized-log analysis: American Association of Petroleum Geologists Bulletin, v. 54, no. 5, pp. 719–729.

Gill, D., Shomrony, A., and Fligelman, H., 1993, Numerical zonation of log suites and log-facies recognition by multivariate clustering: American Association of Petroleum Geologists Bulletin, v. 77, no. 10, pp. 1781–1791.

Hagens, A. and Doveton, J.H., 1991, Application of a simple cerebellar model to geologic surface mapping: Computers & Geosciences, v. 17, no. 4, pp. 561–567.

Haslett, J., 1985, Maximum likelihood discriminant analysis on the plane using a Markovian model of spatial context: Pattern Recognition, v. 18, no. 3–4, pp. 287–296.

Kohonen, T., 1984, Self-organization and associative memory: New York, Springer-Verlag, 255 p.

Krumbein, W.C., and Dacey, M.F., 1969, Markov chains and embedded Markov chains in geology: Mathematical Geology, v. 1, no. 2, pp. 79–96.

Lucia, F.J., 1999, Carbonate reservoir characterization: Springer, Berlin, 226 p.

Luczaj, J.A., 1998, Regional and stratigraphic distribution of uranium in the Lower Permian Chase Group carbonates of southwest Kansas: Log Analyst, v. 39, no. 4, pp. 1–9.

Pettijohn, F.J., 1957, Sedimentary rocks: Harper, New York, 718 p.

Pettijohn, F.J., Potter, P.E., and Siever, R., 1972, Sand and sandstone: New York, Springer-Verlag, 618 p.

Potter, P.E., and Siever, R., 1955, A comparative study of Upper Chester and Lower Pennsylvanian stratigraphic variability: Journal of Geology, v. 63, no. 5, pp. 429–451.

Quirein, J., Kimminau, S., Lavigne, J., Singer, J., and Wendel, F., 1986, A coherent framework for developing and applying multiple formation evaluation models: Transactions of the Society of Professional Well Log Analysts, 27th Annual Logging Symposium, Paper DD, 16 p.

Reading, H.G., 1978, Facies, in H. G. Reading, ed., Sedimentary environments and facies: Elsevier, New York, pp. 4–14.

Rogers, S.J., Fang, J.H., Karr, C.L., and Stanley, D.A., 1992, Determination of lithology from well logs using a neural network: American Association of Petroleum Geologists Bulletin, v. 76, no. 5, pp. 731–739.

Scott, D.W., 1992, Multivariate density estimation: Theory, practice, and visualization: John Wiley & Sons, New York, 317 p.

Selley, R.C., 1976, An introduction to sedimentology: Academic Press, London, 408 p.

Serra, O., and Abbott, H.T., 1980, The contribution of logging data to sedimentology and stratigraphy: Society of Petroleum Engineers, SPE 9270-MS, 19 p.

Serra, O., Delfiner, P., and Levert, J.C., 1985, Lithology determination from well logs: Case studies: Transactions of the Society of Professional Well Log Analysts, 26th Annual Logging Symposium, Paper WW, 19 p.

Siemers, W.T., and Ahr, W.M., 1990, Reservoir facies, pore characteristics, and flow units: Lower Permian Chase Group, Guymon-Hugoton Field, Oklahoma: Soc. Petroleum Engineers, SPE 20757, pp. 417–428.

Smith, M., 1993, Neural networks for statistical modeling: Van Nostrand Reinhold, New York, 235 p.

Stowe, I., and Hock, M., 1988, Facies analysis and diagenesis from well logs in the Zechstein carbonates of northern Germany: Transactions of the Society of Professional Well Log Analysts, 29th Annual Logging Symposium, Paper HH, 24 p.

Tetzlaff, D.M., Rodriguez, E., and Anderson, R.L., 1989, Estimating facies and petrophysical parameters from integrated well data: Transactions of the Society of Professional Well Log Analysts, Log Analysis Software Evaluation and Review (LASER) Symposium, Paper 8, 22 p.

Walther, J., 1894, Einleitung in die Geologie als historische Wissenschaft: Verlag von Gustav Fischer, Jena, 1055 p.

Winters, N.D., 2007, Depositional model, sequence stratigraphy, and distribution of the tan very fine-grained sandstone facies in the upper Chase Group, Hugoton Gas Field, southwest Kansas and Oklahoma Panhandle: MS thesis, University of Kansas, 147 p.

Wolff, M., and Pelissier-Combescure, J., 1982, FACIOLOG—automatic electrofacies determination: Transactions of the Society of Professional Well Log Analysts, 23rd Annual Logging Symposium, Paper FF, 22 p.

CHAPTER 6

Pore-System Facies: Pore Throats and Pore Bodies

THE PETROFACIES CONCEPT

When Archie (1950) first introduced the term "petrophysics," he outlined a tentative petrophysical system "...which revolves mainly around pore-size distribution which defines the capillary pressure curve, permeability, and porosity." As such, a pore distribution does not necessarily coincide with a specific rock type. Different lithologies might contain similar pore distributions and a single lithology might be characterized by several distinctive pore distributions. In the latter case, these differences could be used as the basis for a lithofacies subdivision, where the criteria were defined by pore-network properties rather than more conventional fabric observations. Often, there will be a substantial commonality between the two approaches, because the pore network and rock framework are complementary.

The term "petrofacies" (which comes from "petrophysical facies") extends the facies concept to pore networks. Although this name is commonly (but not exclusively) used for this purpose, the range of published definitions is fairly broad, as pointed out by Sullivan et al. (2003). Some authors intermingle notions of petrofacies with electrofacies and lithofacies, which is understandable, because in many reservoirs there are strong intercorrelations between them. In this text, we distinguish between electrofacies, either seemingly natural petrophysical log associations found by unsupervised methods, or those determined from lithofacies by supervised methods. Lithofacies are generally recognized by standard visual observations of a core, although they may be defined by reference to distinctive porosity-permeability associations (petrofacies) in core measurements.

The two fundamental reservoir components of pore microarchitecture are essentially the same as the spatial elements of conventional architecture: the relative sizes and arrangement of the pore bodies (rooms) and the pore throats (doors between rooms). In an oil or gas reservoir, the volume of pore space contained in the pore bodies dictates the total storage capacity, while the access of hydrocarbon to the pore bodies is regulated by the size of the linking pore throats. Realistic pore-network models for characterizing hydrocarbon recovery from reservoirs are, appropriately, labyrinthine in their intricacy. However, the key pore attributes that are the focus of petrophysical applications are the size distributions of the pore bodies and pore throats, together with the aspect ratio of pore-body size to pore-throat size. The interplay between these parameters and the petrophysical measurements of porosity and permeability are the basis for petrofacies distinctions in a hierarchy of models that attempt to accommodate differing degrees of pore complexity.

EQUIVALENT HYDRAULIC RADIUS OF TUBES

The relationship between permeability and porosity can be defined precisely if a rock is modeled by a pore system of uniform tubes. Combining Darcy's law for flow in porous media and Poiseuille's law for flow in tubes and solving for permeability (Bird et al., 1960) yields:

$$k = \left(\frac{r^2}{8}\right) * \Phi$$

where k is permeability (μm^2), r is the tube radius (μm), and Φ is the porosity (V_{pore} / V_{rock}). This model of pore networks as bundles of capillary tubes is tractable mathematically but is a crude representation of reality. However, this simple approximation does show the fundamental relationship between permeability, porosity, and hydraulic radius that can be modified for porous media. When the equation is rearranged and generalized to solve for equivalent hydraulic radius, then:

$$r = a * \sqrt{\frac{k}{\Phi}}$$

where a is a unit conversion factor. The square-root ratio of permeability to porosity has been used for many years as a basic measure of reservoir quality; its intrinsic meaning is shown by the equation to be the equivalent hydraulic radius of the pore element that is the channel for fluid transmission. For reservoir rocks, this element is the distribution of pore-throat sizes, as evaluated by porosimetry determined from capillary pressure measurements.

CAPILLARY PRESSURE EVALUATION OF PORE-THROAT SIZES

Although capillary pressure data have been obtained from core samples by the oil industry for about fifty years, their primary users have been petroleum engineers. Fortunately, several excellent review papers have been written for geologists that relate capillary pressure to rock type and reservoir structure, as well as applications to both reservoir analysis and exploration (Arps, 1964; Stout, 1964; Berg, 1975; Jennings, 1987; Vavra et al., 1992). The saturation profiles of reservoir rocks can be replicated by laboratory measurements of capillary pressure. Most commonly, mercury is injected into a core plug sample at increasing pressures, and the mercury saturation of the core is recorded at different pressure levels. The laboratory process simulates the intrusion of hydrocarbons into a water-wet rock under the increasing buoyancy pressure that would be experienced by the rock at successively higher levels in a hydrocarbon column. Resistance to the introduction of mercury into the core is provided by the capillary forces within the pore system. In a hydrocarbon system, the wetting fluid (conventionally thought to be water) adheres to the internal surfaces and resists the introduction of the hydrocarbon nonwetting phase. The relationship between capillary forces and pore-throat size is described by the Washburn equation (Washburn, 1921):

$$P_c = \frac{2\sigma\cos\theta}{r}$$

where P_c is the capillary pressure, σ is the surface tension of the wetting fluid, θ is the contact angle between the wetting fluid and the solid surface, and r is the pore-throat radius.

Because the term $2\sigma\cos$ is a constant for any given nonwetting/wetting fluid (or gas) couplet, the capillary pressure is controlled by the pore-throat radius of the rock pores. If the pore throats had a unique radius, then the pore network would be impenetrable up to a critical pressure, at which the entire pore system would be breached. In reality, rock pore systems have a range of pore-throat sizes, so the capillary pressure curve records the saturation of the pore throats at successively smaller sizes with increasing pressure.

Capillary pressure-saturation data for five core samples of medium- to very fine-grained Atokan sandstone facies in the Norcan East field of Kansas are shown as capillary curves in Figure 6.1. These curves represent the cumulative relative frequencies of pore-throat sizes in each core that are penetrated at increasing capillary pressure. Using the Washburn equation, the capillary pressure scale can be converted into an equivalent pore-throat radius scale. Then, by taking the first difference of the curve of saturations, the frequency distribution of pore-throat radii can be displayed and scaled in microns (Figure 6.2). A commonly used subdivision of pore-throat sizes in clastic rocks (e.g., Hartmann and Coalson, 1990) is:

Pore-throat radius, μm	Pore-throat type	Entry pressure, psi (kPa)
<0.1	Nano	1076 (7419)
0.1–0.5	Micro	215 (1482)
0.5–2	Meso	54 (372)
2–10	Macro	11 (76)
>10	Mega	0 (0)

These terms serve as useful general descriptors of pore-throat classes. When a sample is characterized by a principal pore-throat size, it can be assigned to a pore-throat type and equated to a pore-throat petrofacies. Using this approach, four

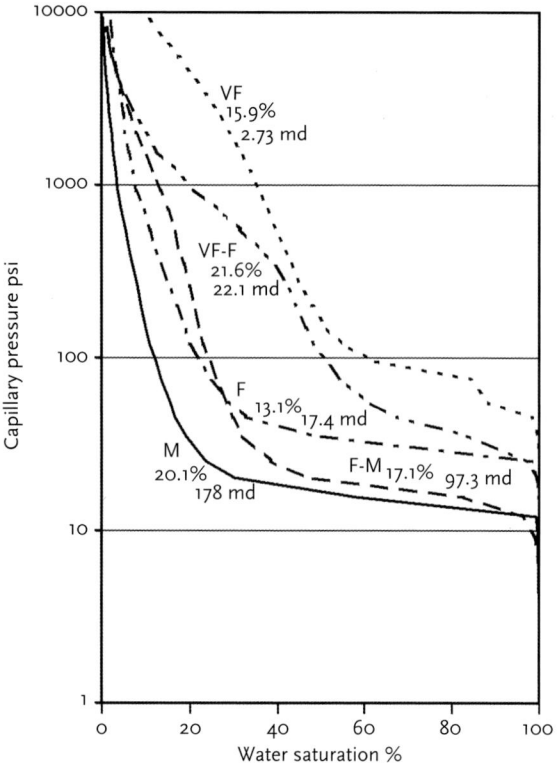

Figure 6.1: Capillary pressure curves of five representative Atoka sandstone cores from the east Norcan field, Kansas.

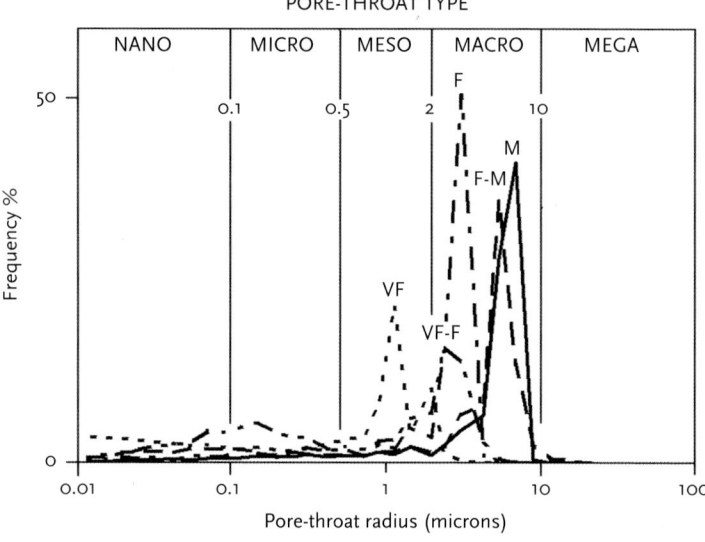

Figure 6.2: Pore-throat radii from capillary pressure measurements of five representative Atoka sandstone cores from the east Norcan field, Kansas.

of the Atokan sandstone core samples can be classified as macropore-throat petrofacies, and the fifth as a mesopore-throat petrofacies. As might be expected, the ordering of dominant pore-throat classes by size is matched by the grain size classes observed in the core.

Because a first-difference plot, such as that shown in Figure 6.2, graphs the slope between successive pore-throat sizes accessed by the nonwetting fluid, the radius of the principal pore throat can be estimated as the geometric average of the two throat sizes that link the maximum difference. A logarithmic crossplot of principal pore-throat radius versus permeability of the core samples (Figure 6.3) shows the strong linear trend that has been observed in Atokan and Morrowan sandstones deposited in marine, estuarine, and fluvial environments (Byrnes et al., 2001).

This pore-throat size and permeability relationship has been reported in many other core capillary studies (e.g., Swanson, 1981; Thompson et al., 1987; Basan et al., 1997). Although the principal pore throat is equated here with the mode of the pore-throat distribution, its high correlation with permeability elevates it to the status of a "characteristic pore throat," because it emulates the hydraulic radius of a simple tube model. In sandstones having a broader range of pore-throat sizes, the location of a characteristic pore throat loses its meaning, and predictions of permeability must be based on additional pore-throat measures. However, although the criteria used for their recognition varies somewhat, principal pore throats are common in intergranular pore systems, and their radius is related to the logarithm of permeability.

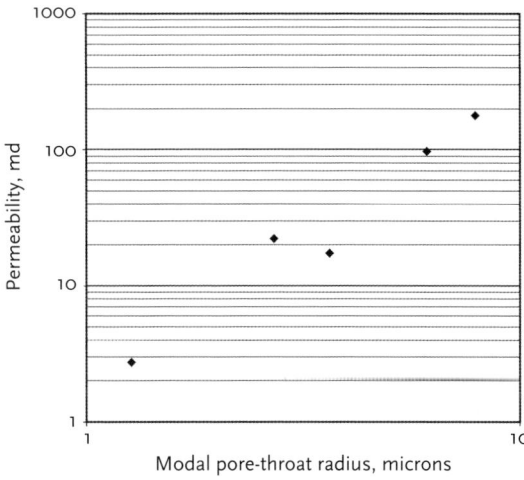

Figure 6.3: Logarithmic crossplot of modal pore-throat radii calculated from capillary pressure measurements for Atoka sandstone cores from the east Norcan field versus core permeabilities.

THE WINLAND EQUATION

The expense involved in core measurements of capillary pressure prompted industry researchers to investigate methods to estimate the characteristic pore-throat size based on porosity and permeability measurements alone. The earliest documented study along these lines was that of Dale Winland, who used capillary measurements of cores from the Spindle field in Colorado. Winland made a regression analysis of pore-throat radii in fifty-six sandstones and twenty-six carbonates, predicted from porosity and permeability measurements over a range of saturation levels; he obtained the highest correlation at a mercury saturation of 35 percent. Kolodzie (1980) published Winland's results, including the Winland equation:

$$\log r_{35} = 0.732 + 0.588 \cdot \log k_{air} - 0.864 \cdot \log \Phi$$

where r_{35} is the pore-throat radius in microns (μ) at 35 percent mercury saturation, k is the absolute permeability to air (md), and Φ is the porosity (%).

Because core measurements are far more common than capillary pressure curves in reservoir databases, the r_{35} estimate of the principal pore-throat radius by the Winland equation has wide applicability. Examples of reservoir studies that have used Winland's equation to discriminate pore type and assess reservoir quality include Hartmann and Coalson (1990) and Martin et al. (1997). Practitioners of the r_{35} method point out that results are valid for rocks whose pore space is dominantly intergranular or intercrystalline.

An instructive demonstration of the estimation of principal pore-throat size and its application to the delineation of flow units is provided by the Atokan sandstone reservoir in the Norcan east field. A comparison between principal pore-throat radii and the Winland r_{35} estimates calculated from permeability and porosity measurements of Atokan sandstone core samples is shown in Figure 6.4. In this field, there is a good concordance between the Winland equation r_{35} predictions and the principal pore-throat sizes based on capillary pressure measurements. In other clastic reservoirs where the Winland equation fails to match core observations, custom-made equations can be developed by regression analyses of capillary curve data related to core permeabilities and porosities using the method described in detail by Pittman (1992).

If the Winland equation is considered to be a useful model for a reservoir, then r_{35} pore-throat sizes can be estimated across the field. A profile of r_{35} pore-throat size estimates is shown in Figure 6.5 for the S2 Atokan sandstone unit in the Murfin #1-3 Patton, which is the type well for the Norcan east field. The Winland equation was applied both to core measurements of porosity and permeability and to log porosity and estimated permeability based on gamma-ray and porosity logs. The estimated permeabilities were provided by the regression relationship developed and described in Chapter 3.

The match between the two r_{35} estimates is good, as judged by their common distinction of three flow units and their assignment to meso- (upper flow unit), micro- (middle flow unit), and macro- (lower flow unit) pore-throat petrofacies. If the log

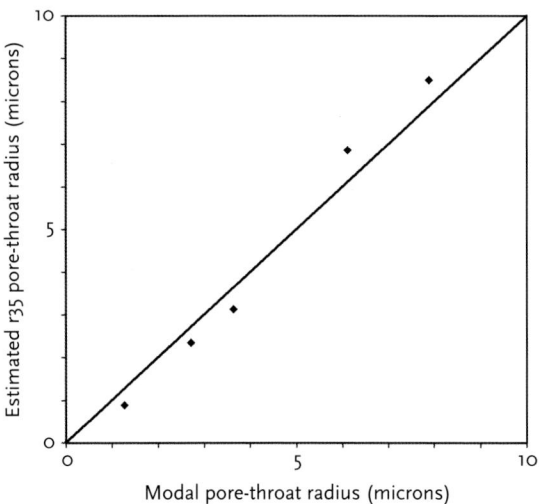

Figure 6.4: Crossplot of modal pore-throat radii calculated from capillary pressure measurements for Atoka sandstone cores from the east Norcan field versus estimates of r_{35} characteristic pore-throat radii calculated by the Winland equation.

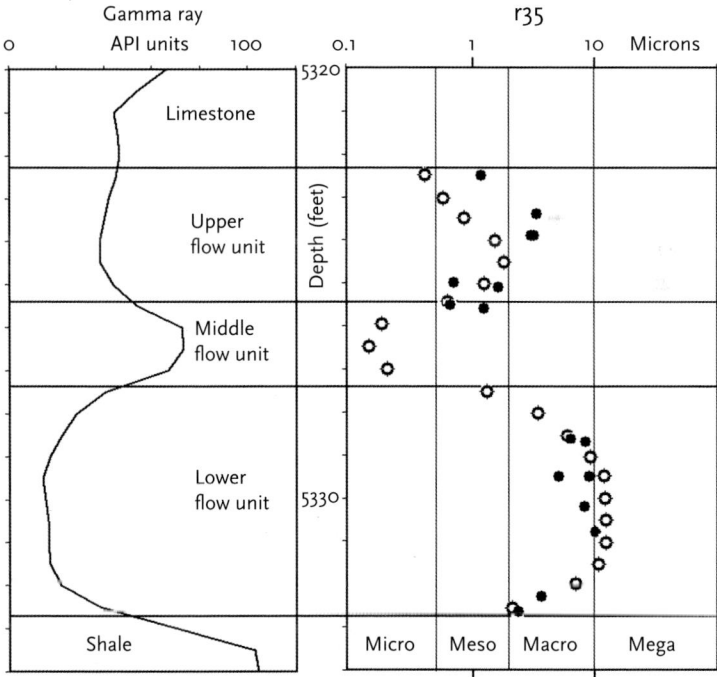

Figure 6.5: Profile of r_{35} pore-throat size in the Atoka sandstone in well Murfin #1-3 Patton estimated from core porosities and permeabilities (solid circles) and well log porosities and estimated permeabilities (open circles). The sandstone is subdivided into three flow units based on estimated r_{35} pore-throat size.

estimates are considered to be validated by the core measurements, then Winland equation r_{35} estimates can be made with some confidence in uncored wells across the field. The log-based results are calculated at the digital log-sampling interval, so estimates can be interpolated between irregularly spaced core measurements. In this sense, the calibrated relationships between core and logs can be leveraged to populate a reservoir model in a more comprehensive manner.

The three distinctive flow units in the Murfin #1-3 Patton are laterally extensive across the field and were related to lithofacies observed in the core by Bhattacharya et al. (2008). The lower flow unit is a fine-grained, cross-bedded marine sandstone overlain by a thin middle flow unit of heterolithic, flaser-bedded, very fine-grained sandstone and shale, interpreted to have been deposited in a distal tidal environment. The upper flow unit of very fine-grained, bidirectional-rippled sandstone is interpreted as tidal rhythmites.

THE FLOW-UNIT CONCEPT

The differentiation of flow units and their correlation between wells has been recognized as a fundamental step in reservoir characterization and modeling since the introduction of the flow-unit concept by Hearn et al. (1984). A "flow unit" is defined as a reservoir zone with lateral continuity between wells and internally consistent properties that control fluid flow and are distinct from those of adjacent flow units. Flow units should not be confused with reservoir compartments, which are not hydraulically connected with one another. Differences in pressure or free-water level (FWL), as discussed in Chapter 7, demonstrate the existence of separate compartments, each of which may contain a number of distinctive flow units.

Flow units commonly are not exactly matched by equivalent lithofacies, although there may be a strong congruence between the two classifications because of the interplay between flow properties and rock-texture descriptors. Criteria that a geologist might use for facies discrimination may subdivide rocks based on interpretations of their depositional environments and/or diagenetic histories. The boundaries of genetic units described by a geologist may be useful for the reservoir engineer if they coincide with distinctive changes in flow properties. However, there may be instances where distinctive flow units are either observed within a single geological facies or flow properties are essentially the same across a facies boundary. Consequently, if the purpose of a reservoir-layer subdivision is the identification of flow units, attention should be focused on the hydraulic properties of porosity ("storage") and permeability ("speed").

An example of the distinction between lithofacies subdivisions and flow units in the Hartzog Draw field of the Powder River Basin, Wyoming was described by Hearn et al. (1984). The reservoir unit is in the Shannon sandstone that was deposited as a complex of sand ridges in the Late Cretaceous seaway. The internal structure of the sand ridge of the Hartzog Draw field was subdivided stratigraphically into three lenses of bar, bar margin, and interbar facies. Distinctions between these facies in cored wells were made on the basis of grain size, bedding structure, and bioturbation

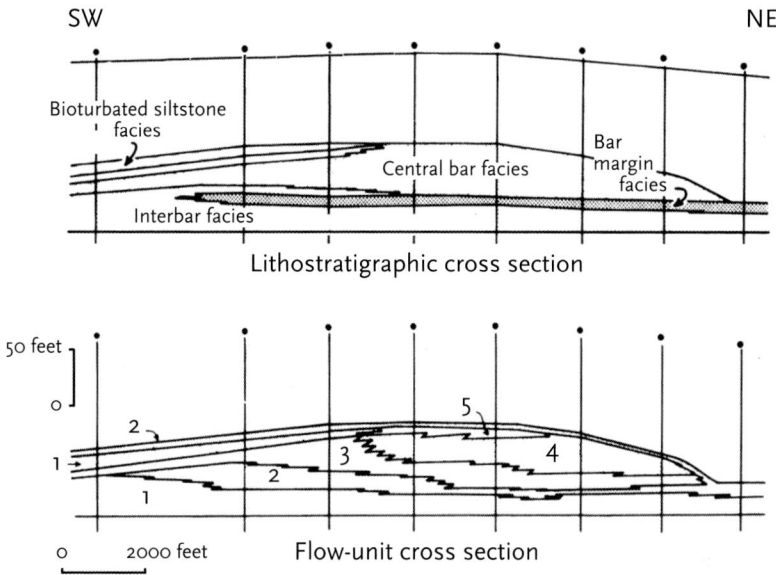

Figure 6.6: Stratigraphic subdivision (above) and flow-unit subdivision (below) of the Shannon sandstone in the Hartzog draw field, Powder River Basin, Wyoming. Adapted from Hearn et al. (1984), Used with permission © 1984 Society of Petroleum Engineers.

features. A lithostratigraphic cross-section of the field is shown in Figure 6.6. Hearn et al. (1984) noted that there was considerable variability in reservoir properties within each lithofacies. However, they were able to subdivide the reservoir into five distinctive flow units on the basis of porosity and permeability and to map these across the field (Figure 6.6).

The two methods of subdivision complement one another, so that standard procedures of reservoir characterization begin with the identification of lithofacies, sequences, and depositional environments from the core and logs. These are correlated across the field in the development of a geomodel interpretation. As part of the process to allow interpolation to uncored wells, attention is also paid to the interplay between lithofacies and flow units that is suggested by porosity and permeability measurements and their common link with petrophysical log properties. The subdivision of flow units is keyed to porosity and permeability associations, as illustrated in Figure 6.7, utilizing capillary pressure data, where available. Because fluid flow is determined by permeability, the most diagnostic rock property is pore-throat size. Consequently, a flow unit is a zone that has a relatively uniform, dominant pore-throat size, and hence a consistent flow behavior.

PETROFACIES CASE-STUDY APPLICATIONS OF THE WINLAND EQUATION

Capillary pressure curves measured on core samples provide the most comprehensive information on flow units because the curves express the entire pore-throat

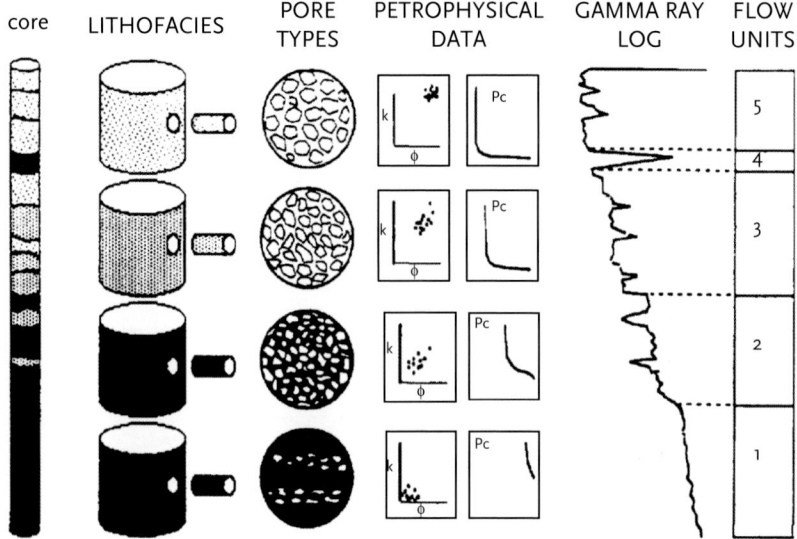

Figure 6.7: Geological and petrophysical data used to delineate flow units (including permeability barrier of flow unit 4), which must correlate between wells. From Ebanks et al. (1992), © 1992 American Association of Petroleum Geologists (AAPG), reprinted by permission of the AAPG, whose permission is required for further use.

distribution. However, in rocks with intergranular porosity, flow units can be approximated using Winland r_{35} estimates of the dominant pore-throat radii from core measurements of porosity and permeability. Porras (1998) applied the Winland equation estimate of r_{35} to Cretaceous and Tertiary clastic rocks of the Carito Norte field in Venezuela. Core measurements of porosity and permeability are shown as a crossplot in Figure 6.8, referenced with contours of the Winland r_{35} pore-throat size and pore-throat classes.

By using this crossplot in conjunction with capillary pressure measurements, Porras (1998) was able to identify five distinctive petrofacies bounded by the Winland r_{35} values (Figure 6.8). Type curves of capillary pressure for these petrofacies are shown in Figure 6.9. Earlier studies of the cores from this field defined nine lithofacies of shaly sandstones and coal, with four sandstone facies distinguished primarily on the basis of grain size. A comparison (Figure 6.10) of the lithofacies classification with the Winland r_{35} pore-throat sizes in a well shows a good overall match between sandstone lithofacies, which is to be expected because the subdivision is keyed so strongly to granulometry. However, Porras (1998) cautioned that there were many instances where different petrofacies occurred within a single lithofacies. In addition, r_{35} estimates are continuous data, in contrast with discrete lithofacies categories, so their information content is higher and linked more directly with reservoir properties. This can be seen in the well profile in Figure 6.10, where the general trend of upward declining porosity is matched by decreases in the r_{35} sizes, but indicated only vaguely by the lithofacies classification.

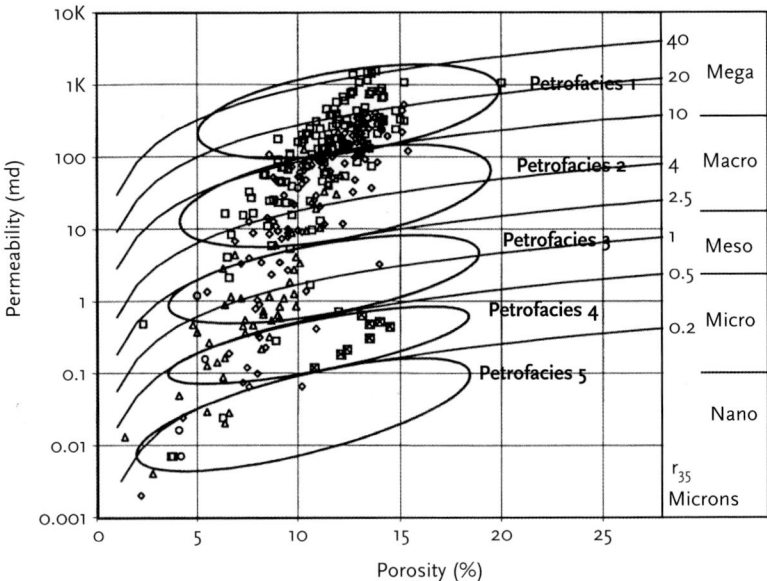

Figure 6.8: Crossplot of core measurements of porosity and permeability from Cretaceous and Tertiary clastic rocks of the Carito Norte field in Venezuela. The plot is overlaid with contours of Winland r_{35} pore-throat sizes (given as pore-throat classes), and subdivided by petrofacies. From Porras (1998), courtesy Society of Petrophysicists and Well Log Analysts.

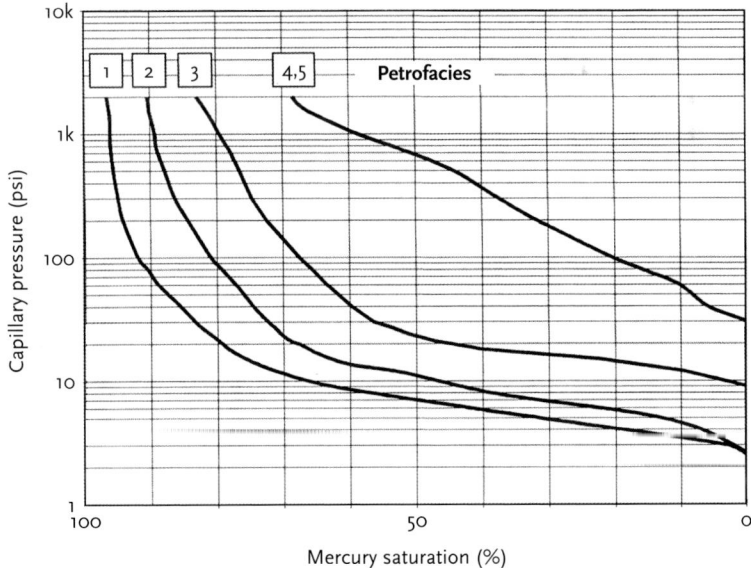

Figure 6.9: Capillary pressure curves for five petrofacies identified in Cretaceous and Tertiary clastic rocks of the Carito Norte field in Venezuela. From Porras (1998), courtesy Society of Petrophysicists and Well Log Analysts.

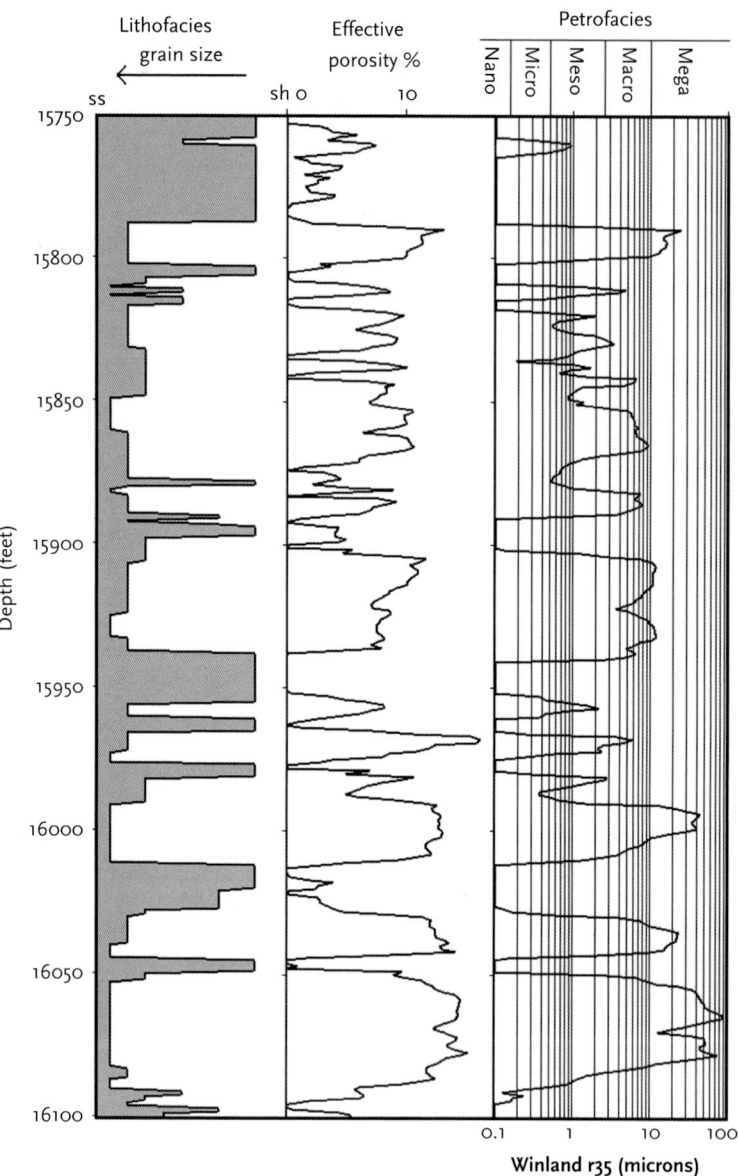

Figure 6.10: Lithofacies, porosity, estimated Winland r_{35} pore-throat radii, and petrofacies in a well in the Carito Norte field in Venezuela. From Porras (1998), courtesy Society of Petrophysicists and Well Log Analysts.

A number of published laboratory studies have critically compared the predictions of the Winland equation with capillary pressure data from a variety of reservoir rock types. In particular, the Winland-equation prediction that the principal pore throat will be breached at a saturation of 35 percent has been examined carefully, leading to alternative conclusions. So, for example, Pittman (1992) extended Winland's work using a larger data set of 202 sandstone samples from fourteen formations ranging from Ordovician to Tertiary in age. Rather than producing a single equation, he published a set of equations that relate the radius of the pore throat to porosity and permeability over a range of different saturations of the nonwetting fluid (see review in Chapter 7). From his analysis he concluded that the best correlation of pore-throat size with permeability occurred at a mercury saturation of 25 percent.

In a detailed core study of the highly porous and permeable Nubia sandstone of Egypt, Nabawy et al. (2010) decided that the best correlation occurred at a saturation of 10 percent, while Sujuan et al. (2011) demonstrated that a 20 percent saturation provided the best estimator in tight Triassic gas sandstones in China. Spearing et al. (2001) examined capillary curves for the tight-gas Sherwood sandstone in England and observed that, while principal pore throats could be discriminated, they were not associated with any specific mercury saturation. They concluded that the Winland equation had poor predictive power. On a more positive note, Leal et al. (2001) found an excellent correlation between the estimates of pore-throat radius by the Winland equation with the pore-throat radius at 35 percent mercury saturation. In spite of the bimodal character of the pore-throat distribution, they obtained good predictions of permeabilty, from which they concluded that the r_{35} calculation had captured the pore-throat size that most contributed to permeability.

The extension of the principal pore-throat estimation to carbonate rocks must take into account the wide range of potential pore shapes and sizes and their degree of connectivity. In nonvugular limestones or dolomites, interparticle pore spaces (either intergranular or intercrystalline) appear to behave in a manner broadly similar to the intergranular network of clastic rocks. In these instances, the Winland equation for estimation of r_{35} may be applied cautiously in reservoir modeling; however, the predictions should be examined carefully. For example, Martin et al. (1997) described the use of Winland-equation estimates of r_{35} to characterize petrophysical flow units in a variety of case studies of carbonate reservoirs. An example of r_{35} pore-throat estimation in a San Andres formation well is shown in Figure 6.11, matched with a water-saturation profile (Martin et al., 1997). They noted that maximum flushing, as shown by high water saturations, coincided with the best flow units discriminated by the r_{35} values. Publication of their 1997 paper resulted in a spirited but gracious discussion by Lucia (1999) and a reply by Martin et al. (1999). Lucia's basic contention was that the Winland equation had been successful in clastic reservoirs, but any realistic reservoir characterization must be rooted in rock fabric distinctions based on geological observations from core and outcrop. In their reply, Martin et al. (1999) once again acknowledged the difficulties posed by vuggy pore space, but emphasized that pore throats control permeability and are difficult to assess in rock-fabric observations, which are more closely keyed to pore-body properties. Whatever their viewpoint, all investigators agree that complex pore morphologies in carbonates frequently create multimodal pore-throat distributions with a broad range of aspect ratios that defy simple characterization. In these instances, their heterogeneous character conflicts with the central idea of the Winland equation as an estimator of principal pore-throat radii.

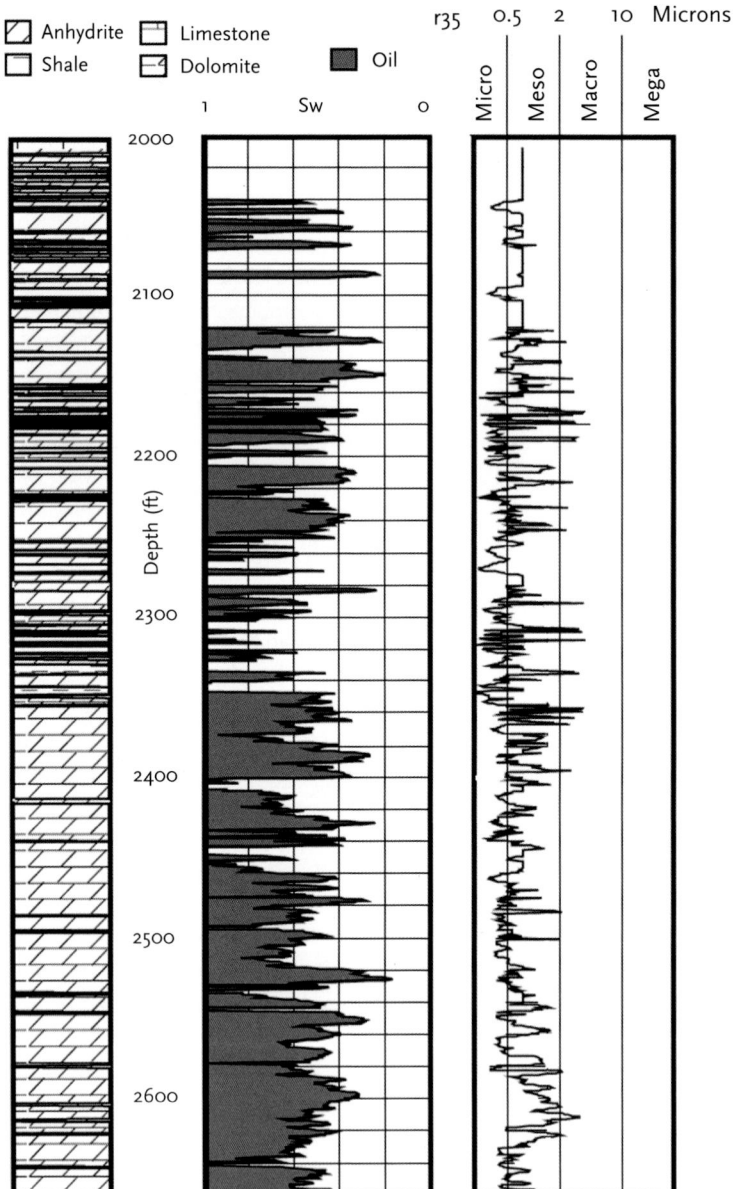

Figure 6.11: Core lithology, water saturation, and Winland r_{35} pore-throat radii in dolomites of the San Andres formation from the W.R. Settles a-92 well in the Howard Glassock field, Permian basin. Modified from Martin et al. (1997), © 1997 American Association of Petroleum Geologists (AAPG), reprinted by permission of the AAPG, whose permission is required for further use.

In a study of the San Andres formation in the Vacuum feld of New Mexico, Pranter (1999) applied the Winland equation, but revised the coefficients to honor capillary pressure measurements from San Andres formation core. Pranter found that estimates of r_{35} pore-throat size from the San Andres core-based equation were generally larger than those computed from the original Winland equation, and speculated that the differences might be caused by higher proportions of bimoldic and intercrystalline pore types.

CARBONATE PETROFACIES PORE-THROAT SIZE DISTRIBUTIONS

Extensive work on the giant Hugoton gas field by the Hugoton Asset Management Project team created a large database that included many capillary pressure measurements of cores from a variety of lithologies in the Lower Permian Chase Group and Council Grove group. For our purposes, a subset of all limestone capillary pressure curves was extracted to examine the pore-throat size characterization and its relationship to limestone textural classes. Capillary pressure curves that typify grainstone (GR), packstone-grainstone (PK-GR), packstone (PK), wackestone-packstone (WK-PK), and wackestone (WK) limestones are shown in Figure 6.12. When replotted

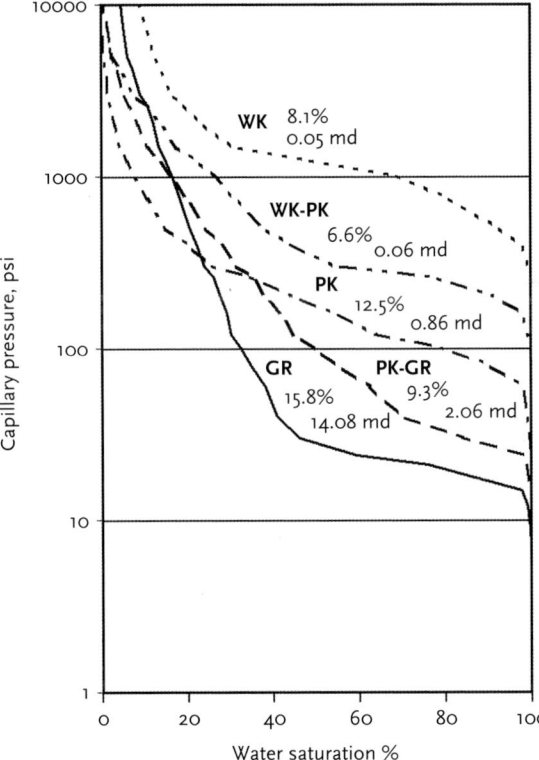

Figure 6.12: Capillary pressure curves that typify grainstone (GR), packstone-grainstone (PK-GR), packstone (PK), wackestone-packstone (WK-PK), and wackestone (WK) limestones from the Lower Permian Chase and Council Grove groups in the Hugoton gas field, Kansas.

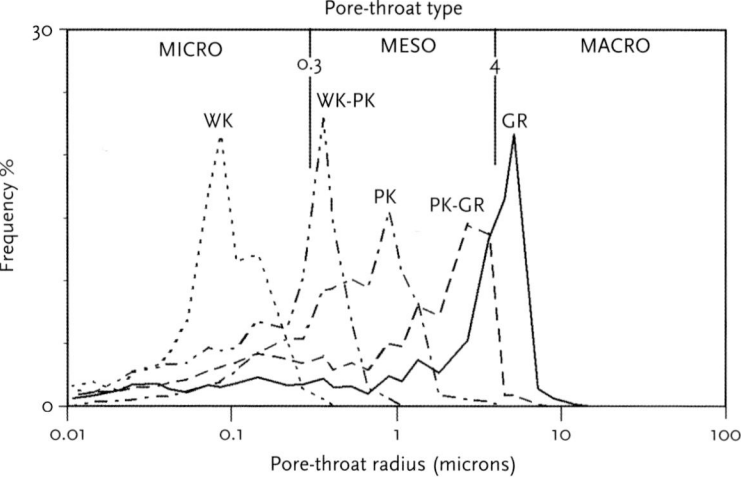

Figure 6.13: Pore-throat radii from capillary pressure measurements of representative grainstone (GR), packstone-grainstone (PK-GR), packstone (PK), wackestone-packstone (WK-PK), and wackestone (WK) limestones from cores of the Chase and Council Grove groups in the Hugoton gas field, Kansas.

as saturation first differences versus capillary pressures rescaled as pore-throat radii (Figure 6.13), the Dunham textural classes are well differentiated by the dominant pore-throat sizes. Notice, however, that although distinctive, the modes of the limestones are not as pronounced as those of the Atokan sandstones (Figure 6.2) when expressed on a saturation-difference scale. However, the fact that the limestones can be distinguished by a single mode rather than a more complex measure is encouragement that a characteristic pore-throat size may be a useful descriptor for these limestones.

On the basis of his laboratory work on hundreds of carbonates in the Thamama and Arab reservoirs of Abu Dhabi, Marzouk (1998) proposed a tripartite classification of micropores (less than 0.3 microns), mesopores (between 0.3 and 4 microns), and macropores (greater than 4 microns) in pore-throat radii. He also observed that curves of the distributions of pore-throat sizes for typical rock types were strongly unimodal with subsidiary low peak responses in the microporosity range. These class boundaries are marked on the size distribution plot in Figure 6.13 and show that the grainstones are macropores, the packstones are mesopores, and the wackestones are micropores. This relationship provides a useful link between visual core descriptions and principal pore-throat sizes.

A crossplot of the modal pore-throat size versus permeability (Figure 6.14) shows a strong relationship (R-squared of 84.6 percent), which validates modal pore-throat sizes as appropriate petrofacies descriptors. There is also a strong relationship between these modal pore throats and predictions of the r_{35} pore-throat size by the Winland equation (Figure 6.15). However, there is a systematic bias in the relationship: modal pore throats are about double the size of the Winland equation r_{35}

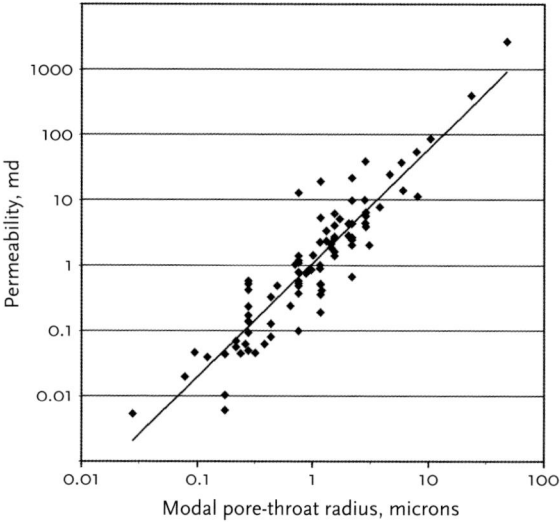

Figure 6.14: Crossplot of modal pore-throat sizes and permeabilities measured in cores of limestones of the Chase and Council Grove groups from the Hugoton gas field, Kansas.

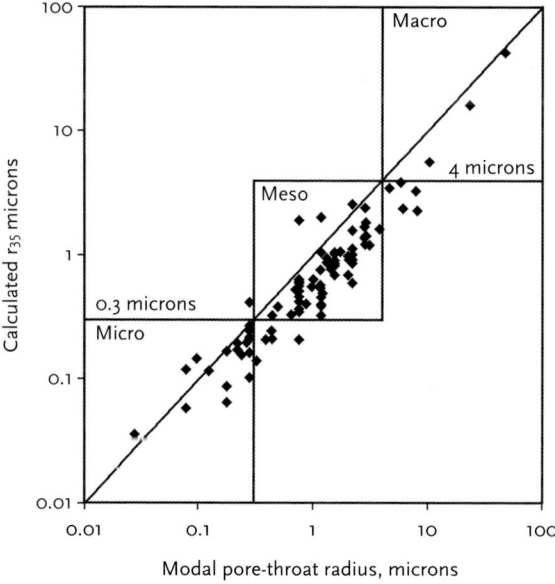

Figure 6.15: Crossplot of modal pore-throat sizes and predictions of the r_{35} pore-throat sizes by the Winland equation. Pore-throat class subdivisions are based on the size boundaries proposed by Marzouk (1998).

estimates, particularly in the mesopore and macropore ranges. This observation echos the experience of Pranter (1999) in his work with the San Andres formation. Consequently, an r_{35} profile computed for these carbonates would be a useful semi-quantitative descriptor for delineating flow units, but would be inadequate for the purpose of estimating the actual size of the principal pore throats.

The distribution of the nonwetting fluid saturation associated with the modal pore throat of each core sample is shown in Figure 6.16. The similar median saturation values for grainstones (28.9 percent), packstones-grainstones (29.6 percent), and packstones (29.2 percent) are contrasted with a median of 49.9 percent for wackestones and an intermediate median of 34.0 percent for wackestone-packstones. These values, together with the bimodal appearance of the distribution, suggest different saturations associated with grain-supported and mud-supported limestones. The distribution was subdivided into two groups, where the partition value was located to minimize the variance within each group and maximize the variance between the two groups (Figure 6.16).

The saturation value of about 30 percent that characterizes the grain-supported limestones probably explains the consistent underestimation of the modal pore-throat size by the Winland r_{35} equation. Recalculation of the equation fitted to an r_{30} estimate would increase the estimates to higher values and more closely match the observations. This would validate approaches that attempt to link a characteristic nonwetting fluid saturation with a principal pore throat. However, the range

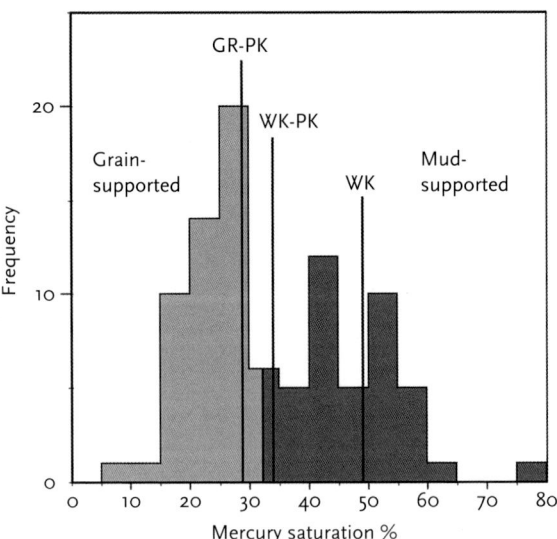

Figure 6.16: Distribution of mercury saturations associated with modal pore throats, indexed by median values for grainstones (28.9 percent), packstone-grainstones (29.6 percent), packstones (29.2 percent), labeled together as GR-PK, wackestone-packstones (34.0 percent) WK-PK, and wackestones (49.9 percent) WK. The distribution is subdivided into two parts that maximize the between-group variance and minimize the within-group variance. The two groups are interpreted as reflecting a differentiation between grain-supported and mud-supported limestones.

of saturation values in the distribution, together with its bimodal aspect, suggest that a single value for nonwetting fluid saturation is only a broad approximation. Instead, an equation that makes a direct prediction of the modal pore-throat radius regardless of saturation is likely to be more efficient and potentially more accurate. A regression analysis of modal pore-throat radii based on core measurements of permeability and porosity (reference?) has an *R-squared* value of 88.5 percent and an equation of:

$$\log(r_{mode}) = 1.179 + 0.616 \log(k) - 1.071 \log(\Phi)$$

where r_{mode} is the modal pore-throat radius (microns), k is the permeability (mD), and Φ is the porosity (%).

The strong correlation between observed modal pore-throat radius and permeability justifies its use for prediction in wells across the Hugoton gas field. Principal pore-throat radii were estimated from core measurements using a regression analysis of data from the Anadarko A-1 Flower "science well," which was drilled into the Chase Group using foam to minimize invasion effects. The neutron-density gas effect shows a good concordance with estimated principal pore-throat variation, which, in turn, shows a good correspondence with the Dunham texture classes observed in the limestone core (Figure 6.17). Finally, gas production from the three drill-stem tests (DST) marked on Figure 6.17 are an excellent match with expectations from the core and log characterizations.

Principal pore-throat radii in the limestone capillary pressure database were subdivided by Dunham textural type and are shown as box plots in Figure 6.18. There is a systematic decline in median pore-throat radii by textural class, implying that a geomodel of lithofacies distributions could be populated by expectations of principal pore-throat radii and, hence, permeabilities. However, the box plots also demonstrate that characterization of the principal pore-throat radii by mean values and standard deviations should be evaluated carefully if the interquartile ranges demonstrate marked asymmetry, as in the grainstones, packstone-grainstones, and packstones. The asymmetric character can be easily accommodated in simulation models by selecting an appropriate nonnormal distribution.

Oomoldic limestones are classic examples of a dual-porosity system, and so their pore-throat characterization poses special challenges. The dual-porosity system of oomolds and interparticle porosity makes permeability difficult to predict because of the great variability in oomold interconnectivity, as discussed in Chapter 3. Textural measures of connectivity reflect both depositional facies and diagenetic processes of dissolution, cementation, and crushing, linked with modification by ground water that have caused either or both occlusion of pore space and improvement in oomold connectivity. An instructive example of this complexity is shown by the capillary pressure curves of five Pennsylvanian oomoldic limestone cores from Kansas (Figure 6.19) and their pore-throat radii frequency distributions (Figure 6.20). The core samples have similar, high porosities, but a wide range of permeability that matches changes in the pore-throat size distribution. The two higher permeability samples show a strong but broad macrosized mode, as contrasted with the lower permeability

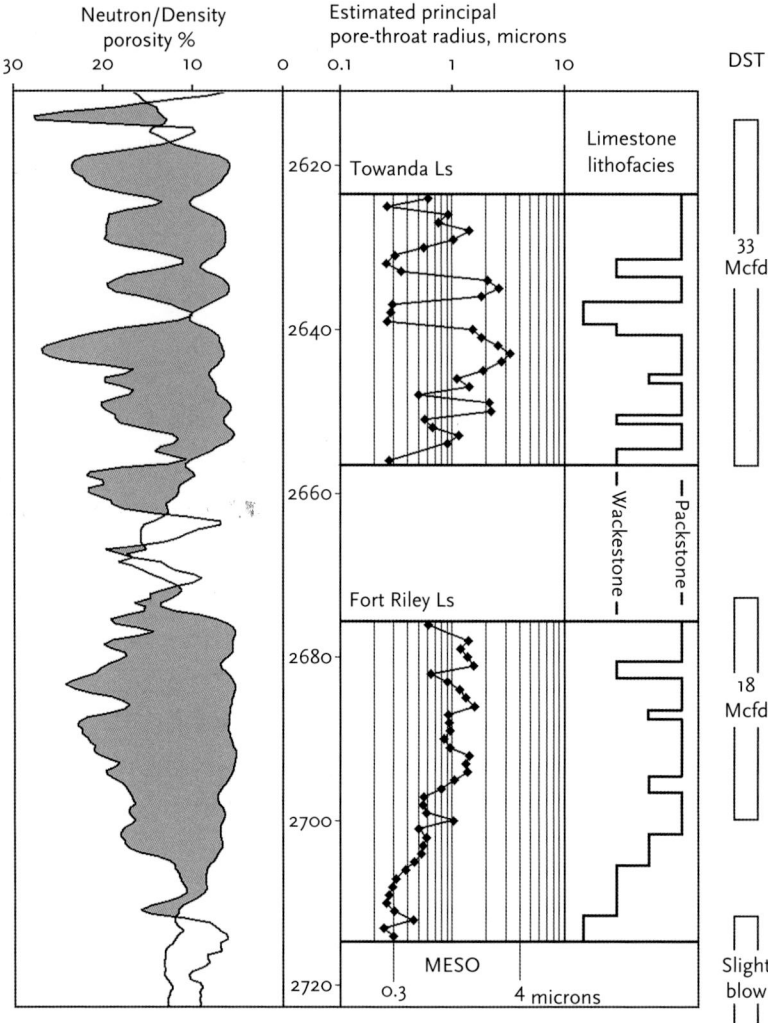

Figure 6.17: Overlaid neutron and density porosity logs with shaded gas effect, predicted principal pore-throat radii, lithofacies descriptions, and DST results for the Towanda and Fort Riley limestone members in the Anadarko #A-1 Flower well in the Hugoton gas field, Kansas.

samples that show a progressive dispersion in pore throats and the loss of a distinctive mode. However, a regression analysis of pore-throat radii at different nonwetting phase saturations shows a distinctive and strong *R-squared* of 91 percent at a saturation of 20 percent (Figure 6.21). The statistics imply that the pore-throat radii that contribute effectively to flow have higher values and that a principal pore throat can be identified with the r_{20} value in these facies. This finding matches the conclusion of Wardlaw and Taylor (1976), who observed that the twentieth percentile represents mercury saturation of the maximum connected pore-throat radius, from their extensive study of carbonate reservoir rocks. The relationship between the r_{20}

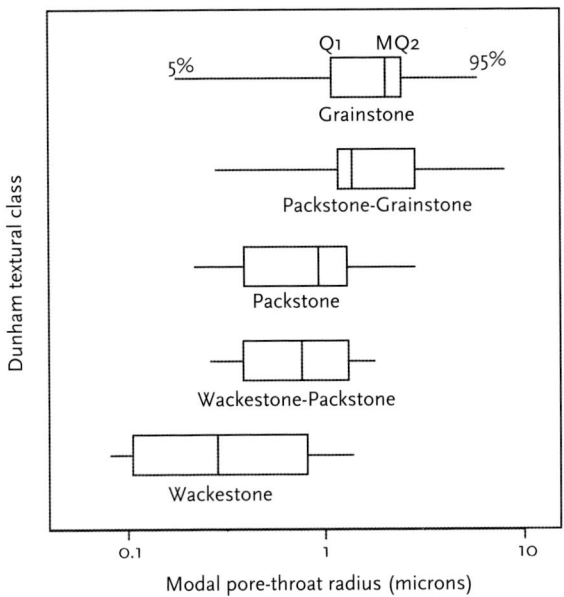

Figure 6.18: Box plots of modal pore-throat radii within textural classes of limestones in the Chase and Council Grove groups of the Hugoton gas field, Kansas.

pore-throat radius and permeability in these Pennsylvanian oomoldic limestone core samples is shown in Figure 6.22, together with the associated regression-analysis prediction function for estimation of the r_{20} pore-throat radius.

Pennsylvanian oolitic reservoirs in Kansas have geometries and architectures that are similar to modern oolite bodies in the Bahamas (Watney et al., 2006), and usually consist of multiple shoals, either stacked or *en echelon*, that formed in response to sea-level fluctuations. The diagenetically modified reservoir properties are linked with the original depositional facies when considered as architectural elements of shoal development (Byrnes et al., 2003). Grain-size variation, coupled with location on oolite buildups and interbedded carbonate mud deposits influenced the character and degree of diagenetic overprinting. An example of a Pennsylvanian oomoldic limestone section from Skelly #16 Colliver is shown in Figure 6.22. Correlation of this unit with neighboring wells showed that the section represents a single shoal body, and core examination identified two upward-coarsening bedsets. The upper bedset contains medium- to coarse-grained oomoldic grainstones overlying a bedset of fine to medium oomoldic grainstones. The grain-size character of these bedsets is duplicated by r_{20} pore-throat radii predicted from the core permeability relationship where mesosize throats are overlain by macrosize throats in this shoal.

The correlation between log-scaled permeability and porosity in core samples from this section is weakly positive (0.27), which is not surprising when considering the great range in permeability but similar porosities in the five oomoldic limestone cores with capillary pressure measurements. More intriguing is the higher correlation between the gamma-ray log measurement and the log-scaled permeability of −0.53.

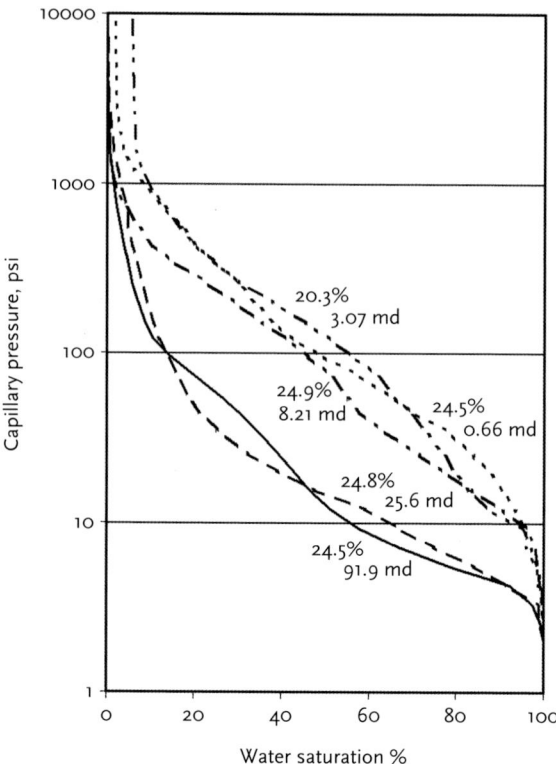

Figure 6.19: Capillary pressure curves of five Pennsylvanian oomoldic limestones from Kansas with similar high porosities.

This inverse relationship was observed by Watney et al. (2006) and interpreted to reflect that coarser-grained and better-sorted ooid facies have lower clay content, resulting in larger and better-connected oomolds. The gamma-ray and grain-size relationship is consistent with general observations that have related limestone textural variation to computed gamma-ray log variation (c.f. Lucia, 2007). However, a more detailed interpretation of oomoldic connectivity should incorporate diagenetic processes of dissolution and cementation in addition to the depositional facies. Because the diagenesis of these oolitic shoals was controlled by processes linked to exposure, the recognition of the architectural elements of the units is an important consideration.

The development (and destruction) of secondary porosity systems by multiple diagenetic episodes often creates pore-throat distributions that are difficult to represent adequately by a single principal pore throat. An instructive study of a wide range of pore-throat variability within a single formation was provided by Luo and Machel (1995). They selected the Grosmont formation in Alberta as a suitable subject for the study of carbonate reservoir properties at all scales because the unit was well known to be highly heterogeneous. This Upper Devonian formation consists of

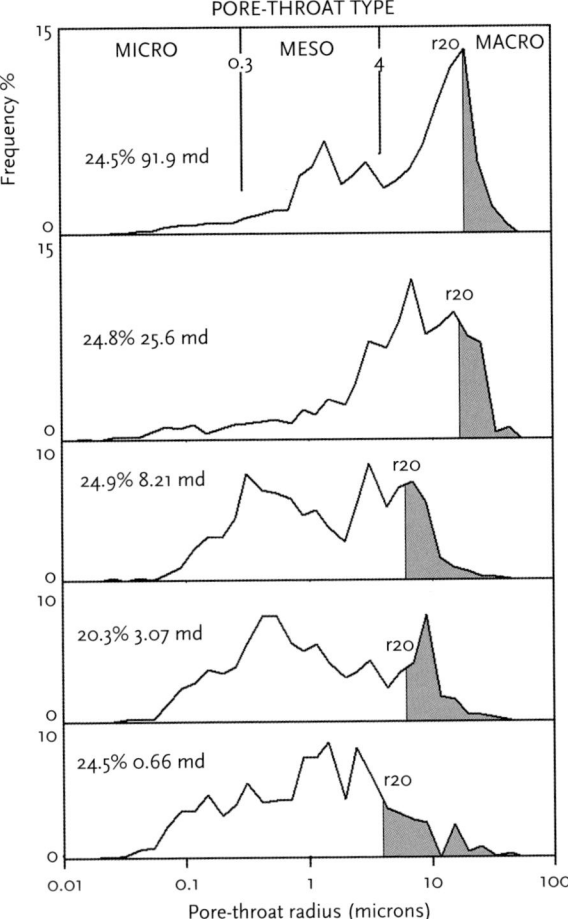

Figure 6.20: Pore-throat radii distributions from capillary pressure measurements of five Pennsylvanian oomoldic limestones from Kansas indexed with the pore-throat radius at 20 percent nonwetting phase saturation (r_{20}).

extensively dolomitized and karstified platform and ramp carbonates with complex associations of homogeneous, multilayered, dual- and triple-porosity systems characterized by fractured, channeled, and pressured solution features. The Grosmont formation consists of shallowing-upward cycles composed of lithofacies designated by a letter convention as *a* to *g* (Figure 6.23). The best reservoir rocks are lithofacies *e* (vuggy, massive mudstones to grainstones) and lithofacies *f* (laminated mudstones). Both have abundant primary porosity with additional secondary porosity created by diagenetic dissolution.

Luo and Machel (1995) made capillary pressure measurements of Grosmont formation core samples and tabulated the statistics of r_{20}, average, median, mode, and standard deviation of pore-throat radii. A crossplot of r_{20} pore-throat radius versus permeability for lithofacies *e* and *f* (Figure 6.24) shows only weak positive trends,

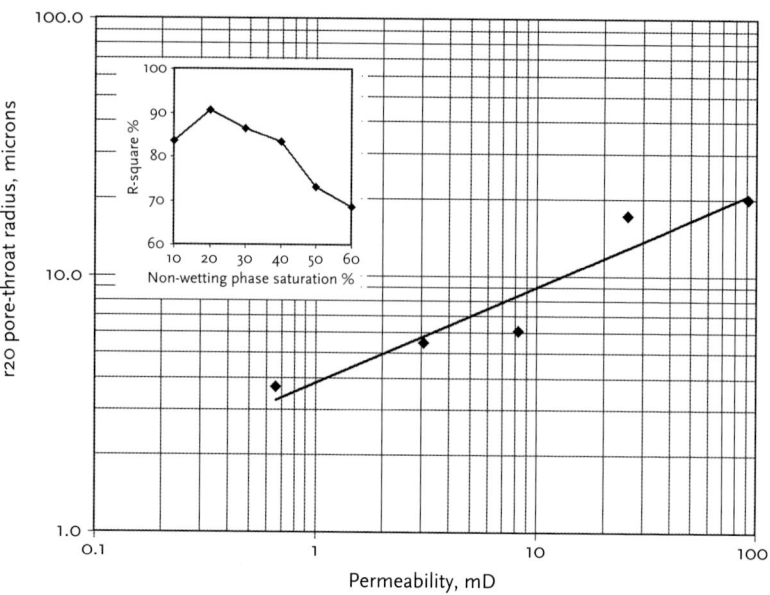

Figure 6.21: Predictive relationship for estimating the principal pore-throat radius at 20 percent nonwetting phase saturation (r_{20}) based on core permeability. Inset shows *R-squared* values for different saturation values.

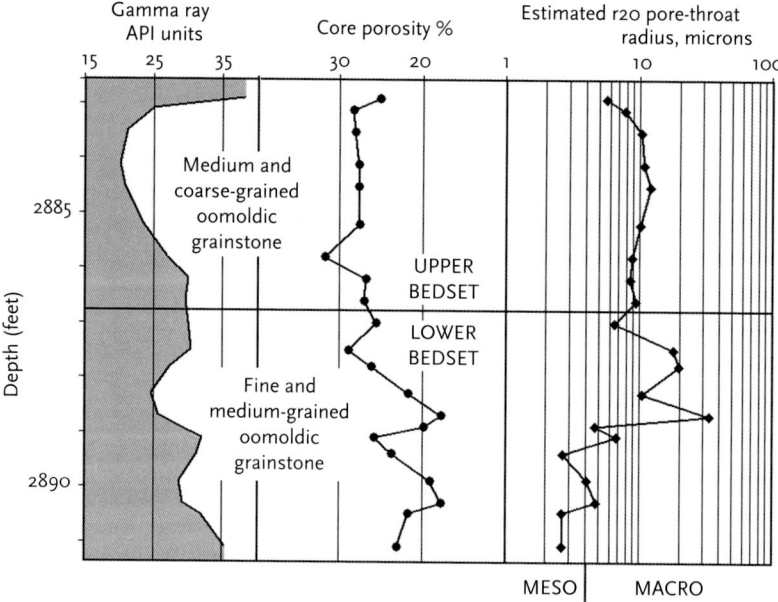

Figure 6.22: Gamma-ray log, core porosities, and estimated principal pore-throat radii (r_{20}) in the upper shoal unit of a Pennsylvanian oomoldic section in the Skelly #16 Colliver well.

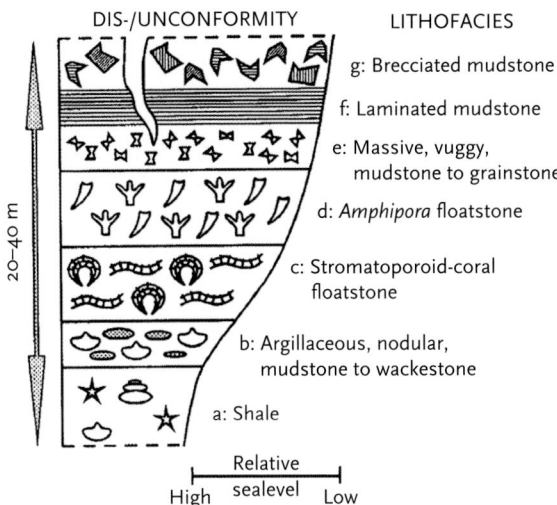

Figure 6.23: Lithofacies classification for the Upper Devonian Grosmont formation of Alberta, Canada. Modified from Luo et al. (1995), © 1995 American Association of Petroleum Geologists (AAPG), reprinted by permission of the AAPG, whose permission is required for further use.

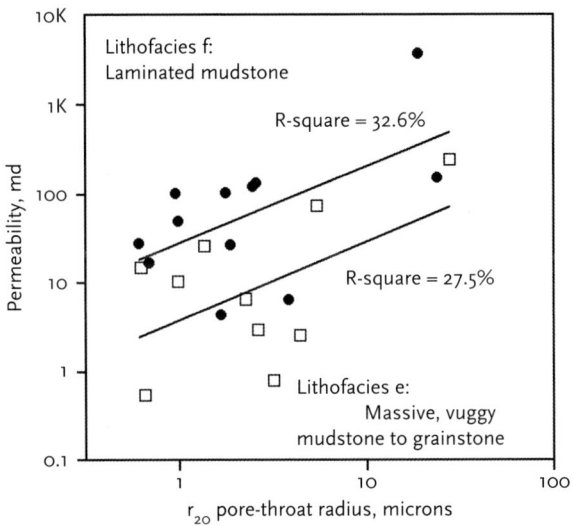

Figure 6.24: Crossplot of permeabilities versus R20 pore-throat radii in the Upper Devonian Grosmont formation lithofacies e and f.

which is to be expected for carbonates whose pore-throat distribution has been modified by a variety of diagenetic processes. However, the two lithofacies show distinctive differences, in that permeabilities in lithofacies f are higher in permeability over a similar porosity range to lithofacies e, where vug development may account for poorer connectivity and heterogeneity. An improved analysis can be made by recognizing once again that pore-throat petrofacies are not necessarily concordant with lithofacies, but can show cross-cutting relationships that differentiate flow units from rock-type units.

Luo and Machel (1995) noted that multiple reservoir types occurred within each lithofacies and so identified six types of pore-throat textures in the Grosmont formation for which the representative capillary pressure curves and pore-throat size frequencies are shown in Figure 6.25. These textures exhibit unimodal, bimodal, and polymodal pore-throat distributions that require multiple statistics for their description. Importantly, these statistical descriptors should capture pore-throat properties

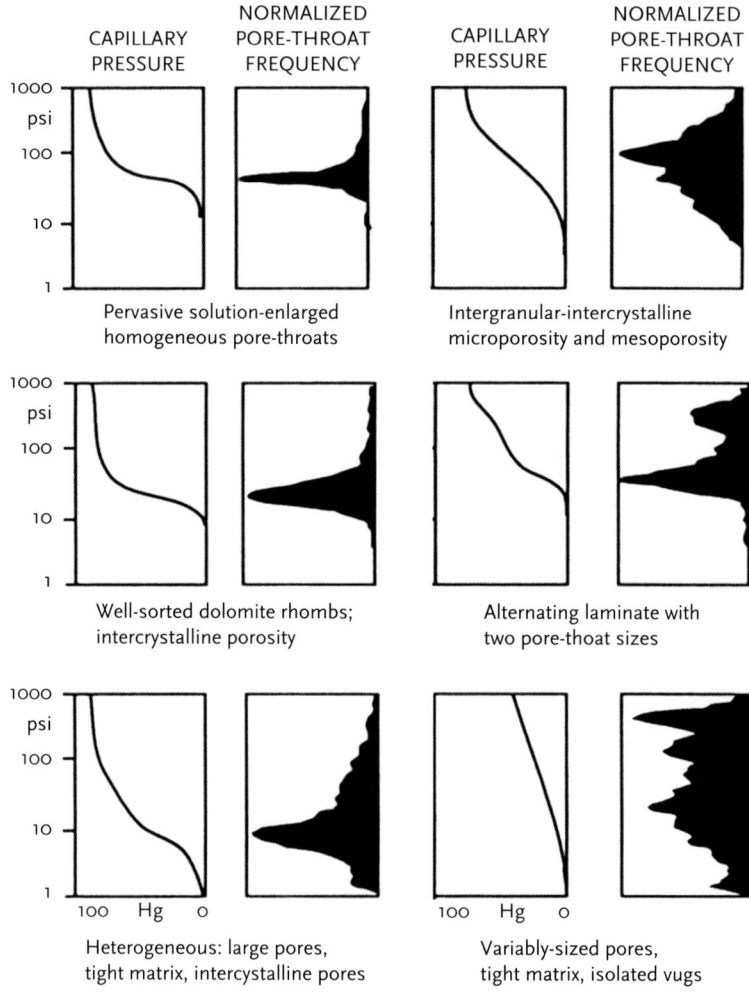

Figure 6.25: Six pore-throat texture classes identified in the Upper Devonian Grosmont formation of Alberta, Canada. From Luo and Machel (1995), © 1995 American Association of Petroleum Geologists (AAPG), reprinted by permission of the AAPG, whose permission is required for further use.

that can be related to reservoir quality. The relationships between the average, mode, and median pore-throat radii were used in the interpretation of the effects of dissolution on matrix heterogeneity. For a normal distribution, these statistics should be approximately equal. However, the mode of the Grosmont data takes progressively higher values with increasing average pore-throat radius, which implies that dissolution enlarged the pore throats. At smaller pore-throat sizes, the median, mode, and average pore-throat radii are similar, suggesting that dissolution did not substantially contribute to the reservoir heterogeneity at this scale. Dolomite dissolution appears to have created relatively large pores and pore throats, rather than microporosity. The average pore-throat radius is positively correlated with standard deviation, and this confirms the interpretation of a higher degree of heterogeneity at greater pore-throat sizes. It was concluded that the single most important pore-throat radius statistic was the twentieth percentile r_{20} value, which matches the assessment by Wardlaw and Taylor (1976) that r_{20} represents the maximum connected pore-throat radius in many carbonate reservoir rocks with either single- or dual-porosity systems.

In summary, the key descriptor of pore-throat petrofacies is the principal pore throat, which shows the maximum correlation with permeability and so represents the equivalent mean hydraulic radius of the rock. Injection of a nonwetting phase (typically mercury) into a core sample at increasing pressure generates a cumulative record of access to successively smaller pore throats. Core petrophysicists have sought to distinguish a characteristic saturation value at which the major connected pore throats are breached and to link this with the associated pore-throat radius in a predictive equation based on porosity and permeability. The benefit of this equation is that conclusions from porosimetry can be applied to routine core measurements of porosity and permeability in the definition of flow units keyed to specific principal pore-throat radii measured in microns.

The widely-used Winland equation sets the characteristic nonwetting phase saturation value at 35 percent, although others use the Pittman equation estimate at a saturation of 25 percent. The pore-throat radius linked with these characteristic saturations shows a close match with the pore-throat radii modes that are usually strongly developed in the intergranular porosity systems of sandstones. The application of the Winland equation to single-porosity carbonate systems appears to underestimate principal pore-throat sizes, and if there is a characteristic principal pore-throat saturation, it may be higher in mud-supported carbonates and lower than 35 percent in grain-supported carbonates. Larger pore throats are developed in carbonates with secondary porosity dissolution, and an r_{20} estimate of principal pore throat size is often a useable estimate. However, diagenetic features in carbonates often introduces extra complexity that requires careful integration of geological history, and the principal pore-throat radius may not be uniquely determined by a single statistic of the pore-throat radii distribution.

Finally, estimates of the principal pore-throat radius can be assigned to size classes whose boundaries have been linked with textural classes of grain or crystal size. These resulting pore-throat facies can be used as flow-unit architectural elements and mapped laterally in the development of dynamic models of flow. The disposition of these petrofacies elements will show a strong concordance with the hydraulic flow units (HFU) described in Chapter 3, because they are defined by the same input measurements. However, their development and interpretation is more closely linked with the geological aspects of rock-matrix texture, while hydraulic flow units use an engineering perspective.

PORE-BODY SIZE DISTRIBUTIONS FROM NUCLEAR MAGNETIC RESONANCE MEASUREMENTS

Although pore-throat sizes control flow properties of rocks, pore-body sizes determine which part of the pore volume can be considered as effective for fluid storage and mobility. Pore bodies are easier to evaluate visually than pore throats, so many petrographic studies have been made to characterize their volumes and size distributions. Traditional methods of point counting the pore space in thin sections gave statistically weak estimates of porosity, while evaluating pore sizes and shapes were both extremely tedious and challenging.

The introduction of computer processing of rock images (e.g., Ehrlich et al., 1984) dramatically resolved the statistical problem of small sample sizes and removed questions of bias in visual estimations. Individual pores could be distinguished and their perimeters and areas measured. In addition to characterizing pore size, the perimeter-to-area ratio is a measure of pore shape. The shape aspects of pores have important implications for understanding the meaning of pore "sizes" estimated from nuclear resonance logging, especially in the complex pore systems of carbonate rocks. So, for example, Anselmettii et al. (1998) reported from their image analyses of carbonates from the Great Bahama Bank that moldic pores and micropores are well rounded with low shape ratios. In contrast, interparticle pores, which have more complex geometry, have high shape ratios. The major limitation of conventional image analyses stems from their use of two-dimensional representations of three-dimensional solids and the inherent distortions in estimates of three-dimensional reservoir pore attributes. Recent advances in computed tomography (CT) scans have overcome this dimensional limitation, but their relatively high cost, coupled with the large quantity of low-cost legacy core image data, has tended to focus efforts on understanding the "meaning" and potential usefulness of traditional measurements.

The application of nuclear magnetic resonance (NMR) log responses to pore-size measurement in both core and borehole logging applications has been the subject of extensive research in recent decades (e.g., Kenyon, 1992). The physical basis for the NMR tool measurement and its use in permeability estimation has already been reviewed in Chapter 3 of this book. In this chapter, we are concerned with pore facies characterization and will focus our attention on the pore-size implications of NMR relaxation-time distributions obtained by borehole logging. Kleinberg et al. (1993) showed that T2 times measured at low frequency and short pulse spacing have almost the same information about pore sizes as T1 longitudinal relaxation times. Consequently, pore-size conclusions from T1 measurements, which are more common in core studies, can be related directly to subsurface measurements of T2 acquired by logging tools (Kleinberg, 2001).

The T2 time reflects the relaxation of protons in terms of three components, as expressed by the equation:

$$\frac{1}{T2} = \frac{1}{T2b} + \frac{1}{T2s} + \frac{1}{T2d}$$

Where *T2b*, *T2s*, and *T2d* are the transverse relaxation times for bulk fluid, surface effects, and diffusion effects. The bulk-fluid relaxation time corresponds to an infinitely large container, and its value of about three seconds makes it so small that it can be eliminated from consideration. The effect of diffusion is also generally very small, so that its relaxation time can be ignored unless the rock contains vugs. Consequently, the T2 relaxation time is dominated by surface effects in simple, homogeneous pore systems that are characterized by interparticle porosity. The surface relaxation time is controlled by the surface-to-volume ratio as given by the equation:

$$\frac{1}{T2s} = \rho 2 \frac{S}{V}$$

where $\rho 2$ is the surface relaxivity, measured in microns per second, S is the pore surface area, and V is the pore volume. The surface relaxivity measures the ability of the pore wall to promote proton relaxation and is especially sensitive to paramagnetic ions such as iron and manganese (Kleinberg et al., 1994). Surface relaxivity can be measured from core samples and shows a wide range of values dictated by compositional variability. However, Chang et al. (1994) suggested a surface relaxivity of fifteen microns/second for sandstones (attributed to iron content) and a value of five microns/second for carbonates, reflecting their lower content of impurities. In computer simulations, Toumelin and Torres-Verdin (2004) chose a range of ten to thirty microns/second as reasonably representative of sandstones, and one to seven microns/second as representative of carbonates. It is important to remember that the surface relaxivity is estimated from core measurements of surface and volume, so the value represents a calibration transform that may differ according to the method of core measurement; it is not an independent physical property. In addition, relaxivity will vary according to the nature of the wetting fluid, so oil relaxivity will be different from water relaxivity and will vary according to the wettability of the grain surface. In this text, we will assume a water-wet surface as our baseline model for pore-size evaluation, keeping in mind that hydrocarbon effects may affect pore-size evaluations by causing a shift of T2 distributions.

Because area has two dimensions and volume has three, the ratio (S/V) is a measure of length and can be expressed as the pore radius of a standard pore shape. So,

$$\frac{1}{T2s} = \rho 2 \frac{G}{r}$$

where r is the pore radius and G is a geometric shape factor. If a pore is spherical, G has a value of three; for a tubular pore, G is two; and for a planar (fracture) pore, G is one. Real pores have more complex and diverse shapes and will show differences with respect to pore size as demonstrated in image analysis studies (e.g., Anselmetti et al., 1998). Consequently, a simple rescaling of T2 relaxation times to an equivalent pore "radius" is an unnecessary and potentially misleading oversimplification. Instead, the relaxation time is a direct function of a pore length measure, whose meaning must be interpreted in the context of the formation pore geometry that was created during its depositional and diagenetic history. Any given T2 distribution therefore represents an expression of pore sizes that are best considered in terms of facies.

NMR FACIES IN SANDSTONES

The term "NMR facies" was probably first introduced by Lowden (1996) and further discussed by Walsgrove et al. (1997), who defined it as "a set of similar NMR distributions that summarize the characteristics of the rock" in their study of a magnetic resonance log of a clastic succession in a North Sea well. They partitioned the facies by a supervised classification using criteria that included porosity, permeability, bulk volume water, capillary pressure data, pore geometry models, and lithological descriptions. In this chapter, we will consider both supervised and unsupervised methods of NMR facies analysis.

The relationship between the distribution of T2 relaxation times and pore sizes in sandstones should be simpler than in the complex pore network of carbonates, which will be reviewed below. Relaxation times for sandstones generally range between ten and five-hundred milliseconds (Delhomme, 2007). The pores are primarily interparticle, and their size distribution closely mirrors the size distribution of the grains, but modified by sorting, cementation, and other processes.

Sedimentologists have used a variety of numerical methods to characterize grain size distributions, ranging from arbitrarily chosen percentiles to more systematic descriptive statistics. Although the sizes of sedimentary particles occur on a continuum, grain size is almost always measured, displayed, and analyzed in discrete form as histograms composed of intervals or bins. Krumbein (1936) proposed using the negative logarithm to the base two as a transform of the grain diameter measured in millimeters, which results in classes that correspond to those in the Wentworth grade scale (Wentworth, 1922). Grain size distributions are commonly represented by a lognormal model, which reflects a multiplicative process of grain breakage that is applicable to particulate matter. On a logarithmic scale, the arithmetic average corresponds to the geometric average of the sizes on an arithmetic measurement scale. This statistic is the first moment and locates the centroid of the distribution. Krumbein (1936) described the computation of higher-order moments; the second moment expresses the dispersion, and the third moment expresses the skewness of a distribution. The "method of moments" presented a way for sedimentologists to condense grain size distributions to a set of statistics for purposes of classification and genetic studies.

The same concept of moment calculation and interpretation is applicable to T2 distributions of clastics for characterizing their pore sizes, and, by implication, the properties of their associated grain size distributions. The first moment of the T2 distribution is widely used as the pore-size term in the SDR equation to estimate permeability (Kenyon, 1992). The geometric average in the SDR equation is equivalent to the first moment, given by the equation:

$$m_1 = \frac{\sum \phi_i \log T2_i}{\sum \phi_i}$$

where ϕ_i is the porosity assigned to the ith bin and $T2_i$ is the transverse relaxation time of the ith bin. Similarly, the second moment is calculated by:

$$m_2 = \frac{\sum \Phi_i (\log T2_i)^2}{\sum \Phi_i}$$

The second moment expresses the dispersion about the origin, but can be referenced to the centroid: $n_2 = m_2 - m_1^2$. From this, the standard deviation $s = \sqrt{n_2}$ can be calculated, which expresses the dispersion about the centroid in units of relaxation time. Finally, the third moment is calculated by the equation,

$$m_3 = \frac{\sum \Phi_i (\log T2_i)^3}{\sum \Phi_i}$$

This moment may be standardized by $n_3 = m_3 - 3m_2 m_1 + 2m_1^3$ and converted to a dimensionless measure of skewness,

$$Sk = \frac{\sqrt{n_3}}{\sqrt{n_2}}$$

Higher moments also can be computed, such as the fourth moment, which measures kurtosis or "peakedness." Although kurtosis is easy to calculate, it is difficult to interpret its meaning as related to size distributions, unlike the first three moments, which represent location, dispersion, and degree of asymmetry.

The NMR log of a Pennsylvanian Stalnaker sandstone section measured in a well in southern Kansas (Figure 6.26) is used as an example for the interpretation of the moments of the distribution of the T2 relaxation times. The depositional environment of the Stalnaker sandstone in this area has been interpreted as deltaic, with burrowed, coarsening-upwards fine-grained sandstone sequences produced by prograding delta lobes succeeded by fining-upward sequences with abundant plant material, which are suggestive of channels in a delta-plain setting (Walton and Griffith, 1985). The image log of the T2 distribution in the Stalnaker sandstone (Figure 6.26) shows a strong mode throughout the unit, with a secondary mode that indicates a patchy development of finer pore sizes. This suggests that the Stalnaker sandstone consists of distinctive zones rather than being homogeneous. At a large scale,

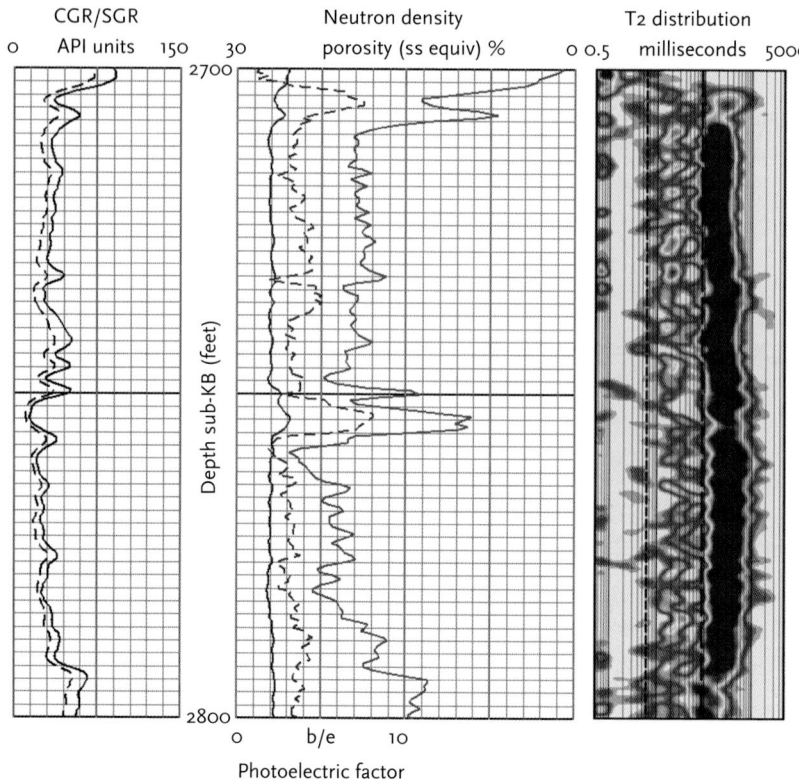

Figure 6.26: Computed gamma-ray (CGR), standard gamma-ray (SGR), photoelectric-factor, neutron and density porosity (sandstone equivalent units), and T2-relaxation-time image logs for a Stalnaker sandstone section in a southern Kansas well.

the sandstone interval is clearly subdivided into two parts that are separated by a calcite-cemented sandstone.

A log of the first three moments of the distribution of the T2 relaxation times is shown in Figure 6.27. In the lower sandstone, the first moment shows a systematic upward increase in relaxation time that reflects increasing pore size, and, by implication, increasing grain size. The pattern is consistent with an interpretation of prograding delta lobes from a core in a nearby well (Walton and Griffith, 1984). Three zones can be discriminated within this lower sandstone by matching clay content inferred from the computed gamma-ray log with values of the second (dispersion) and third (skewness) moments. Zones at a finer scale could be made from a more detailed subdivision of the second- and third-moment curves. Above the calcite-cemented sandstone that terminates the lower sandstone section, the first moment of the T2 relaxation time shows a distinctive decrease, representing finer pore and grain sizes. This upper sandstone appears to be zoned at different scales, with indications of both coarsening- and fining-upward pore- and grain-size distributions. From comparisons with nearby wells, the upper

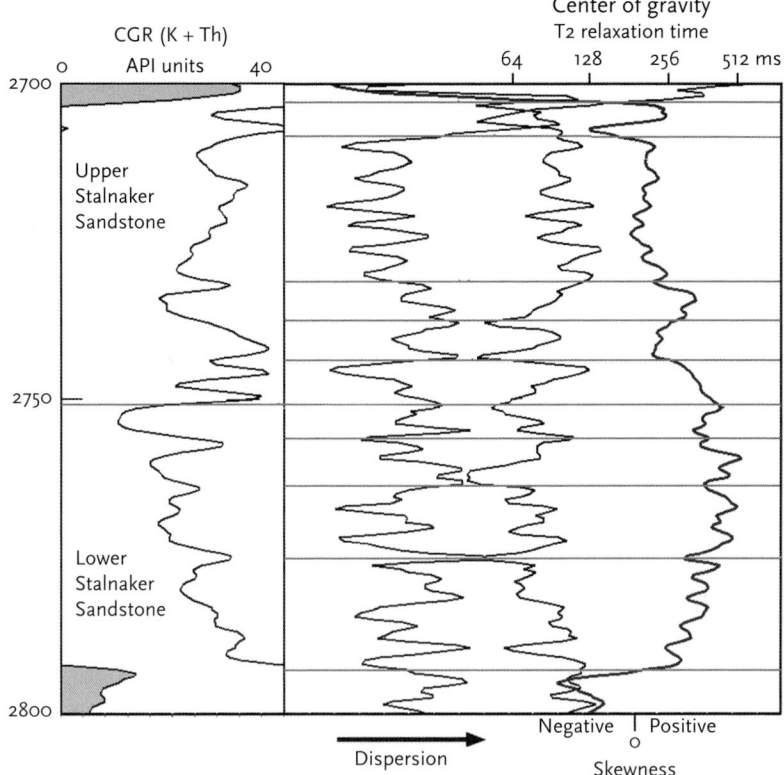

Figure 6.27: Computed gamma-ray (CGR), standard gamma-ray (SGR), T2-relaxation-time centroid, dispersion, and skewness for a Stalnaker sandstone section subdivided into moment facies zones in a southern Kansas well.

sandstone section can be classified as a delta-plain deposit representing a variety of depositional environments.

The strong primary mode of the T2 relaxation times throughout the entire Stalnaker sandstone means that the first moment is the most important descriptor of changes in gross pore size. The dispersion and skewness moments are more useful for defining zonal subdivisions that can be interpreted as individual beds that differ internally in their pore and grain sizes, which are responses to short-term depositional processes. A clear distinction between the upper and lower sandstone subdivisions is seen on the crossplot of the computed gamma-ray log versus the first moments of the T2 relaxation time (Figure 6.28). The lower sandstone has a coarsening-upward trend of coarser pores and lower clay content, while the upper sandstone becomes finer upward, is more clay-rich, and has smaller pores. These patterns are consistent with interpretations of the lower sandstone as the product of prograding delta lobes and the upper sandstone as delta plain deposits.

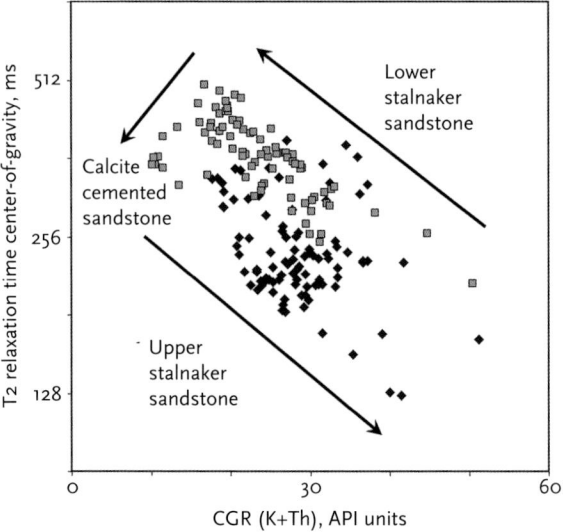

Figure 6.28: Crossplot of computed gamma-ray (CGR) against T2-relaxation-time centroid that differentiates coarsening-upward trend of prograding delta lower Stalnaker sandstone with fining-upward delta-plain upper Stalnaker sandstone.

NMR PORE-SIZE INTERPRETATION IN CARBONATES

The interpretation of T2 relaxation times in carbonates is more complex than the interpretation in sandstones, which generally have relatively simple and homogeneous distributions of interparticle pores. The radically different sizes and shapes of pores within carbonates are the result of the multiplicity of diagenetic processes of dissolution, cementation, and mineral transformation. Multimodal pore distributions are the norm rather than the exception. As a consequence, conclusions drawn from the study of one carbonate unit may be misleading when applied to another. Every carbonate interval should be considered on a case-by-case basis. However, it is possible to relate a carbonate reservoir to a catalog of published magnetic resonance data, provided there is an acceptable match with an analog that has similar pore types and diagenetic history.

As noted earlier, the T2-relaxation-time measurement has three components: the bulk, surface, and diffusion effects. For interparticle porosity, bulk and diffusion effects are very small compared to the surface relaxivity and can be neglected. As a result, the measured T2 relaxation times are dominated by the pore surface-to-volume ratio and are proportional to a length measure of the pore bodies. Sandstone surfaces are more efficient than carbonates in relaxing nuclear magnetism, and the variability in carbonate composition and texture creates a wider range of values for surface relaxivity. So, for example, Quintero et al. (1999) reported that mud-supported carbonates have lower relaxivities than grain-supported carbonates. This makes the conversion from relaxation time to pore size more difficult to ascertain. More fundamentally, the enormous variation in carbonate pore shape calls into question the very meaning of the "radius" of a pore.

By comparing vugs observed in the core with the T2 relaxation times, Chang et al. (1994) concluded that the porosities of carbonates whose relaxation times exceeded 750 milliseconds showed a good match with vug volumes estimated by image point counting. In vuggy carbonates, protons within the larger pores have a much lower likelihood of interacting with pore surfaces, so the bulk diffusion term can be significant. More importantly, diffusion of protons between the larger and smaller pores can cause a decrease in the proportion of measured finer pores. This will shift the proportion of larger pore relaxation times downward to lower values (Ramakhrisna, 1999). The net effect may be the merging of the two peaks of a dual-porosity system into a single mode in the T2-relaxation-time distribution (Delhomme, 2007). These circumstances explain some of the anomalies that puzzled early workers in magnetic resonance studies of carbonates, when compared to the more successful applications in sandstones. These explanations are helpful guides to a realistic interpretation of carbonate pore distributions but suggest that strategies expressed in terms of facies are more likely to be fruitful than attempts at numerical transformations.

In working with magnetic resonance data from the Shuaiba Formation, Lodola (2004) computed the mean, median, variance, skewness, and ninetieth percentile statistics to characterize the shape of the T2-relaxation-time distributions. He concluded that the ninetieth percentile was the best discriminator of pore types for separating vugs from smaller pores. The other statistics were more difficult to interpret because of the compounding effect of multiple pore-type distributions. Lodola (2004) decomposed the T2 distributions by partitioning them successively into as many as three lognormal distributions. The means and variances of the lognormal distributions were estimated through a best-fit Newton-Raphson procedure that minimized the sum of the squared errors. He concluded that samples dominated by depositional pore types showed no increase in the quality of the fit of several distributions over the fit of a single lognormal distribution. In contrast, samples with facies-selective pores and especially diagenetic porosity showed better fits when partitioned into multiple lognormal distributions.

NMR-PARTITIONED POROSITY AND DUNHAM TEXTURAL CLASSES

The issue of the discrimination of vugs and their partitioning into connected and nonconnected pores has been discussed in Chapters 2 and 3, so we will focus here on the integration of NMR pore facies with Dunham carbonate textural classes. Computed gamma-ray, photoelectric factor, neutron and density porosity (limestone equivalent), and magnetic resonance image logs are shown in Figure 6.29 for a section of Cambro-Ordovician Arbuckle limestone in a southern Kansas well. This section consists of dolomitic mudstones, packstones, and grainstones deposited as cyclic peritidal units that have minimal vug porosity. The T2 distribution shows a generalized bimodal character that reflects pore-size variability that is potentially related to carbonate lithofacies. Histograms of T2 relaxation times for a larger set of core observations of Dunham textural classes in the Arbuckle limestone are shown in Figure 6.30. The histograms clearly show a bimodal distinction between longer

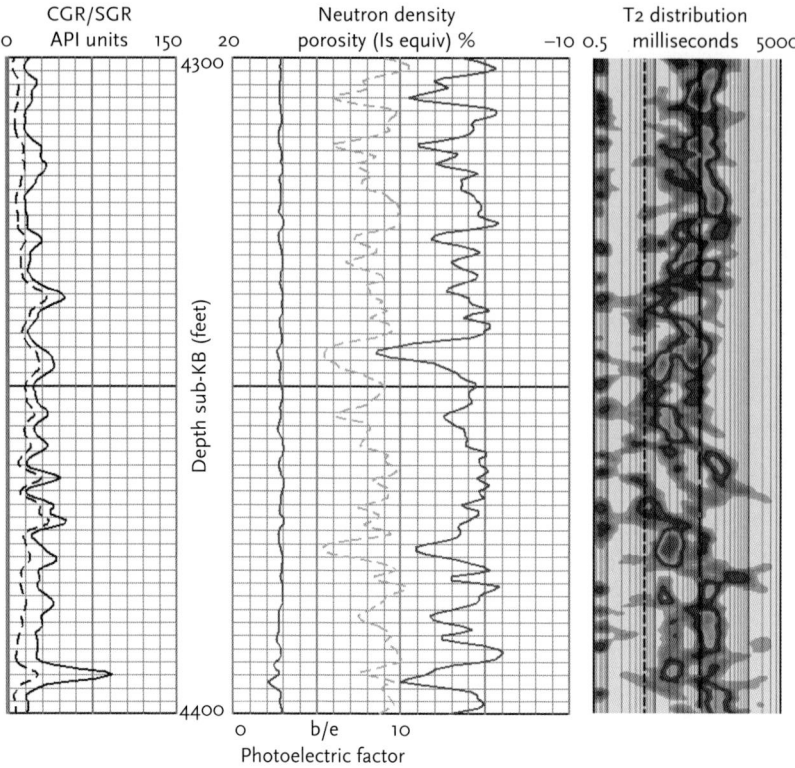

Figure 6.29: Computed gamma-ray (CGR), standard gamma-ray (SGR), photoelectric-factor, neutron and density porosity (limestone equivalent units), and T2-relaxation-time image logs for an Arbuckle limestone section in a southern Kansas well.

and shorter relaxation times. There is a clear differentiation between the larger pore mode in grain-supported grainstones and packstones, contrasted with the secondary mode of smaller pore sizes associated with mudstones. Consequently, the distribution can be subdivided between "microporosity" and "macroporosity."

A supervised classification of mudstones, packstones, and grainstones was made by a discriminant function analysis on the basis of the percentage of macroporosity and the computed gamma-ray (CGR) log. The results of classifying the Arbuckle limestone section (Figure 6.29) are shown in Figure 6.31, with log tracks of the computed gamma-ray curve, proportion of pore space based on T2 relaxation times, the Dunham lithofacies as observed in the core, and the "Punham facies" classified by petrophysics. In Figure 6.31, the T2 relaxation times have been shaded as three intervals that represent the primary and secondary modes (microporosity and macroporosity) and the intervening time interval of the Dunham-class T2 histograms. Deflections of the computed gamma-ray curve to higher values suggest that increased clay content differentiates mudstones from grain-supported grainstones and packstones and is matched by complementary shifts in the T2 relaxation times. Cleaner carbonates have higher proportions of larger pores and more shaly carbonates have

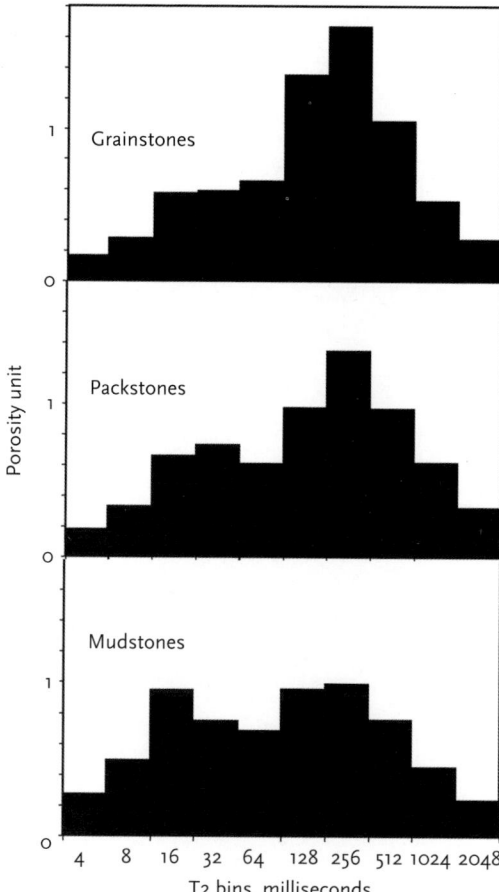

Figure 6.30: Histograms of average T2-relaxation-time bin porosities for core intervals described as mudstones, packstones, and grainstones in the Arbuckle limestone in a southern Kansas well.

higher proportions of smaller pores. The informal terminology of "Punham facies" is applied here to emphasize the distinction between Dunham grain-size textural facies and NMR pore-size facies. While there appears to be a good overall correlation, there is no implicit physical reason why a pore-size driven classification should give an exact prediction of a Dunham texture.

This supervised classification case study demonstrates the complementary nature of Dunham lithofacies classes and pore-size distributions not only as a conceptual model but also a model with quantitative descriptive and predictive power. The prediction model could be extended to uncored wells whose lithofacies can be classified based on T2-relaxation-time distributions. Alternatively, a derived magnetic resonance log could be approximated by plotting the expectations of T2 relaxation times based on the Dunham textural classes observed in the core.

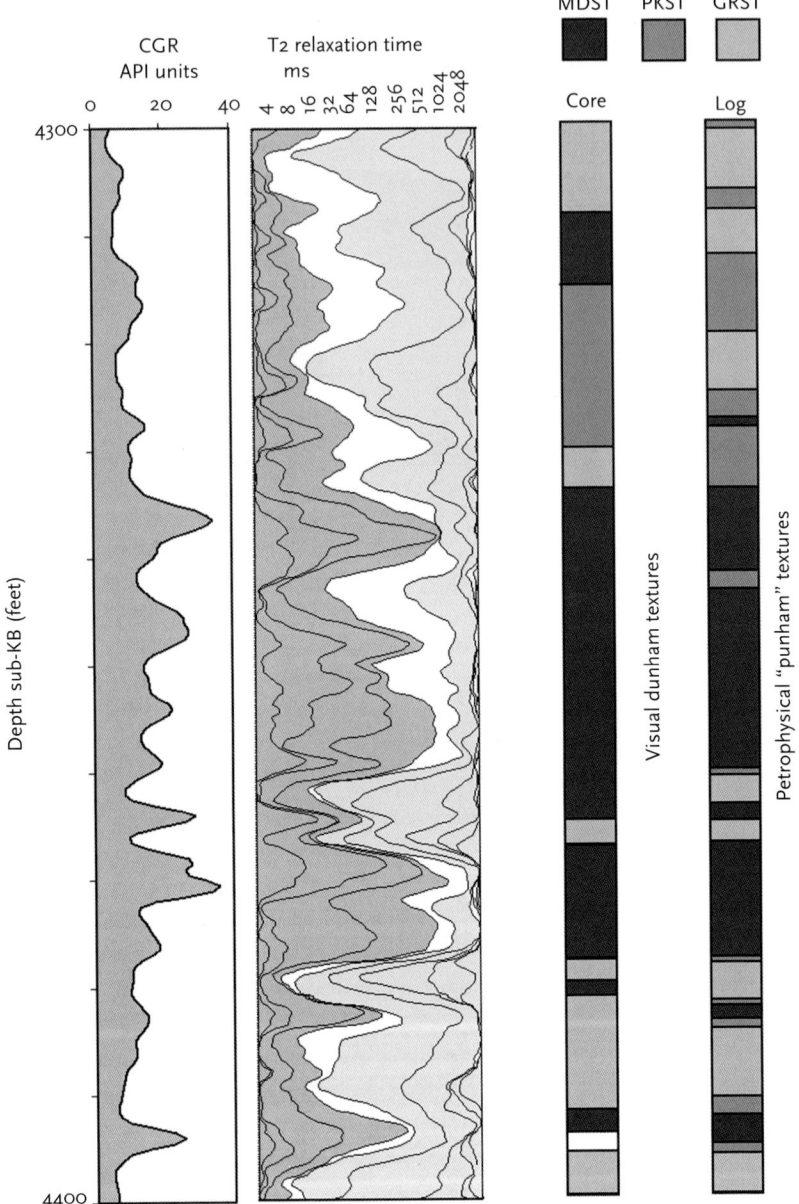

Figure 6.31: Computed gamma-ray (*CGR*), binned T2 relaxation times, core observations of Dunham textural facies, and discriminant function classification for an Arbuckle limestone section in a southern Kansas well.

NMR FACIES IN CARBONATES

In an alternative approach, T2-relaxation-time logs can be analyzed in an unsupervised manner to extract intrinsic associations for interpretation and application. A principal components analysis was made of the T2 relaxation times of the entire Arbuckle lmestone formation. In doing this, the data set was expanded beyond subtidal and peritidal mudstones, packstones, and grainstones to include multiple brecciated intervals that were created by periods of extensive karstic weathering that led to the development of vug-size pores over a range of scales. Principal components reveal the intrinsic structure of the T2 relaxation times within associations of intervals or bins that are related by distinctive pore-size groupings. Table 1 gives the percent contribution by each porosity interval to each principal component. There are ten T2 porosity intervals and the analysis indicates that there are five pore-size types. The contributions to the first three principal components are shown in Figure 6.32 and can be informally labeled the macro-, meso-, and micropore subdivisions of the matrix porosity; this classification is consistent with the T2-relaxation-time range. The range encompassed by macro- and mesopores coincides with the principal mode of the lithofacies histogram shown in Figure 6.30. This substantiates the idea that this mode is actually the merger of two distinctive pore-size types. Consequently, the bimodal character of the matrix-porosity T2-relaxation-time distributions observed in the lithofacies should be attributed to a triple pore system.

Bins representing longer relaxation times discriminate vugs (in this case study, they are called "megapores") and their contributions are restricted to several higher-order principal components. Bins that represent the shortest relaxation times (nanopores) also contribute to higher components; their values are consistent

TABLE 6.1. PERCENT CONTRIBUTIONS OF T2 BIN POROSITIES TO EACH PRINCIPAL COMPONENT, WITH PERCENT TOTAL VARIABILITY ACCOUNTED FOR BY EACH PRINCIPAL COMPONENT INDIVIDUALLY AND CUMULATIVELY

	PC1	PC2	PC3	PC4	PC5	PC6	PC7	PC8	PC9	PC10
4ms	0.2	0.1	0.0	0.2	0.2	5.9	18.8	22.0	49.5	3.3
8ms	0.8	0.0	2.6	0.0	15.7	26.8	23.2	0.8	26.6	3.5
16ms	1.7	0.3	47.4	4.7	25.0	0.0	8.3	4.5	6.6	1.4
32ms	1.1	0.2	39.8	2.5	19.4	16.0	6.6	9.0	3.7	1.7
64ms	5.9	12.0	0.6	48.9	1.8	5.8	0.4	21.4	1.3	1.8
128ms	1.8	63.1	3.1	0.0	7.1	3.0	2.8	16.5	0.3	2.3
256ms	29.2	23.3	1.8	18.9	8.1	2.7	5.8	0.2	0.0	0.0
512ms	49.0	0.3	2.9	16.9	0.9	3.0	7.3	2.8	1.0	15.9
1024ms	10.1	0.5	0.4	9.5	17.1	20.1	5.8	0.1	3.6	32.7
2048ms	0.3	0.1	1.4	0.3	4.6	16.6	21.2	16.7	7.5	31.4
Percent contribution of the principal components to the total variability:										
Individual	38.9	15.7	15.1	11.8	8.1	4.9	2.7	1.6	0.7	0.5
Cumulative	38.9	54.6	69.7	81.5	89.6	94.5	97.3	98.8	99.5	100.0

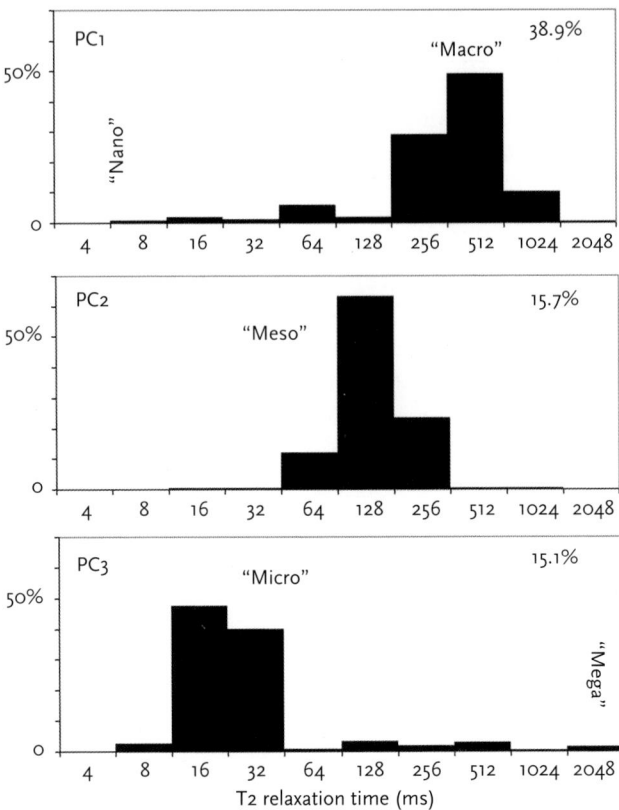

Figure 6.32: Percent contributions of T2 relaxation times to the first three principal components in the Arbuckle limestone in a southern Kansas well.

with clay-bound water. It is interesting to note that mega- and nanopores jointly contribute to higher-order principal components, suggesting they are associated in clay-rich brecciated zones. Following this interpretation, the first principal component discriminates matrix porosity associations in contrast to vuggy and clay-rich karstic breccias in the remaining principal components.

A crossplot of the first two principal component scores from T2-relaxation-time distributions of the Arbuckle limestone is shown in Figure 6.33; symbols indicate mudstones, packstones, and grainstones. The crossplot represents 54.6 percent of the total variability in the T2-relaxation-time distributions, with the first principal component representing macropores and the second component dominated by mesopores. The locations of the centroids of the three lithofacies match our expectations for the pore-size associations seen in supervised lithofacies analyses. In this plot, the principal mode of the T2-relaxation-time distributions of the observed lithofacies is mapped in terms of the larger matrix pores. The third principal component captures the microporosity and would be plotted on an axis orthogonal to this plot.

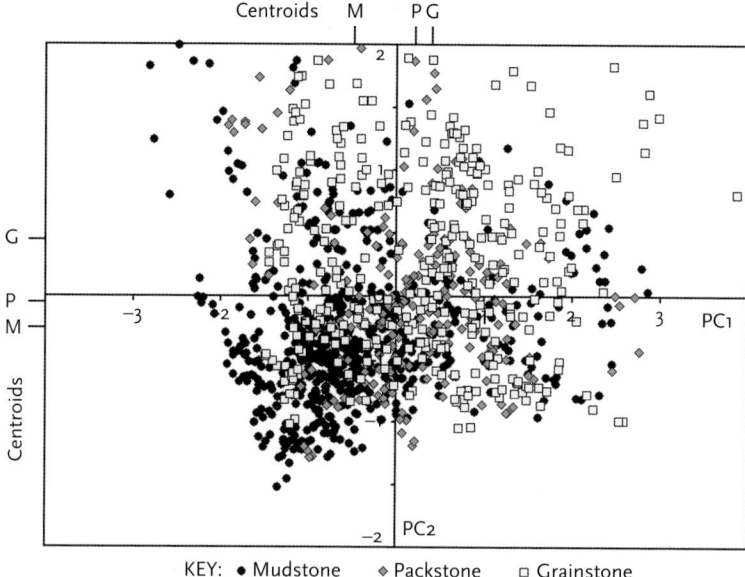

Figure 6.33: Crossplot of first two principal-component scores calculated from T2-relaxation-time porosities indexed by core observations of mudstones, packstones, and grainstones in the Arbuckle limestone in a southern Kansas well.

All three matrix porosity pore-size types can be extracted from the first three principal components and plotted on ternary diagrams in order to clarify the interrelationships between pore-size and lithofacies. In Figure 6.34, mudstones, packstones, and grainstones for the Arbuckle limestone formation are shown on composition triangles in terms of macro-, meso-, and microporosity. The mudstones show a broad scatter of pore sizes, but microporosity is relatively higher than in grainstones and packstones. Packstones and grainstones both tend towards macroporosity, but they overlap, reflecting their definition as members of a continuum of grain-supported texture classes. The plot of grainstones is interesting because it indicates that there are two types, distinguished by their relative proportions of meso- and macropores. The majority are coarser grainstones with high macropore content. These are distinct from finer grainstones dominated by micro- and mesopores; these were identified in a subsequent examination of core as fine-grained peloidal grainstones.

Similar approaches to carbonate categorization have been made for major reservoirs in the Middle East, using the magnetic resonance relaxation time as one facet of an integrated log and core analysis (Ramamoorthy et al., 2008). The identification of mineralogy and lithology was followed by the partitioning of porosity between pore geometry components using magnetic resonance logs in conjunction with borehole imagery and full waveform acoustic logs. An example of a ternary pore-size diagram resulting from this integrated procedure is shown in Figure 6.35, where the triangle is subdivided between pore-system classes for the purpose of improving permeability estimates (Hassall et al., 2004). This approach is more petrophysically

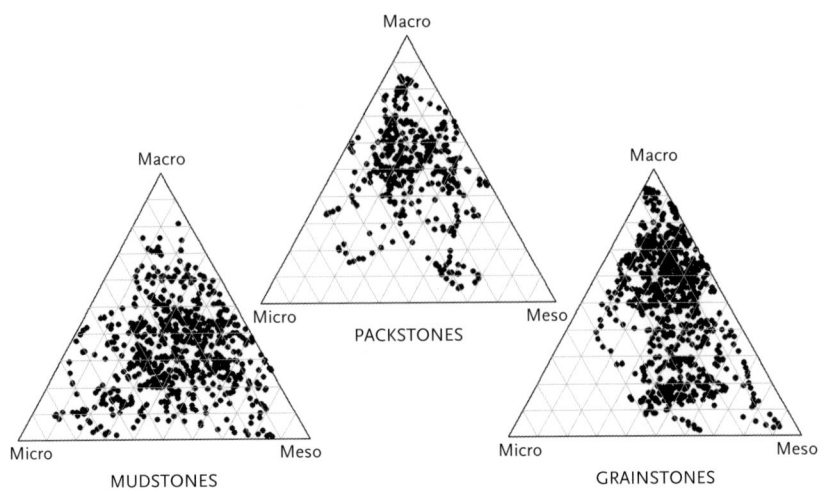

Figure 6.34: Ternary composition plots of mudstones, packstones, and grainstones in the Arbuckle limestone in a southern Kansas well. End members are micro-, meso-, and macroporosities identified by principal component analysis of T2-relaxation-time porosities.

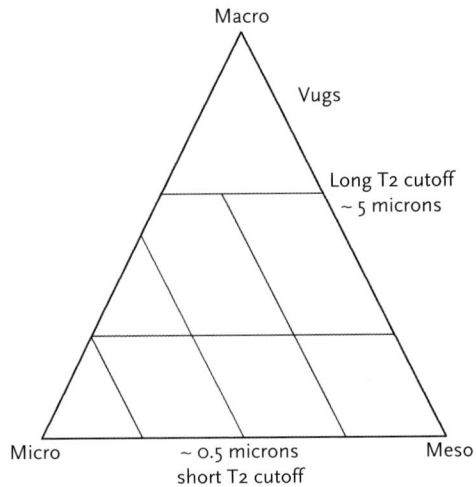

Figure 6.35: Ternary diagram with micro-, meso-, and macroporosity endmembers partitioned into eight petrophysical rock types based on NMR logs in conjunction with borehole imagery. Modified from Hassall et al. (2004), Used with permission © 2004 Society of Petroleum Engineers.

driven, so that the classes are petrofacies in nature, although they are closely linked with associated lithofacies.

Conclusions from Cretaceous limestones of the Middle East will show both similarities and differences from Cambro-Ordovician dolomites of the American Midwest. Similar patterns of pore-size development result from the same processes

[212]   *Principles of Mathematical Petrophysics*

of deposition and diagenesis, but their complex interplay results in characteristics that are peculiar to every stratigraphic interval. Each carbonate formation must be considered independently, using all available logs to augment the information from magnetic resonance logging. Suites of modern well logs provide an extraordinary amount of information. Current research is focused on the use of this information and the development of systematic strategies to clarify carbonate pore systems. This requires geometrical measures of pore characteristics that are meaningful conceptually and can be used in the emerging discipline of digital petrophysics.

REFERENCES

Anselmetti, F.S., Lüthi, S., and Eberli, G.P., 1998, Quantitative characterization of carbonate porosity by digital image analyses: American Association of Petroleum Geologists Bulletin, v. 82, no. 10, pp. 1815–1836.

Archie, G.E., 1950, The electrical resistivity log as an aid in determining some reservoir characteristics: Transactions of the American Institute of Mechanical Engineers, v. 146, no. 1, pp. 54–62.

Arps, J.J., 1964, Engineering concepts useful in oil finding: American Association of Petroleum Geologists Bulletin, v. 48, no. 2, pp. 157–165.

Basan, P.B., Lowden, B.D., Whattler, P.R., and Attrad, J.J., 1997, Pore-size data in petrophysics: A perspective on the measurement of pore geometry, in M.A. Lowell and P.K. Harvey, eds., Developments in Petrophysics: Geological Society Special Publication 122, London, pp. 47–67.

Berg, R.R., 1975, Capillary pressure in stratigraphic traps: American Association of Petroleum Geologists Bulletin, v. 59, no. 6, pp. 939–956.

Bhattacharya, S., Byrnes, A.P., Watney, W.L., and Doveton, J.H., 2008, Flow unit modeling and fine-scale predicted permeability validation in Atokan sandstones: Norcan East field, Kansas: American Association of Petroleum Geologists Bulletin, v. 92, no. 6, pp. 709–732.

Bird, R.B., Stewart, W.E., and Lightfoot, E.N., 1960, Transport Phenomena: John Wiley and Sons, New York, 780 p.

Byrnes, A.P., Buatois, L.A., Mángano, M.G., and Carr, T.R., 2001, Integration of lithofacies and petrophysics in marine and estuarine Morrow sandstone, southwest Kansas: A midcontinent rock catalog: Oklahoma Geological Survey Circular, v. 104, pp. 59–65.

Byrnes, A.P., Franseen, E.K.,Watney, W.L., and Dubois, M.K., 2003, The role of moldic porosity in Paleozoic Kansas reservoirs and the association of original depositional facies and early diagenesis with reservoir properties (abs): American Association of Petroleum Geologists Convention Abstracts.

Chang, D., Vinegar, H., Morriss, C., and Straley, C., 1994, Effective porosity, producible fluid and permeability in carbonates from NMR logging: Transactions of the Society of Professional Well Log Analysts, 35th Annual Logging Symposium, Paper A, 21 p. Later published in 1997 in Log Analyst, v. 38, no. 2, pp. 60–72.

Delhomme, J.P., 2007, The quest for permeability evaluation in wireline logging, in L. Chery and G. de Marsily, eds., Aquifer systems management: Darcy's legacy in a world of impending water shortage: Hydrology 10, Taylor and Francis, New York, pp. 55–70.

Ebanks, W.J., Jr., Scheihing, M.H., and Atkinson, C.D., 1992, Flow units for reservoir characterization: Part 6, Geological Methods, Volume ME 10: Development Geology Reference Manual: American Association of Petroleum Geologists, Houston, Texas, pp. 282–285.

Ehrlich, R., Crabtree, S.J., Kennedy, S.K., and Cannon, R.L., 1984, Petrographic image analysis I: Analyses of reservoir pore complexes: Journal of Sedimentary Petrology, v. 54, no. 4, pp. 1365–1376.

Hartmann, D.J., and Coalson, E.B., 1990, Evaluation of the Morrow sandstone in Sorrento field, Cheyenne County, Colorado, in S.A. Sonnenberg, L.T. Shannon, K. Rader, W.F. von Drehle, and G.W. Martin, eds., Morrow sandstones of southeast Colorado and adjacent areas: Rocky Mountain Association of Geologists, Denver, Colorado, pp. 91–100.

Hassall, J.K., Ferraris, P., Al-Raisi, M., Hurley, N.F., Boyd, A., and Allen, D.F., 2004, Comparison of permeability predictors from NMR, formation image, and other logs in a carbonate reservoir: Society of Petroleum Engineers, SPE 88683-MS, 13 p.

Hearn, C.L., Ebanks, W.J., Jr., Tye, R.S., and Ranganathan, V., 1984, Geological factors influencing reservoir performance of the Hartzog draw field: Journal of Petroleum Technology, v. 36, pp. 1335–1344.

Jennings, J.B., 1987, Capillary pressure techniques: application to exploration and development geology: American Association of Petroleum Geologists Bulletin, v. 71, no. 10, pp.1196–1209.

Kenyon, W.E., 1992, Nuclear magnetic resonance as a petrophysical measurement: Nuclear Geophysics, v. 6, no. 2, pp. 153–171.

Kleinberg, R.L., Straley, C., Kenyon, W.E., Akkurt, R., and Farooqui, S.A., 1993, Nuclear magnetic resonance of rocks: T 1 vs T 2: Society of Petroleum Engineers, SPE 26470-MS, 11 p.

Kleinberg, R.L., Kenyon, W.E., and Mitra, P.P., 1994, Mechanism of NMR relaxation of fluids in rocks: Journal of Magnetic Resonance, Series A, v. 108, no. 2, pp. 206–214.

Kleinberg, R.L., 2001, NMR measurement of petrophysical properties: Concepts in Magnetic Resonance, v. 13, no. 6, pp. 404–406.

Kolodzie, S., Jr., 1980, Analysis of pore-throat size and use of the Waxman-Smits equation to determine OOIP in Spindle Field, Colorado: Society of Petroleum Engineers, SPE-9382-MS, 10 p.

Krumbein, W.C., 1936, Application of logarithmic moments to size frequency distribution of sediments: Journal of Sedimentary Petrology, v. 6, no. 1, pp. 35–47.

Leal, L., Barbato, R., Quaglia, A., Porras, J., and Lazarde, H., 2001, Bimodal behavior of mercury-injection capillary pressure curve and its relationship to pore geometry, rock quality, and production performance in a laminated and heterogeneous reservoir: Society of Petroleum Engineers, SPE 69457-MS, 10 p.

Lodola, D.D., 2004, Identification of pore type and origin in a Lower Cretaceous carbonate reservoir using NMR T2 relaxation times: MS thesis, Texas A&M University, 67 p.

Lowden, B., 1996, NMR facies analysis: SPWLA workshop: Improving NMR log interpretation using core data: Society of Professional Well Log Analysts, 37th Annual Meeting, Paper A, 22 p.

Lucia, F.J., 1999, Characterization of petrophysical petrophysical flow units in carbonate reservoirs: Discussion: American Association of Petroleum Geologists Bulletin, v. 83, no. 7, pp. 1161–1163.

Lucia, F.J., 2007, Carbonate reservoir characterization: An integrated approach: Springer-Verlag, Berlin, 336 p.

Luo, P., and Machel, H.G., 1995, Pore size and pore throat types in a heterogeneous dolostone reservoir, Devonian Grosmont formation, western Canada sedimentary basin: American Association of Petroleum Geologists Bulletin, v. 79, no. 11, pp. 1698–1719.

Martin, A.J., Solomon, S.T., and Hartmann, D.J., 1997, Characterization of petrophysical flow units in carbonate reservoirs: American Association of Petroleum Geologists Bulletin, v. 81, no. 5, pp. 734–759.

Martin, A.J., Solomon, S.T., and Hartmann, D.J., 1997, Characterization of petrophysical flow units in carbonate reservoirs: Reply: American Association of Petroleum Geologists Bulletin, v. 83, no. 7, pp. 1164–1173.

Marzouk, I., Takezaki, H., and Suzuki, M., 1998, New classification of carbonate rocks for reservoir characterization: Society of Petroleum Engineers, SPE 49475-MS, 10 p.

Nabawy, B.S., Geraud, Y., Rochette, P., and Bur, N., 2010, Pore-throat characterization in highly porous and permeable sandstones: American Association of Petroleum Geologists Bulletin, v. 93, no. 6, pp. 719–739.

Pittman, E.D., 1992, Relationship of porosity and permeability to various parameters derived from mercury injection-capillary pressure curves for sandstone: American Association of Petroleum Geologists Bulletin, v. 76, no. 2, pp. 191–198.

Porras, J.C., 1998, Determination of rock types from pore throat radius and bulk volume water, and their relation to lithofacies, Carito Norte field, eastern Venezuela basin: Transactions of the Society of Professional Well Log Analysts, 39th Annual Logging Symposium, Paper OO, 13 p.

Pranter, M.J., 1999, Use of a petrophysical-based reservoir zonation and multicomponent seismic attributes for improved geologic modeling, Vacuum Field, New Mexico: PhD dissertation, Colorado School of Mines, 366 p.

Quintero, L., Boyd, F., al-Wazeer, F., 1999, Comparison of permeability from NMR and production analysis in carbonate reservoirs: Society of Petroleum Engineers, SPE 56798-MS, 8 p.

Ramakhrisna, T.S., 1999, Forward models for nuclear magnetic resonance in carbonate rocks: Log Analyst, v. 40, no. 4, pp. 260–270.

Ramamoorthy, R., Boyd, A., Neville, T., Seleznev, N., Sun, H., Flaum, C, and Ma, J., 2008, A new workflow for petrophysical and textural evaluation of carbonate reservoirs: Transactions of the Society of Professional Well Log Analysts, 49th Annual Logging Symposium, Paper B, 15 p.

Spearing, M., Allen, T., and McAulay, G., 2001, Review of the Winland R35 method for net pay definition and its application in low permeability sands: Special Core Analysis Symposium Volume 2001, Society of Core Analysts, Edinburgh, no. 63, p. 1–7.

Stout, J.L., 1964, Pore geometry as related to carbonate stratigraphic traps: American Association of Petroleum Geologists Bulletin, v. 48, no. 3, pp. 329–337.

Sujuan, Y., Zhengxiang, L., and Li, R., 2011, Petrophysical and capillary pressure properties of the upper Triassic Xujiahe formation tight gas sandstones in western Sichuan, China: Petroleum Science, v. 8, no. 1, pp. 34–42,

Sullivan, K.B., Campbell, J.S., and Dahlberg, K.E., 2003, Petrofacies: Enhancing the deepwater reservoir characterization effort in West Africa: Petrophysics, v. 44, no. 3, pp. 177–189.

Swanson, B.F., 1981, A simple correlation between permeabilities and mercury capillary pressures: Journal of Petroleum Technology, v. 33, no. 12, pp. 2498–2504.

Thompson, A.R., Katz, A.J., and Raschke, R.A., 1987, Estimation of absolute permeability from capillary pressure measurements: Society of Petroleum Engineers, SPE 16794-MS, pp. 475–481.

Toumelin, E. and Torres-Verdin, C., 2004, A numerical assessment of modern borehole NMR interpretation techniques: Society of Petroleum Engineers, SPE 90539, 19 p.

Vavra, C.L., Kaldi, J.G., and Sneider, R.M., 1992, Geological applications of capillary pressure: A review: American Association of Petroleum Geologists Bulletin, v. 76, no. 6, pp. 840–850.

Walsgrove, T., Stromberg, S.G., Lowden, B.D., Basan, P.B., 1997, Integration of nuclear magnetic resonance core analysis and nuclear magnetic resonance logs: An example from the North Sea, UK: Transactions of the Society of Professional Well Log Analysts, 38th Annual Logging Symposium, Paper UU, 12 p.

Walton, A.W., and Griffith, G., 1985, Deltaic deposition in the Tonganoxie or "Stalnaker" sandstone (Stranger formation, Virgilian): TXO Robinson C-1, Harper County, Kansas, in M. Adkins-Heljeson, ed., Core studies in Kansas, Subsurface Geology Series, v. 6, Kansas Geological Survey, Lawrence, Kansas, pp. 145–160.

Washburn, E.W. 1921, The dynamics of capillary flow: Physical Review, v. 17, no. 3, pp. 273–283.

Watney, W. L., Franseen, E. K., Byrnes, A. P., Miller, R., Raef, A. E., Reeder, S. L., and Rankey, E. C., 2006, Characterization of seismically-imaged Pennsylvanian ooid shoal geometries and comparison with modern (abs): American Association of Petroleum Geologists Convention Abstracts.

Wentworth, C.K., 1922, A scale of grade and class terms for clastic sediments: Journal of Geology, vol. 30, no. 5, pp. 377–392.

CHAPTER 7
Saturation-Height Functions

INTEGRATION: THE SATURATION-HEIGHT MODEL

As observed by Worthington (2002), "The application of saturation-height functions forms part of the intersection of geologic, petrophysical, and reservoir engineering practices within integrated reservoir description." It is also a critical reference point for mathematical petrophysics; the consequences of deterministic and statistical prediction models are finally evaluated in terms of how closely the estimates conform to physical laws. Saturations within a reservoir are controlled by buoyancy pressure applied to pore-throat size distributions and pore-body storage capacities within a rock unit that varies both laterally and vertically and may be subdivided into compartments that are not in pressure communication. Traditional lithostratigraphic methods describe reservoir architecture as correlative rock units, but the degree to which this partitioning matches flow units must be carefully evaluated to reconcile petrofacies with lithofacies. Stratigraphic correlation provides the fundamental reference framework for surfaces that define structure and isopach maps and usually represent principal reflection events in the seismic record. In some instances, there is a strong conformance between lithofacies and petrofacies, but all too commonly, this is not the case, and petrofacies must be partitioned and evaluated separately. Failure to do this may result in invalid volumetrics and reservoir models that are inadequate for fluid-flow characterization.

A dynamic reservoir model must be history matched to the actual performance of the reservoir; this process often requires adjustments of petrophysical parameters to improve the reconciliation between the model's performance and the history of production. Once established, the reservoir model provides many beneficial outcomes. At the largest scale, the model assesses the volumetrics of hydrocarbons in place. Within the reservoir, the model establishes any partitioning that may exist between compartments on the basis of pressure differences and, therefore, lack of communication. Lateral trends within the model trace changes in rock reservoir quality that control anticipated rates and types of fluids produced in development wells. Because the modeled fluids represent initial reservoir conditions, comparisons can be made between water saturations of the models and those calculated from logs in later wells, helping to ascertain sweep efficiency during production. The model can be used to estimate initial water saturations in older wells that lack resistivity logs or whose resistivity logs are unreliable. A major problem with older resistivity measurements is their coarse vertical resolution as contrasted with porosity measurements, so water saturations in thin sandstone beds are poorly evaluated and the model may provide inaccurate estimates. In some underpressured reservoirs, the resistivity log measurements may be compromised by deep invasion.

There are a number of saturation-height methods used whose relative merits have been discussed in the oil and gas industry (sometimes heatedly). Before evaluating some of the more popular approaches, it is important to review the basic elements of a reservoir in terms of both its rock properties and the distribution of fluids. Fortunately, there is common agreement on this aspect because the reservoir is the result of the operation of physical laws. The debate centers on what are the appropriate reservoir descriptors to be used and in what formulation to generate robust predictors of saturation at any location within the reservoir.

THE BASICS OF RESERVOIR SATURATION PROFILES

A schematic representation of the fluid distribution in a simple, homogeneous reservoir is shown in Figure 7.1. Migrating oil (and/or gas) has completely filled the anticlinal structure above the spill point. The free-water level (FWL) coincides with the spill point in this example because the trap is completely charged. Notice that the FWL does not coincide with the oil-water contact (OWC), which is the deepest zone in which oil will flow (as well as water). A minimum threshold entry pressure is required for the hydrocarbons to penetrate the largest pore throats. The controls for the fluid distribution and fluid production in this reservoir are found in the equilibrium between the buoyancy pressure exerted by lighter hydrocarbons moving upward and the resistance exerted by capillary forces that hold the wetting phase of formation water on the pore surfaces. As hydrocarbons move upward, the obstacles to the movement of the oil globules are the constrictions of the pore throats, rather than the pores themselves.

A conventional rock has a distribution of progressively smaller pore-throat sizes that will be penetrated at different heights above the FWL as the buoyancy pressure increases. Immediately above the FWL, the buoyancy pressure is very low, and at an entry (or displacement) pressure, the largest pore throats are breached by hydrocarbons. The majority of pores are still completely filled with water. At this level, some minor oil staining may be noticed in the drill cuttings. Core samples will also show staining and low oil saturations. However, if a production test is run, the produced sample will be entirely water. The oil remains behind in the rock as isolated globules and constitutes "residual oil saturation," S_{or}. Moving higher above the FWL, the produced fluid changes from water only to water with a small oil cut. This level marks the oil-water contact at the base of the transition zone and occurs where the increased buoyancy pressure causes a greater penetration of the pore throats, to the extent that oil globules begin to form continuous filaments throughout the pore network. The continuity of the oil phase allows it to flow together with the water. At still shallower

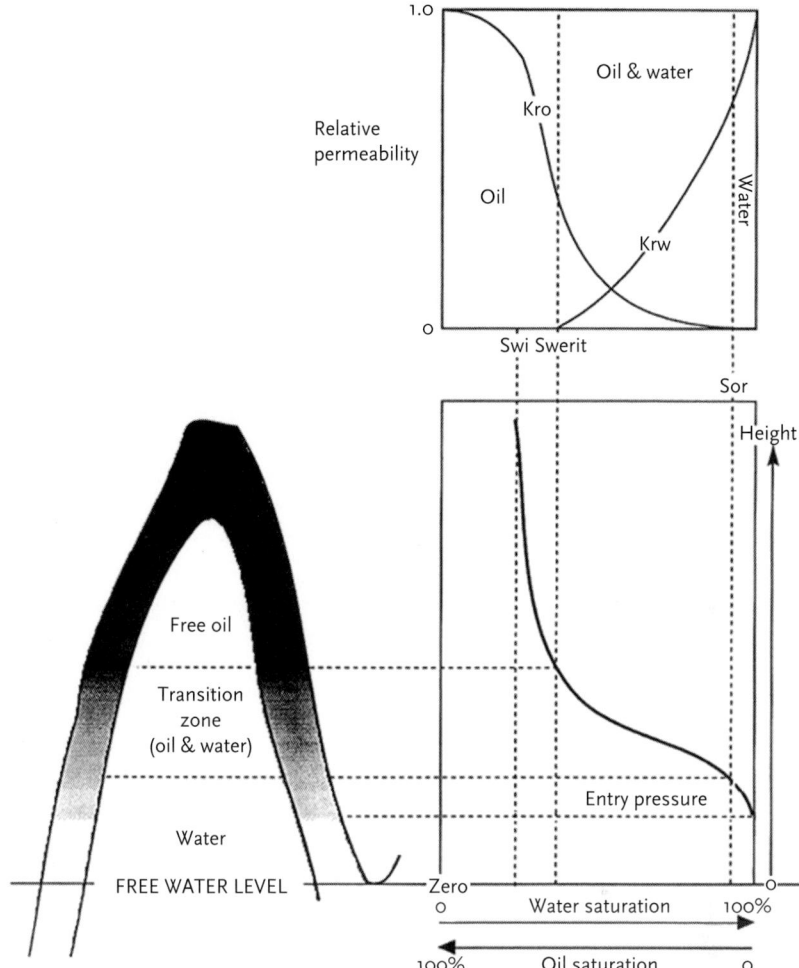

Figure 7.1: Idealized reservoir with corresponding water saturation–height profile and relative permeability plot.

depths, the oil cut becomes greater until the produced fluid becomes water-cut oil. Finally, a depth is reached where only oil is produced. The remaining water in the reservoir is immobile and is found only as thin films on micropore surfaces.

The fluid production performance of this reservoir profile is an expression of the relative permeabilities of the reservoir rock with respect to oil and water as a function of fluid saturations. Simple concepts of permeability consider the situation where the pores of a reservoir rock contain a single fluid, typically water; then the permeability measured is the "absolute permeability," K. If the pores contain two (or more) phases, the ability of a fluid to flow is not only controlled by the pore network, but also by the distribution of the other phases within the pores. For an oil/water system, two relative permeabilities can be defined, symbolized as K_{ro} and K_{rw}. Their values are determined by the ratio of the permeabilities of each phase to the absolute permeability. Between the FWL and the base of the transition zone, the isolated globules of oil are not produced and the relative permeability to oil, K_{ro}, is essentially zero. Above the base of the transition zone, K_{ro} increases progressively; its value asymptotically approaches the absolute permeability as the water phase becomes immobile and confined to smaller and smaller pores.

This description is a highly idealized portrayal of a reservoir composed of a single pore system with a constant porosity and distribution of pore-throat sizes. In reality, there may be lateral changes in facies within a reservoir unit across a field, as well as rapid vertical changes resulting from the interbedding of different lithologies. If there is a lateral gradient from larger to smaller pore-throat sizes, then the oil-water contact will actually rise in the field, producing an apparent tilted contact, because the transition zone has been lengthened. This is shown in Figure 7.2, taken from Arps (1964), together with the saturation profiles that would be expected for three wells in the hypothetical structural trap. This behavior is seen in many fields, where sometimes puzzling shifts in oil-water contacts merely reflect lateral facies changes in the reservoir rock. So, for example, Cuddy et al. (1993) advocated the use of the FWL, rather than the gas-water contact (GWC), as a consistent datum in gas fields in the North Sea, precisely because the GWC shows vertical shifts that reflect rock qualities. In addition, lengthy transition zones have meant that "many equity determinations have foundered over arguments" concerning an appropriate water saturation criterion to determine the GWC.

In the vertical dimension, there is an overall trend of water saturation decreasing upward, but the trend may be broken by excursions to lower or higher saturations

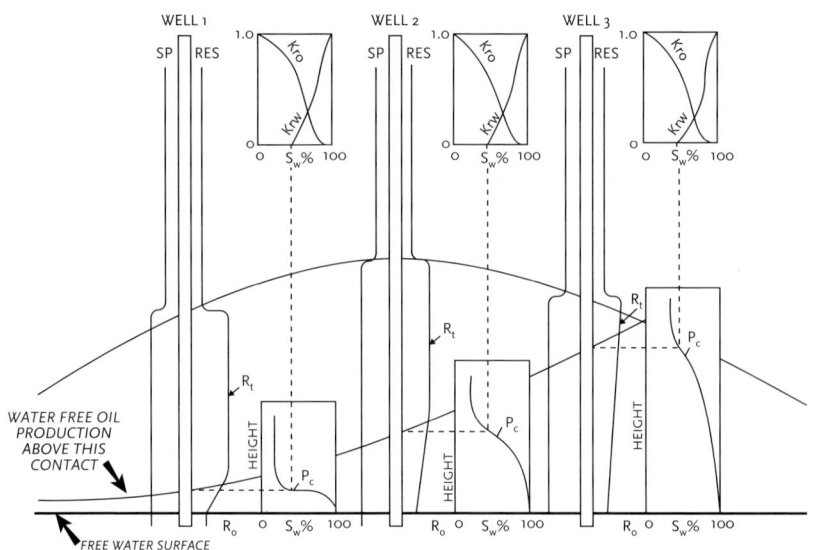

Figure 7.2: Changes in water–free oil contact controlled by reservoir rock quality. From Arps (1964), © 1964 American Association of Petroleum Geologists (AAPG), reprinted by permission of the AAPG, whose permission is required for further use.

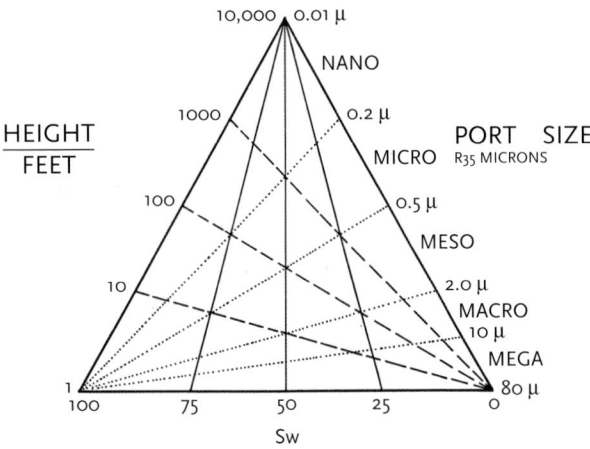

Figure 7.3: Closed system relating height above free-water level, water saturation, and the r35 estimate of pore-throat size penetrated. From Hartmann and MacMillan (1992), courtesy Utah Geological Association.

that reflect changes in permeability and pore-throat sizes. The exact pattern will be determined by the structure of the reservoir, whether it is relatively homogeneous or whether it consists of layers of rock with distinctly different pore-size distributions. The interrelationships between water saturation, r_{35} pore-throat size, and height above the free-water level are shown diagrammatically in Figure 7.3 (Hartmann and MacMillan, 1992). The closed system of the triangle means that any one of these variables can be solved if the values of the other two variables are known. The numerical solution will vary according to the relationship between the buoyancy pressure controlled by the hydrocarbon density and the height above free-water level. Also, measures of permeability can be substituted for the r_{35} pore-radius variable because of its strong relationship with the principal pore-throat size.

The interrelationships can also be seen on Figure 7.4, where a hypothetical reservoir consisting of two petrofacies that differ in their permeabilities is shown in terms of depth and pressure. If two of the three variables (height above the free-water level, permeability, and water saturation) are known, then the remaining variable is determined. The system provides a model for the reconciliation of saturations calculated from logs with those predicted from capillary pressure measurements and a determination of the depth to the free-water level.

SATURATION-HEIGHT MODELING IN SANDSTONES FROM CAPILLARY PRESSURE MEASUREMENTS

An application of capillary measurements to free-water level estimation was described by Bhattacharya et al. (2008) for the Atokan sandstone reservoir in the Norcan East field of Kansas. Methods for predicting permeability from porosity and

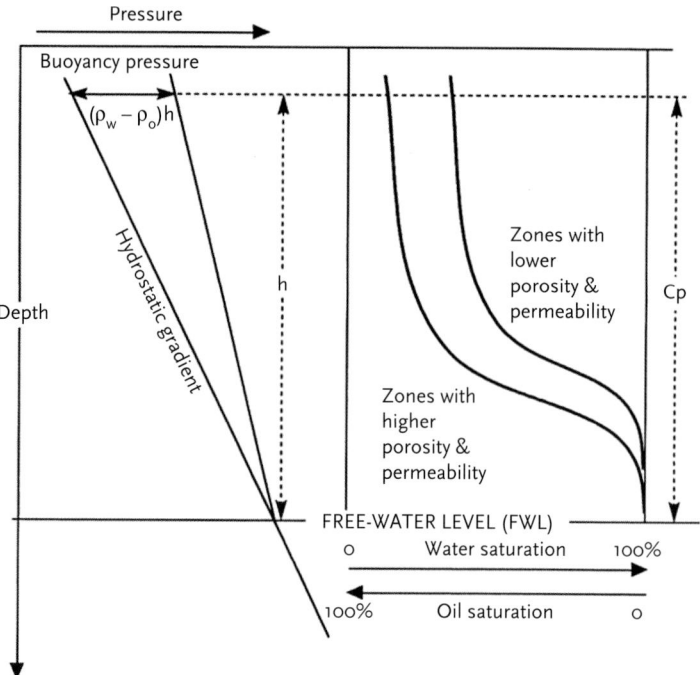

Figure 7.4: Hypothetical reservoir consisting of two petrofacies determined by permeability, shown in terms of depth and pressure.

gamma-ray logs of this unit were described in Chapter 3, and capillary pressure and pore-throat characterizations of representative cores were discussed in Chapter 6. Bhattacharya et al. (2008) modeled capillary pressure curves for various reference permeabilities to produce a generalized match to capillary pressure data from Atokan sandstone cores (Figure 7.5). The laboratory measurements of the mercury-air system were first converted to an oil-brine system representative of the Norcan East field through the equation:

$$P_{cR} = \frac{P_{cL}(\sigma\cos\theta)_R}{(\sigma\cos\theta)_L}$$

where P_{cR} and P_{cL} are capillary pressures in the oil-brine reservoir and mercury-air systems, and $\sigma\cos\theta$ is the product of the interfacial tension and the cosine of the contact angle in each system. As shown on Figure 7.5, the oil-water capillary pressure scale was also converted into an equivalent scale of height (H) above free-water level, using density measurements of oil (ρ_o = 0.704 gm/cc) and water (ρ_w = 1.08 gm/cc) in this field:

$$H = \frac{P_{cR}}{g(\rho_w - \rho_o)}$$

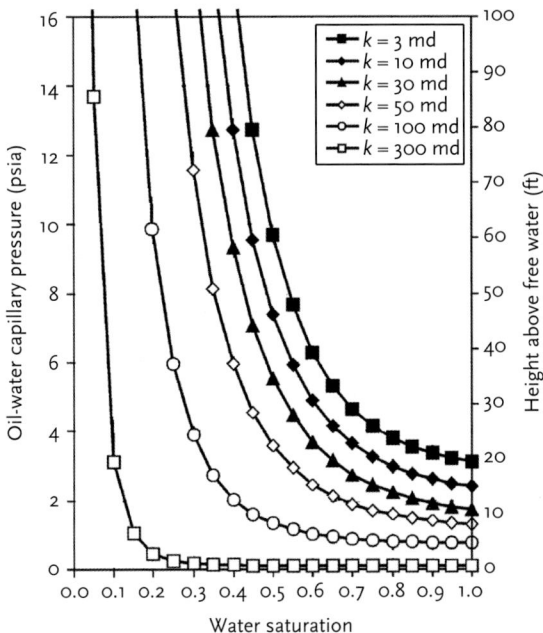

Figure 7.5: Generalized capillary pressure curves for Atokan Sandstone reservoir permeabilities in the Norcan East field, Kansas. From Bhattacharya et al. (2008), © 2008 American Association of Petroleum Geologists (AAPG), reprinted by permission of the AAPG, whose permission is required for further use.

where g is the acceleration due to gravity.

Using this equation, multiple predictions of the depth of the free-water level can be made using estimates of permeability and water saturation from log measurements. Each prediction is referenced to a sandstone zone, based on its subsea depth, water saturation calculated from the Archie equation, and permeability estimated by the regression analysis based on porosity and gamma-ray logs. The predictions are summarized as a frequency plot and cumulative frequency plot (Figure 7.6) to determine a best estimate of the free-water level. Bhattacharya et al. (2008) concluded that the FWL was probably located between 2,730 and 2,760 feet subsea; the spread in depths is attributable to insensitivities in the water saturation values, statistical error in permeability estimates, and scaling issues. A major problem is the lack of data from the transition zone; such data would provide more reliable projections than those from water-free zones with immobile water saturations. Perforations with no associated water production occur as deep as 2,753 feet subsea, which limits the range of the FWL to depths between 2,754 and 2,760 feet subsea. In further work, multiple predictions of water saturations based on the capillary pressure model for different values of the FWL within this range were compared to water saturations calculated from porosity and resistivity logs. The best fit was found when the free-water level was set at 2,758 feet subsea (Figure 7.7). As a consequence of this validation of the model, the free-water level, water

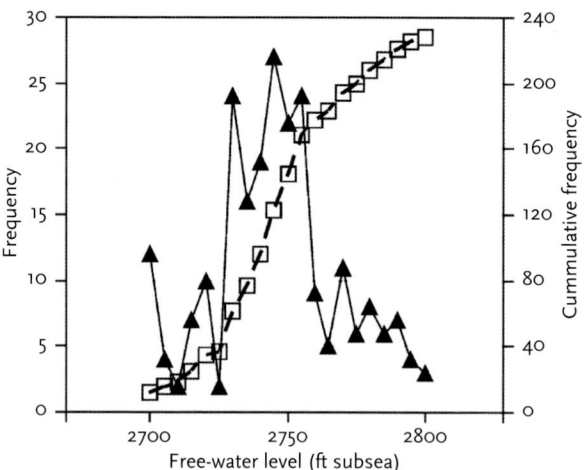

Figure 7.6: Estimation of the free-water level in the Norcan East field, Kansas, based on permeability estimated from porosity and gamma-ray logs and log-calculated water saturation. From Bhattacharya et al. (2008), ©2008 American Association of Petroleum Geologists (AAPG), reprinted by permission of the AAPG, whose permission is required for further use.

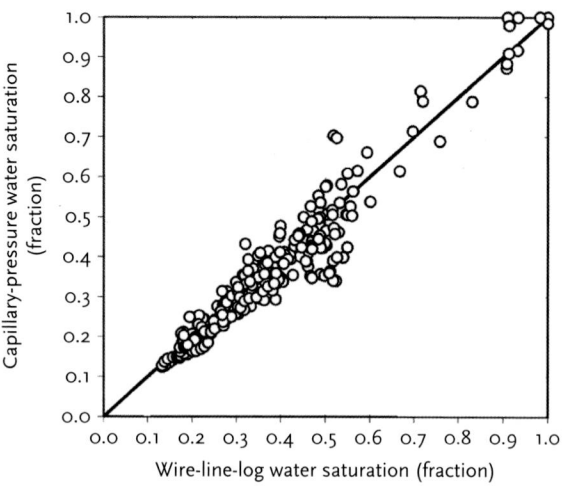

Figure 7.7: An optimum best-fit relationship between water saturations estimated from capillary pressure measurements and water saturations calculated from logs is reached when a free-water level of 2,758 feet subsea is selected. From Bhattacharya et al. (2008), © 2008 American Association of Petroleum Geologists (AAPG), reprinted by permission of the AAPG, whose permission is required for further use.

saturation distribution, and permeability characterization captured the reservoir parameters for the entire field. When used for reservoir simulation, there was a successful match with the field performance history, both in terms of primary and secondary production.

HEIGHT FUNCTIONS FOR BULK-VOLUME WATER

There are many situations where capillary pressure measurements from core data do not exist and even conventional core measurements of porosity and permeability are not available. In these circumstances, the only recourse is to work with log measurements of porosity and calculations of water saturations to establish saturation-height functions. The oldest and simplest relationship is the "power function" which takes the form:

$$S_w = aH^b$$

Water saturations (S_w) are fitted to height above free-water level (H) for discrete ranges of porosity, using regression analysis to estimate the parameters a and b (Skelt and Harrison, 1995). The "lambda function" is a power function variant with an added constraint that it is asymptotic to the "irreducible" water saturation.

Even if successful, the application of these functions introduces artificial discontinuities at the boundaries of each porosity class. At a more fundamental level it is recognized that pore-throat sizes control the saturation function through permeability rather than porosity, which simply measures pore volume. Porosity is used for pragmatic reasons because, unlike permeability, porosity is a log measurement. If there is a useable relationship between porosity and permeability, then porosity can be used as a proxy for permeability. In this case, saturation-height functions should incorporate petrofacies classes in their development.

Cuddy et al. (1993) recognized the intrinsic problem of porosity bundling and proposed that saturation-height modeling should be cast in terms of bulk-volume water (BVW) using the function:

$$BVW = aH^b$$

With a logarithmic transformation, this function becomes:

$$\log BVW = \log a + b \cdot \log H$$

The constants a and b can be determined by least-squares regression, with BVW as the dependent (predicted) variable. A single function that successfully fits the log data can project an estimate of the free-water level and predict water saturation at any height, given porosity values from a log. Cuddy et al. (1993) pointed out that the function was not merely a useful empirical formula for saturation-height modeling but was a mathematical consequence of the Leverett J-function (Leverett, 1941):

$$J(S_w) = \frac{P_c}{\sigma \cos \theta} \sqrt{\frac{k}{\phi}}$$

where J is a dimensionless function of water saturation (S_w), P_c is capillary pressure, $\sigma \cos \theta$ is the product of the interfacial tension and the cosine of the contact angle, and k and ϕ are, respectively, the permeability and porosity. When the J-function was first proposed, it was considered a universal method to average a set of capillary curves into a single coherent, dimensionless function. However, subsequent work has demonstrated that *J*-function averaging "*only applies* [emphasis in original] if the porous rock types have similar pore-size distributions or pore geometry" (Harrison and Jing, 2001). This inherent constraint must be borne in mind when evaluating the success of a *BVW*-height function. The success depends on the degree to which the fitted reservoir zones are drawn from a common petrofacies.

Kay and Cuddy (2002) applied this function to locating free-water levels within separate blocks of the Heather field and modeling fluid saturation–height functions. The Heather oil field is located in the northern North Sea, where oil is produced from sandstones of the Middle Jurassic Brent group. As one example, the *BVW*-height function and FWL were estimated in the North Terrace satellite field using data from two wells (Figure 7.8). In order to establish the position of the FWL, *BVW* data were first plotted against the true vertical depth below sea level (TVDSS). A porosity cut-off of 10 percent was applied to exclude data that were not pay, particularly because many of these intervals were almost completely water saturated. The FWL is located where the plotted data project to an asymptotic value on the depth axis. The computational procedure used to solve for the free-water level is iterative; a range of potential depth values for the FWL are successively evaluated by fitting the *BVW* values to each potential datum. The best choice for the FWL is then selected as the level that produces the minimum summed least-square difference between the measured and estimated *BVW* values. In the North Terrace of the Heather field, the FWL was estimated to be approximately 10,730 feet below sea-level and closely matches the structural spill point. If the free-water level elevation is known, the water saturation can be calculated using the *BVW*-height function. In the Heather field, there is a generally good agreement between the water saturation computed from resistivity logs and that predicted by the *BVW* function.

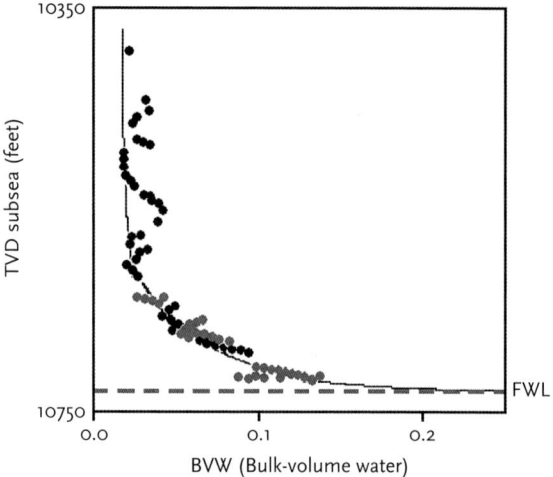

Figure 7.8: Bulk-volume water (BVW) versus true vertical depth (TVD) feet subsea in two wells in the North Sea Heather field fitted with a BVW-height function to estimate the free-water level (FWL). Adapted from Kay and Cuddy (2002), courtesy the Geological Society of London.

The *BVW*-height function can be justified mathematically, because it can be derived from the Leverett model, and also empirically, because it can provide viable predictions of water saturation. In addition, the function can be justified because it conforms to physical principles. When written in logarithmic form, the *BVW*-height function is:

$$\log BVW = \log a + b \cdot \log H$$

Because capillary pressure is a linear function of height, it follows that a log-log crossplot of *BVW* and capillary pressure should be linear in form. Gagnon et al. (2008) provided a useful illustration of this in their study of North Sea fields. Two crossplots of capillary pressure versus *BVW* for representative core plugs from six North Sea clastic reservoirs (Figure 7.9) show strong linear trends that appear to confirm a relationship between *BVW* and capillary pressure. Because each trend is from a single plug, the relationship is supported at the petrofacies level but lacks descriptive power in reservoirs that are composed of a range of petrofacies. This behavior is similar to that of the Leverett *J*-function to which the *BVW*-height function is linked. Notice how the trends in the *BVW*–capillary pressure crossplot in the left diagram of Figure 7.9 appear to reflect relative grain sizes (and pore sizes), so that the lowest *BVW* occurs within the Jurassic conglomerate. However, plugs taken as representative of two reservoirs interpreted as Jurassic marine sandstones (Figure 7.9, right) are markedly different, implying that "Reservoir A" has larger pores than "Reservoir B." The trend in the *BVW*–capillary pressure crossplot for the "Reservoir A" marine sandstone is virtually indistinguishable from the trends for the Permian eolian sandstone and the Triassic distal delta sandstone reservoirs. This further demonstrates that generalizations based only on depositional environments have little value unless they are linked to a specific grain-size distribution and, therefore,

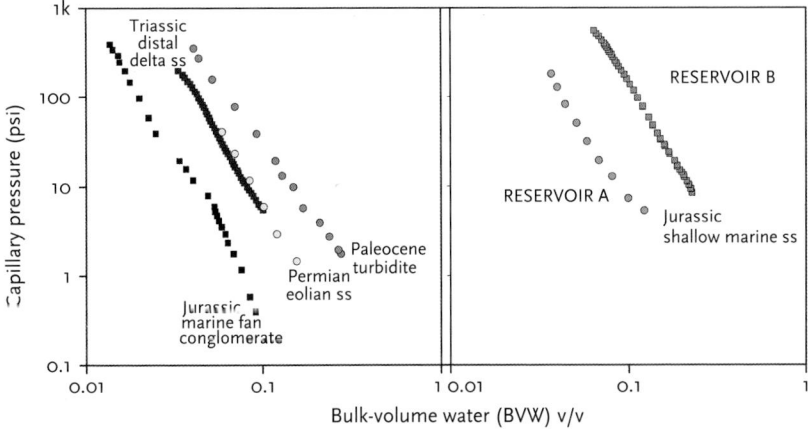

Figure 7.9: Capillary pressure versus bulk-volume water from representative core plugs from six North Sea reservoirs, plotted and referenced to depositional environments (left) and within a depositional environment, but from two different reservoirs (right). Adapted from Gagnon et al. (2008), courtesy Society of Petrophysicists and Well Log Analysts.

to petrofacies. Of course, any diagenetic cementation or dissolution will introduce yet another variable to be considered.

Gagnon et al. (2008) computed representative *BVW*-height functions based on log analyses for eleven North Sea clastic reservoirs deposited in a wide range of environments (eolian, fluvial, lacustrine, deltaic, and turbidite) and ages (Devonian to Paleocene). Their results are shown in Figure 7.10, both on linear (left) and logarithmic (right) scales of *BVW* and height. The log-log plot clearly demonstrates the relative role of parameter *a*, which expresses the intercept, and parameter *b*, which controls the slope. Gagnon et al. (2008) pointed out that the rock quality and reservoir fluid parameters of the Leverett *J*-function are all associated with the *a* term, and that the *b* term is a dimensionless scaling factor with little variability between fields. Consequently, the *a* term can be considered a fundamental measure of rock quality. The trends are consistent with earlier observations that lower *BVW* values are anticipated for reservoir units having larger pore throats and occurring at greater heights above the free-water level, where the water fraction is immobile.

Although Gagnon et al. (2008) confirmed earlier conclusions by Cuddy et al. (1993) in reporting that "*BVW* is largely independent of porosity and permeability for the typical porosity range" of the North Sea fields in their study, Harrison and Jing (2001) had a more nuanced view of this claim. They concluded that by ignoring variations in rock type, the method of fitting a height function to *BVW* values was biased toward the fit to water saturation in better quality reservoir rock. This often resulted in poor predictions in intervals having lower porosities and broke down in the transition zone. However, Harrison and Jing conceded that the procedure was the simplest height-saturation method available and that Cuddy et al. (1993) held an opposing view—that the *BVW* method was rooted in physics in contrast to other methods they regarded as arbitrary data-fitting techniques.

The performance of the *BVW*-height method compared to capillary pressure modeling is explored by returning to the Atokan sandstones of the Norcan East field.

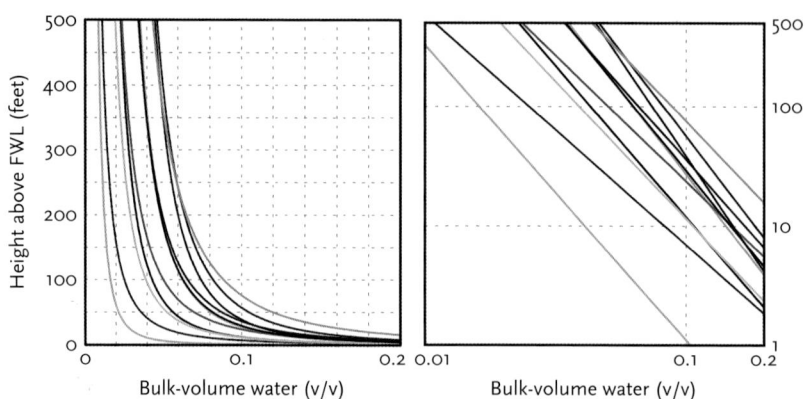

Figure 7.10: Bulk-volume water versus height above free-water level functions for eleven North Sea clastic reservoirs plotted on linear scales (left) and logarithmic scales (right). From Gagnon et al. (2008), courtesy Society of Professional Well Log Analysts.

A composite depth plot of *BVW*s from twenty wells in the Norcan East field is shown in Figure 7.11. The sandstone zones are subdivided between those with gamma-ray log values of less than thirty API units and those with values between thrity and forty API units. This discriminates between two petrofacies of better and poorer reservoir rock quality. An iterative procedure was used to establish the free-water level by successively changing the potential FWL values in the range of 2,752 to 2,765 feet depth subsea and fitting a regression model of the form:

$$\log BVW = \log a + b \cdot \log H$$

The highest *R-squared* was associated with a free-water level at 2,768 feet depth subsea; the fitted function is shown in Figure 7.11.

If the regression model had a better statistical fit to the observations, then the function could be used with some confidence to predict water saturations at any height above the FWL. The water saturation at a specific depth is estimated by

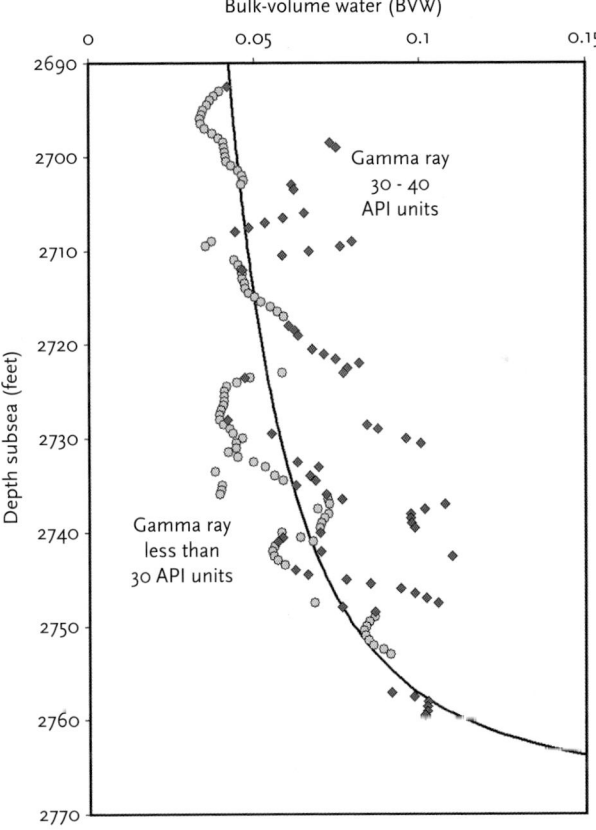

Figure 7.11: Water saturations for zones subdivided between those with gamma-ray log values of less than thirty API units (circles) and those between thirty and forty API units (diamonds), plotted against depth subsea in wells of the Norcan East field, Kansas. The curve shows the best-fit BVW-HAFL function with a free-water level of 2,756.5 feet subsea.

dividing the *BVW* by the porosity log reading at the same depth. However, not only are there large deviations about the fitted trend, but there also is a significant bias caused by overestimation of the *BVW* in better-quality reservoir rock and underestimation of the *BVW* in poorer-quality zones. This bias results from zones with lower permeabilities having higher *BVW* values than zones with higher permeabilities, at any given height above the free-water level. As a consequence, the *BVW*-height function predicts increased water saturations in units with higher permeability and lower water saturations in low-permeability zones.

The *BVW*-height function was reevaluated using only data from zones with gamma-ray values less than thrity API units; these zones would be considered the best candidates for pay intervals. An iterative search for a solution with the best fit resulted in a revised estimate of the free-water level at 2,756.5 feet subsea. This depth is very close to the final estimate of 2,758 feet, which was found by Bhattacharya et al. (2008) in the reconciliation of their saturation model with capillary pressure data. The revised *BVW*-height function provides better estimates of water saturations, but only in sandstone intervals with gamma-ray values of less than thirty API units. A bias towards lower water saturations would result if the revised function were to be applied to sandstone zones with higher gamma-ray values.

The conclusions from this case study confirm that the *BVW*-height function may be applicable to a single petrofacies identified as pay, giving a viable method for predicting water saturation in pay zones. Unfortunately, this precludes water saturation estimation in other petrofacies when a complete characterization of saturation is needed for a comprehensive reservoir model. The better the fit of the *BVW*-height function in the pay petrofacies, the greater the confidence that can be placed on the estimate of the free-water level. All these conclusions are consistent with the notion that the *BVW*-height model is derived from the Leverett *J*-function. If the *J*-function fails to average separable petrofacies, the *BVW*-height function will also fail. However, with this caveat in mind, *BVW*-height analyses can provide some useful predictive outcomes, particularly considering that its modest data requirements can be met using logs alone.

PERMEABILITY-HEIGHT FUNCTIONS

To model all sandstones within a reservoir consisting of multiple petrofacies in a single comprehensive model, a saturation-height function should be keyed to both permeability (k) and capillary pressure (Pc) as the fundamental physical controls. In its simplest implementation, the predictive relationship takes the form:

$$S_w = c \cdot Pc^d \cdot k^e$$

This equation was successfully applied by Mitchell et al. (1999) to North Sea field data. Slightly more complex variants of this equation were used by Johnson (1987) and Søndenå (1992), although Mitchell et al. (1999) claimed that their simpler form outperformed these alternatives and was less tedious to use. The linear relation between height above FWL (h) and capillary pressure allows the simple substitution:

$$S_w = c \cdot h^d k^e$$

When solved as a linear regression, the model becomes:

$$\log(S_w) = \log(c) + d \cdot \log(h) + e \cdot \log(k)$$

As an example, a water saturation-height function was developed for the Atokan sandstones in the Norcan East field. Water saturations computed from logs were regressed against height above the FWL, and the permeability was estimated from porosity and gamma-ray logs. The free-water level depth was set to the value producing the highest *R-squared* in the associated regression analysis. It should be noted that the logarithmic scaling of water saturation gives higher weightings to the lower water-saturation values. In order to avoid this bias, the regression was applied to arithmetic-scale water saturations; a crossplot of the model predictions versus values calculated from logs is shown in Figure 7.12. The regression model resulted in minor changes of the predicted values and moved the estimate of the FWL from 2,765 to 2,761 feet depth subsea. The use of regression analysis to estimate reservoir parameters should be considered in terms of the implicit assumptions in regression and the degree to which these assumptions might conflict with the aims of the reservoir evaluation. It is assumed that the deviations of the dependent variable from the fitted regression model are normally distributed and that the means and dispersions of these deviations do not change for different values of the predictor variables. However, water saturation is a proportional measure with finite bounds, and its distribution varies

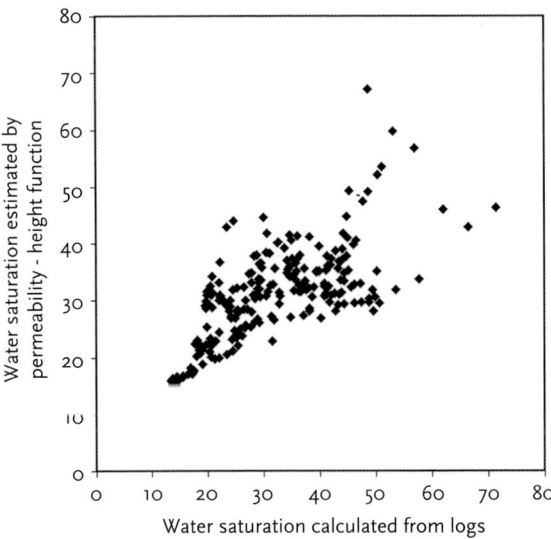

Figure 7.12: Crossplot of water saturations calculated from porosity and resistivity logs against water saturations estimated by a function of permeability and height above free-water level.

according to the ratio of the length of the reservoir to the length of the transition zone. If the reservoir section is long compared to the transition zone, then most water saturations take on their "irreducible" values (S_{wi}) and the values are distributed between the component petrofacies modified by a gradual drift downwards at increasing heights above the free-water level. If the transition zone is lengthy, then the wide range of water saturations results in a distribution with a strong positive skew.

The potential problems associated with parametric statistical analysis can be addressed by a variety of different approaches, but more fundamental issues should be considered. What is the purpose of the analysis? Is it to predict water saturations with equal errors, regardless of whether the saturations are high or low? If this requirement is made to satisfy the demands of a simulation model, then what are the consequences for volumetrics? If saturations are honored equally, regardless of their associated porosities, then small volumes of water are matched to the same degree as large volumes of water. Perhaps a better saturation-height model would rely on the fraction of the rock that is water-filled (*BVW*) rather than the fraction of the pore space (S_w). At the largest scale, if a major goal of the analysis is to assess the total hydrocarbons in place, then perhaps saturation height should be analyzed in terms of bulk-volume hydrocarbon (*BVH*), where:

$$BVH = \Phi \cdot (1 - S_w)$$

If *BVH* is chosen as the critical variable, then water saturation values used for simulation should be derived from *BVH* in order to maintain a consistent and coherent model with no internal conflicts. Both Heseldin (1974) and Alger et al. (1989) chose *BVH*-height models, primarily because the choice of *BVH* resulted in more robust averaging of capillary pressure data and a way to integrate log and core data that was more amenable to regression analysis. Finally, the criterion of "fit-for-purpose" is widely used in the petroleum industry to evaluate competing methods, based on their ability to produce robust estimates within an acceptable range defined by economics. In selecting an appropriate saturation-height model, the purpose of the reservoir characterization should be clearly defined, particularly if there are multiple objectives.

The Leverett *J*-function (Leverett, 1941) continues to be widely used as an initial procedure for averaging capillary pressure curves. It also is applied in a more adaptive fashion by assigning different functions to petrofacies that are distinguished by their porosity-permeability relationships. This is explicitly recognized when the "rock quality index" (*RQI*) equation is applied:

$$RQI = 0.0314 \cdot \sqrt{\frac{k}{\Phi}}$$

because it is derived from the *J*-function. For example, Obeida et al. (2005) applied *RQI* values to develop separate *J*-functions associated with seventeen reservoir rock types identified from the core, logs, and seismic data. These were used to model saturations within a giant Lower Cretaceous carbonate reservoir in the Middle East. This study, in common with all saturation-height models that are built from multiple petrofacies, requires that the reservoir architecture be described by the spatial stacking of the petrofacies elements. These can be identified in cored wells, but they must be predicted from logs calibrated to the core in uncored wells.

As an alternative to the iterative fitting of models to reservoir measurements, forward modeling can be used to create hypothetical reservoir architectures. Predicted log responses of the models are matched against the observed log measurements in an iterative manner that reconciles log responses with reservoir properties. One immediate benefit is that the architecture of the reservoir can be expressed in terms of the spatial arrangement of flow units. The effects of long-term trends in the grain-size/pore-size ratio can be evaluated, as can short-term cycles, such as those in stacked clastic sequences. At the most basic level, simple forward models can be used to evaluate the consequences of different styles of reservoirs and the methodologies used for their analysis. So, for example, Doveton (2008) described a simple forward-model procedure (Figure 7.13) where a sequence of flow units was generated by a petrofacies transition probability matrix. Although stationary, the Markov-chain realization could be modified to incorporate long-term trends. Representative porosities and permeabilities are assigned to each level, and the water saturation is predicted based on the petrofacies and height above the free-water level. Finally, resistivity and porosity logs are modeled for the simulated section and displayed as a Pickett plot for comparison with real sections. More sophisticated forward-modeling procedures can be developed to match real reservoirs. The use of petrofacies results in a hierarchy of simulated reservoirs in which distinctive flow units are preserved; it is thus more likely that there will be a match with the performance of a real reservoir (Maharaja and Journel, 2005).

SATURATION-HEIGHT MODELING IN CARBONATES

As described in Chapter 3, Lucia (1995) subdivided limestones and dolomites into three petrophysical classes based on porosity and permeability descriptors linked to textures observed in the core. The first two classes are grain-dominated fabrics ranging from grainstones to packstones, while the third class consists of mud-dominated packstones, wackestones, and mudstones. Using capillary pressure data, Lucia (1995) also developed an equation for each fabric class that relates the hydrocarbon column height to porosity and saturation:

$$S_w = a \cdot H^b \Phi^c$$

The coefficients a, b, and c for each of the three classes are:

Class 1: $a = 0.02219$, $b = -0.316$, $c = -1.745$
Class 2: $a = 0.1404$, $b = -0.407$, $c = -1.440$
Class 3: $a = 0.6110$, $b = -0.505$, $c = -1.210$

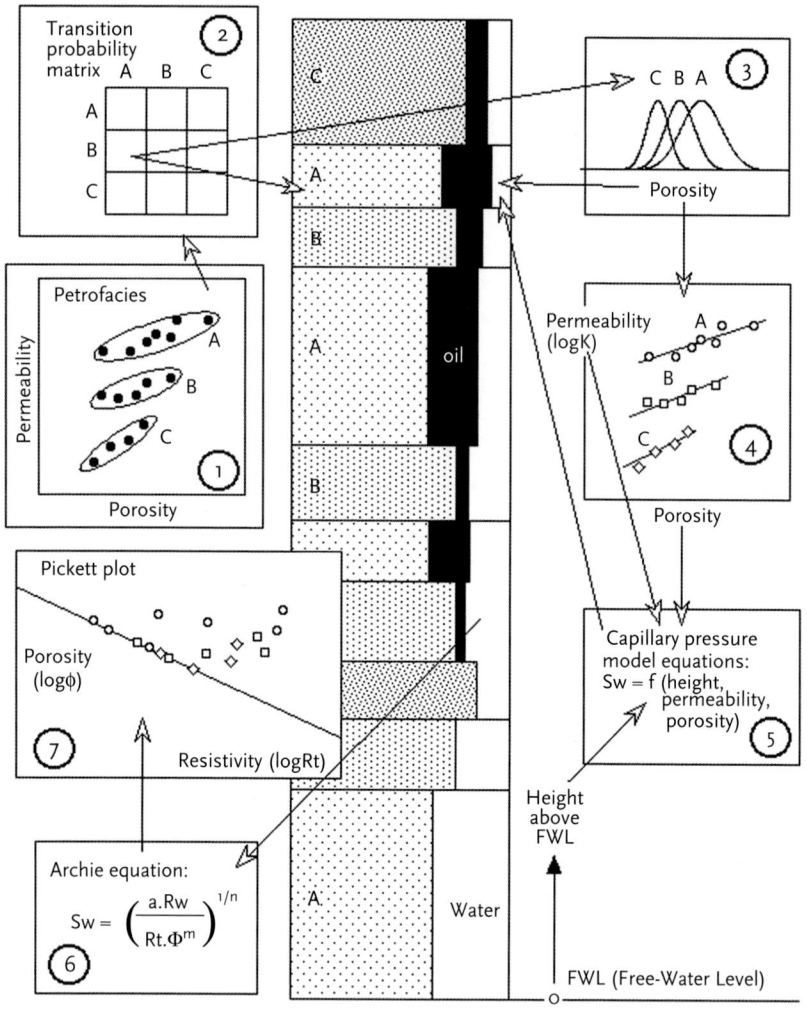

Figure 7.13: Flow diagram of operational steps (numbered in order) to simulate a sandstone reservoir and compute its wireline log responses. From Doveton (2008), courtesy Society of Professional Well Log Analysts.

These equations can be applied to standard log-analysis computations of porosity and water saturation to model the structure of nonvuggy carbonate reservoirs. An example is shown in Figure 7.14, where anticipated saturation profiles are shown for the three classes in an interval of the San Andres formation in the Seminole field of the Permian basin of Texas (Lucia, 2007). In the upper, cored section, there is an excellent match between the water saturation calculated from logs and the class saturation curve based on the core fabric. If fabric classes can be predicted from logs in uncored sections, then a composite saturation profile can be created that honors both the height above the free-water level and the reservoir architecture.

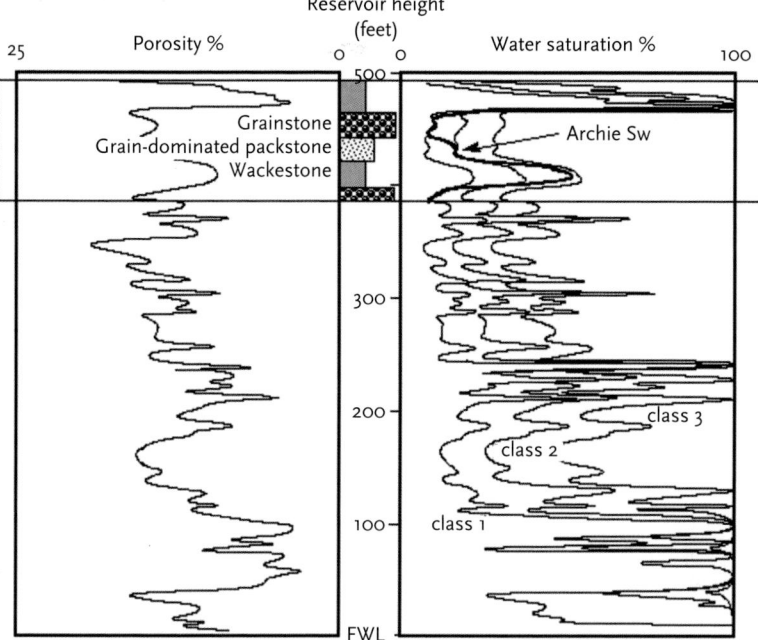

Figure 7.14: Water saturation–height profiles based on equations linked with capillary pressure data for three particle-size classes in a section of the San Andres formation in the Seminole field, Permian basin, Texas. Adapted from Lucia (2007), courtesy Springer Science+Business Media.

The Lucia model expresses carbonate architecture in terms of framework descriptors that commonly are used by geologists and are applicable to most carbonate reservoirs. However, even though the classes are associated with distinctive segments of the porosity-permeability continuum, they are based on depositional lithofacies. They can serve only as first-order descriptors of carbonate microarchitecture, which may exhibit complex variability in pore-body and pore-throat size distributions. Saturation-height analyses of carbonates are much more problematic than analyses of clastics, because carbonate pore systems may be multimodal rather than the unimodal distributions of intergranular porosity. If a pore-throat system is multimodal, then the J-function and related methods will not be adequate predictors because they are tied to an average permeability-porosity ratio.

The Thomeer method has been widely applied to capture the multimodality of carbonate pore systems by superposing multiple hyperbolas, each corresponding to a distinctive modal size of pore throat. Advocates of the method point out that the J-function approach is not "rock-texturally intuitive" (Clerke et al., 2008), in contrast to the fitting of Thomeer hyperbolas, which are explicitly linked to observed pore-throat sizes. The Thomeer model (Thomeer, 1960) is expressed as a hyperbola when data are plotted as the logarithm of capillary pressure versus the logarithm of the bulk volume of the nonwetting phase (mercury in the laboratory). The Thomeer hyperbola is controlled by three parameters that have a simple geometrical interpretation, as shown in Figure 7.15. The parameter P_d is the horizontal asymptote and represents the extrapolated entry (or displacement pressure); $V_{b\infty}$ is the vertical asymptote and represents the bulk volume of mercury extrapolated to infinite pressure; c is a hyberbolic constant that controls the shape of the curve. Expressed as an equation, the Thomeer hyperbola is:

$$\log\left(\frac{V_b}{V_{b\infty}}\right) \cdot \log\left(\frac{P_c}{P_d}\right) = c$$

V_b is the fractional volume occupied by mercury, and P_c is the capillary pressure. The hyperbolic constant, c, is usually recast as G, the "pore geometric factor" by the equation:

$$G = \frac{-C}{\log_e 10} = -0.4343 \cdot c$$

This modification dates from the *precomputer* methodology of Thomeer (1960) and was designed "…to simplify this equation and available mathematical tabulations." In either guise (c or G), the shape parameter reflects the width of the distribution of pore-throat sizes as an expression of sorting. The parameter P_d is controlled by the largest pore throat and $V_{b\infty}$ corresponds to the total porosity volume.

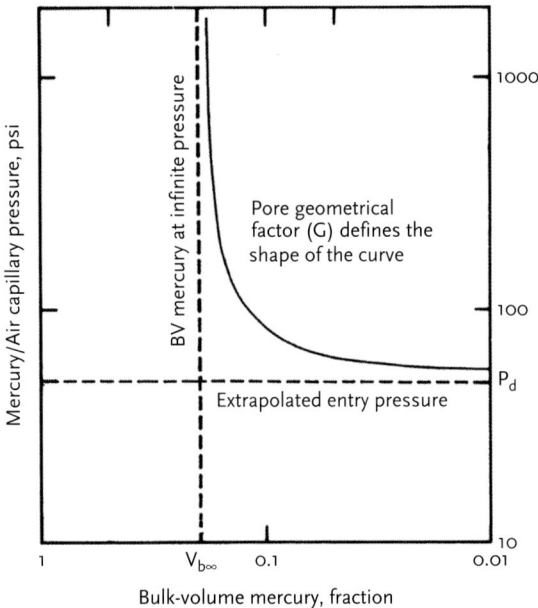

Figure 7.15: Parameters of Thomeer hyperbolic capillary pressure curve. Adapted from Thomeer (1960).

If a single Thomeer hyperbola provides a good fit to the data, then the bulk volume of mercury at infinite pressure should be approximately equal to the porosity of the sample. If this is not the case, then the modeler may choose to apply two hyperbolas for an improved fit; this decision may be obvious if there is a kink in the capillary pressure curve. In a double hyperbola model, there are two sets of Thomeer parameters, whose entry-pressure parameters represent two distinctive modal pore-throat sizes (Figure 7.16).

Clerke and Martin (2004) developed a shareware spreadsheet to interactively fit multiple hyperbolas. Clerke et al. (2008) applied the procedure to 125 samples representing a variety of Dunham textures taken from the Upper Jurassic Arab D limestone in the Ghawar field of Saudi Arabia. The first step of the analysis was to determine the pore-system modality of each sample, because this establishes the number of Thomeer hyperbolas that are required to fit the capillary pressure data. The authors found that 35 percent of the samples were unimodal, 62 percent were bimodal, and 3 percent were trimodal. When differentiated by the Dunham textural classes (Figure 7.17), the modalities follow a distinctive trend that matches the patterns observed in the pore-throat size and pore-body size distributions that were described in Chapter 6. The identification of a microporous mode in the Arab D is important, because this mode does not significantly contribute to permeability. The analysis demonstrates that entry pressure is the most important Thomeer parameter. Because the entry pressure is strongly correlated with permeability, this

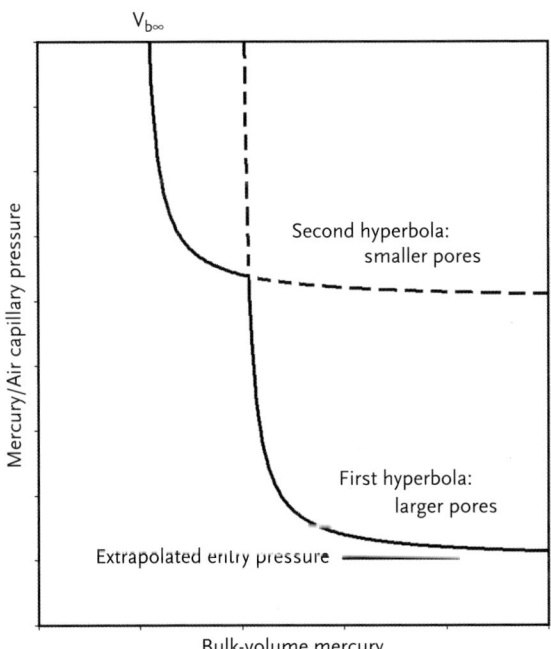

Figure 7.16: Geometry of a double hyperbola Thomeer model with two sets of Thomeer parameters, where the entry pressure parameters are linked with two distinctive modal pore-throat sizes.

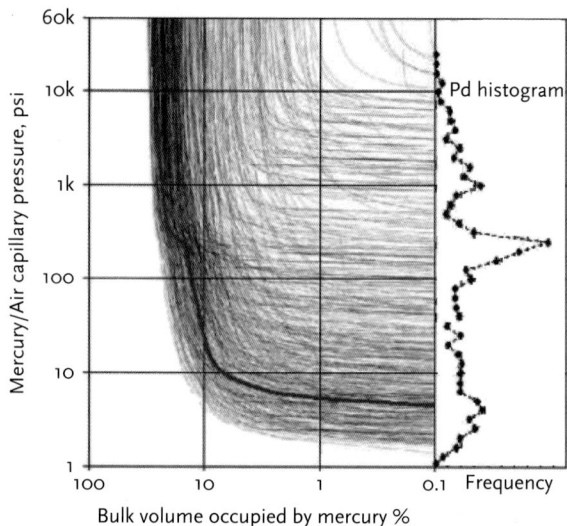

Figure 7.17: Ghawar field Arab D limestone capillary pressure curves fitted with multiple Thomeer hyperbolas. Graph on right plots frequency against Thomeer entry pressures. One of the double Thomeer hyperbolas is marked in bold. Adapted from Clerke (2009), Used with permission © 2009 Society of Petroleum Engineers.

confirms that the Lucia petrophysical classification is a viable first-order model for saturation-height modeling.

A second consequence of modeling multimodal porosity by Thomeer analysis is the isolation of distinctive modes, which Clerke et al. (2008) characterized as "porositons." Four principal porositons were identified, with mean pore-throat diameters of 58.27, 1.05, 0.16, and 0.04 microns. Mutimodal systems were then defined in terms of porositon combinations; these are expressed as distinctive clusters on permeability-porosity crossplots and as associations with textural facies. Clerke et al. (2008) also examined the relationship between T2-relaxation-time spectra and pore-throat size distributions in core samples to see if multimodality in pore throats matched the equivalent multimodality of inferred pore-body sizes. Results from both core (Figure 7.18) and magnetic resonance logging were encouraging; this suggests a possible link between "porobodons" and "porositons" that could be exploited. This would be valuable because of the greater availability and sampling volume of magnetic resonance logs compared to core plugs.

SATURATION-HEIGHT MODELING BASED ON MAGNETIC RESONANCE LOGS

The role of magnetic resonance logging in permeability estimation has been discussed in Chapter 3. In this section, we turn our attention to their potential to develop capillary pressure curves and saturation-height models. The fluid saturation profile is

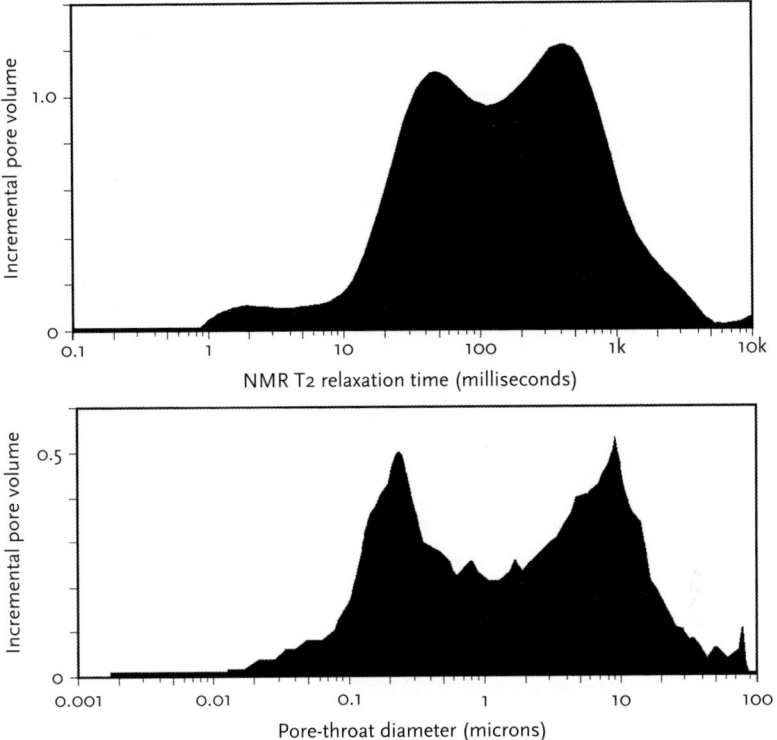

Figure 7.18: Comparison of NMR T2 relaxation time and pore-throat diameter measurements from an Arab D limestone core plug at 100 percent water saturation. Adapted from Clerke et al. (2008), courtesy GeoArabia.

determined by the entry of hydrocarbons through successively smaller pore throats, and the magnetic resonance log records a spectrum of relaxation times that reflects the distribution of pore-body sizes. The fundamental challenge is to find a relationship between pore-body and pore-throat sizes that can be used to convert the T2 relaxation time to capillary pressure. This relationship exists in sandstones (and some carbonates) because the grain size determines both the pore- and pore-throat sizes to some degree (Volokitin et al., 2001). If the ratio of pore-throat radius (r_t) to pore-body radius (r_b) is a constant (A), then the procedure ideally is a simple rescaling:

$$r_t = A \cdot r_b$$

However, the pore-body dimension measured by the T2-relaxation-time distribution is a function of the ratio of the internal surface area (S) to the volume (V), so the estimated "radius" varies with pore shape, as discussed in Chapter 6. If the bulk relaxation and diffusion effects are neglected:

$$\frac{1}{T2} = \rho \cdot \frac{S}{V} = \rho \cdot \frac{G}{r_b}$$

where ρ is the surface relaxivity and G is the geometric shape factor. The relationship between pore-throat radius and capillary pressure is given by the Washburn equation:

$$P_c = \frac{2\sigma \cos\theta}{r_t}$$

These three equations can be combined to transform the T2 relaxation time into capillary pressure:

$$\frac{1}{P_c} = \frac{\rho}{2\cos\theta} \cdot A \cdot G \cdot T2$$

This can be simplified:

$$P_c = C \cdot T2^{-1}$$

Volokitin et al. (2001) calculated C in an extensive study of 380 sandstone core plugs to determine an "optimal conversion constant," which they termed *kappa* (κ). This could be used to transform magnetic resonance logs to equivalent capillary pressure curves. This methodology would be important, if successful, because logs provide continuous records of greater rock volumes compared to limited, discrete, and destructive core-plug measurements. Volokitin et al. (2001) compared saturations from uniformly sampled capillary pressure curves to saturations predicted from T2 distributions by a root-mean-square evaluation of the differences between measurements and predictions. They concluded that below 500 psi, the optimal value for *kappa* in sandstones was three, when the T2 relaxation time was scaled in seconds and the mercury/air capillary pressure was measured in psi units. This prediction model is appropriate for zones that are completely water-saturated. Volokitin et al. (2001) described a method to adjust the T2-relaxation-time spectra of partially saturated zones to the distributions expected at complete saturation by evaluating the "irreducible" water saturation and rescaling the free-fluid distribution.

Volokitin et al. (2001) cautioned that allowance must be made for variability in relaxivity in different sandstone formations when applying this simple relationship and default value to magnetic resonance logs. They also noted that incorporating a variable scaling parameter provides an improved fit, particularly for finer pores, where a more realistic fit to "irreducible" saturations can be made. Altunbay et al. (2001) extended the methodology by considering the variability of the factor, C. There are three components of the equation that define C:

$$C = \frac{\rho}{2\cos\theta} \cdot \frac{r_t}{r_b} \cdot G$$

The first term combines the relaxivity of the pore surface with the surface-tension properties of the wetting fluid, and can be considered a constant for each rock sample. However, the two terms of the ratio of pore-throat radius to pore-body radius and the geometric shape factor are subject to variation. Because C can be calculated by matching a T2 relaxation time with its corresponding capillary pressure at an equivalent water saturation, the conversion factor can be evaluated as a function of pore size. So, the recommendation of Volokitin et al. (2001) to use a higher value of *kappa* for finer pores suggests either a change in the shapes of the pore bodies, a convergence in the sizes of the pore throats and pore bodies, or a combination of the two.

Extending this methodology to carbonates is, as usual, more complex for a variety of reasons. First, the surface relaxivity of carbonates is more variable than that of sandstones, so that formation variability must be taken into account. The distinction between pore bodies and pore throats and their representation by basic geometric shapes can be incorporated into intergranular models. In contrast, the shapes of carbonate pores may be difficult to quantify, as they are irregular voids and the relationship between pore-body and pore-throat size is obscure. Multiple-porosity systems also are common in carbonates, requiring conceptual pore models that are more complex than those of simple intergranular networks.

The consequences of all these factors can be seen in the empirical results from scaling T2 relaxation times with capillary pressures in carbonate samples. Although the conversion factor, C, is unlikely to be a constant across all sizes of pores within a carbonate sample, its variability can be modeled either as a continuous function with respect to water saturation or as values associated with specific pore sizes. The work of Clerke et al. (2008) suggests that distinctive pore-throat sizes ("porositons") occur in carbonates and these can be linked with corresponding sizes of pore bodies ("porobodons"). If this is true, the scaling factor may be a stepped function of characteristic values for discrete pore sizes based on combinations of geometric shapes and ratios of the pore-throat to pore-body radii.

PUTTING IT ALL TOGETHER: THE STATIC RESERVOIR MODEL

In this chapter we have reviewed saturation-height models as a final step in characterizing a reservoir. Earlier sections of this book discussed the analysis of properties that collectively describe the architectural components of the static model. Once the boundaries of a reservoir have been circumscribed by geophysical methods coupled with structural mapping, petrophysical properties estimated from well logs must be interpolated to populate a three-dimensional cellular model of the reservoir's interior. This static model (sometimes referred to as the geological model or geomodel) is the precursor to a dynamic model where reservoir simulation and history matching reconcile the model to the reservoir's production performance. An example of a static reservoir model is shown in Figure 7.19 (Dubois et al., 2012), with three-dimensional cellular renditions of facies, porosities, and water saturations within the Chesterian incised valley sandstone reservoir of the Pleasant Prairie south field in Haskell County, Kansas. The field was discovered by three-dimensional seismic exploration in 1990 and has produced 4.4 million barrels of oil; it was put on waterflood in 2002. Petrophysical data for the model were drawn from logs run in twenty-five wells. The cell dimensions of the model were set at fifty-five feet horizontally and two feet vertically, resulting in 700,000 active cells.

The four key components of a static model are facies, porosity, permeability, and water saturation. Lithofacies recognized in the core can be integrated with petrophysical logs for prediction of lithofacies in uncored wells, as described in Chapters 4 and 5. The interpreted depositional environments associated with these lithofacies are the key to the development of a three-dimensional framework in which appropriate lateral interpolation functions are used to create geobodies with geologically realistic shapes and sizes. At the same time, the distinctions between petrofacies and lithofacies should be recognized, as discussed in Chapter 6, because a failure to identify and discriminate flow units will compromise history matching in the dynamic model. Methods to evaluate porosity were described in Chapter 2; the discrimination of effective porosity, in the engineering sense of a pore space containing movable fluids, is important in the static model. As described in Chapter 3, the realistic estimation of permeability within a static model requires the subdivision of the model into petrofacies, each with distinctive porosity-permeability associations. Finally, Chapter 1 described the calculation of water saturation by the Archie equation and its variants: in this last chapter, the fluid saturations that populate the static model are reconciled with the forces of gravity that control buoyancy pressures and the capillary forces that constrain fluid entry and exit through the pore system.

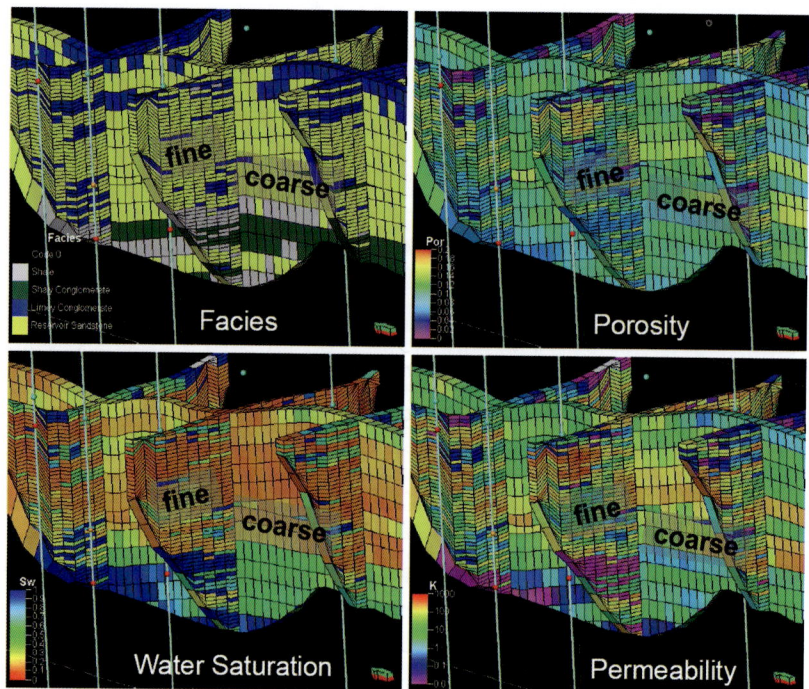

Figure 7.19: Static model of the incised valley Chesterian sandstone reservoir of the Pleasant Prairie South field, Haskell County, Kansas. From Dubois et al. (2012), courtesy Geophysical Society of Oklahoma City.

So, in the final analysis, the static model is the fruition of petrophysics, as first realized by Archie as a concept, and now actualized as a computer construct. In the pages of this book we have seen the mix of empiricism and rival analytical models that are the lifeblood of everyday petrophysical practice. Those who constrain themselves to think of the Archie equation as "Archie's law" will find it to be a chimera and should instead recognize it as a useful model that has been the basis for a variety of successful formation evaluation methods. However, there are indeed physical laws that govern petrophysics, and they control the movement and entrapment of fluids within rock pore spaces. As a popular bumper sticker proclaims: "Gravity. It's the Law." The results of mathematical petrophysics must give results that are consistent with the physical world as we understand it. But rocks are complicated, so we can never fully characterize the subsurface with the limited data that are generally available. Useful models, rather than complete realizations, are the goal, and these should have sufficient power to deliver practical results. The implementation of mathematical petrophysics have evolved from charts, through calculators, to computer applications. Pattern recognition from a reservoir perspective and algebraic solutions of reservoir properties have been the common thread throughout. However, the future is upon us, now that not only static models are improving in scope and sophistication at the large scale, but "digital rocks" are starting to emerge as practical petrophysical solutions at the microscopic scale.

REFERENCES

Alger, R.P., Luffel, D.L., and Truman, R.B., 1989, New unified method of integrating core capillary pressure data with well logs: Society of Petroleum Engineers Formation Evaluation, v. 4, no. 2, pp. 145–152.

Altunbay, M., Martain, R., and Robinson, M., 2001, Capillary pressure data from NMR logs and its implications on field economics: Society of Petroleum Engineers, SPE 71703-MS, 10 p.

Arps, J.J., 1964, Engineering concepts useful in oil finding: American Association of Petroleum Geologists Bulletin, v. 48, no. 2, pp. 157–165.

Bhattacharya, S., Byrnes, A.P., Watney, W.L., and Doveton, J.H., 2008, Flow unit modeling and fine-scale predicted permeability validation in Atokan sandstones: Norcan East field, Kansas: American Associaton of Petroleum Geologists Bulletin, v. 92, no. 6, pp. 709–732.

Clerke, E. A., and Martin, P.R., 2004, Thomeer Swanson Excel spreadsheet, FAQ's and user comments: Presented and distributed at the Society of Professional Well Log Analysts, 2004 Carbonate Workshop, Noordwijk, The Netherlands.

Clerke, E.A., Mueller, H.W. III, Phillips, E.C., Eyvazzadeh, R.Y., Jones, D.H., Ramamoorthy, R., and Srivastava, A., 2008, Application of Thomeer hyperbolas to decode the pore systems, facies, and reservoir properties of the Upper Jurassic Arab D limestone, Ghawar field, Saudi Arabia: A "Rosetta Stone" approach: GeoArabia, v. 13, no. 4, pp. 113–160.

Clerke, E., 2009, Permeability, relative permeability, microscopic displacement efficiency, and pore geometry of M-1 bimodal pore systems in Arab-D limestone. Society of Petroleum Engineers Journal, v. 14, no. 3, pp. 524–531.

Cuddy, S., Allinson, G., and Steele, R., 1993, A simple, convincing model for calculating water saturations in southern North Sea gas fields: Transactions of the Society of Professional Well Log Analysts, 34th Annual Logging Symposium, Paper H, 17 p.

Doveton, J.H., 2008, Enhanced formation evaluation based on forward-modelling of resistivity and porosity logs from flow-unit architecture simulations of carbonate and clastic reservoirs: Transactions of the Society of Professional Well Log Analysts, 49th Annual Logging Symposium, Paper NNN, 6 p.

Dubois, M.K., Senior, P.R., Williams, E., and Hedke, D.E., 2012, Reservoir characterization and modeling of a Chester incised valley-fill reservoir, Pleasant Prairie South field, Haskell County, Kansas: Geophysical Society of Oklahoma City Education 2012, Oklahoma City, Oklahoma, 37 p.

Gagnon, D., Cuddy, S., Conti, F., and Lindsay, C., 2008, The effect of pore geometry on the distribution of reservoir fluids in UK North Sea oil and gas fields: Transactions of the Society of Professional Well Log Analysts, 49th Annual Logging Symposium, Paper CCC, 16 p.

Harrison, B., and Jing, X.D., 2001, Saturation height methods and their impact on volumetric hydrocarbon in place estimates: Society of Petroleum Engineers, SPE 71326-MS, 12 p.

Hartmann, D.J., and MacMillan, L.,1992, Petrophysics of the Wasatch formation and Mesaverde group, Natural Buttes producing area, Uinta Basin, Utah, in T.D. Fouch, V.F. Nuccio, and T.C. Chidsey, Jr, eds., Hydrocarbon and mineral resources of the Uinta Basin, Utah and Colorado: Utah Geological Association Guidebook 20, Salt Lake City, Utah, pp. 175–192.

Heseldin, G.M., 1974, A method of averaging capillary pressure curves: Transactions of the Society of Professional Well Log Analysts, 15th Annual Logging Symposium, Paper E, 8 p.

Johnson, A., 1987, Permeability averaged capillary data: A supplement to log analysis in field studies: Transactions of the Society of Professional Well Log Analysts, 28th Annual Logging Symposium, Paper EE, 11 p.

Kay, S., and Cuddy, S., 2002, Innovative use of petrophysics in field rehabilitation, with examples from the Heather Field: Petroleum Geoscience, v. 8, no. 4, pp. 317–325.

Leverett, M.C., 1941, Capillary behavior in porous solids: Transactions of the American Institute of Mining, Metallurgical, and Petroleum Engineers, v.142, no. 1, pp. 159–172.

Lucia, F.J., 1995, Rock-fabric/petrophysical classification of carbonate pore space for reservoir characterization: American Association of Petroleum Geologists Bulletin, v. 79, no. 9, pp. 1275–1300.

Lucia, F.J., 2007, Carbonate reservoir characterization: An integrated approach: Springer, Berlin, 336 p.

Maharaja, A., and Journel, A., 2005, Hierarchical simulation of multiple-facies reservoirs using multiple-point geostatistics: Society of Petroleum Engineers, SPE 95574-MS, 15 p.

Mitchell, P., Walder, D., and Brown, A.M., 1999, Prediction of formation water saturation from routine core data populations: Proceedings of the 1999 International Symposium, Society of. Core Analysts, SCA-9955, 10 p.

Obeida, T.A., Al-Mehairi, Y.S., Suryanarayana, K., 2005, Calculations of fluid saturations from log-derived J-functions in giant complex Middle-East carbonate reservoir: Society of Petroleum Engineers, SPE 95169-MS, 5 p.

Skelt, C., and Harrison, B., 1995, An integrated approach to saturation height analysis: Transactions of the Society of Professional Well Log Analysts, 36th Annual Logging Symposium, NNN, 10 p.

Søndenå, E., 1992, An empirical method for evaluation of capillary pressure data: Proceedings, Society of Core Analysts, Third European Core Analysis Symposium, Paris, 129 p.

Thomeer, J.H.M., 1960, Introduction of a pore geometrical factor defined by the capillary pressure curve: Journal of Petroleum Technology, v.12, no. 3, pp. 73–77.

Volokitin, Y., Looyestijn, W.J., Slijkerman, W.F.J., and Hofman, J.P., 2001, A practical approach to obtain primary drainage capillary pressure curves from NMR core and log data: Petrophysics, v. 42, no. 4, pp. 334–343.

Worthington, P.F., 2002, Application of saturation-height functions in integrated reservoir description, in M. Lovell and N. Parkinson, eds., Geological application of well logs: American Association of Petroleum Geologists, Methods in Exploration No. 13, p. 75–89.

INDEX

acoustic
    transit time 55–57, 95, 118–119
    pseudo-matrix transit time 56
    Wyllie equation 55–56, 60
    Raymer-Hunt-Gardner (RHG)
        transform 56
    evaluation of vuggy porosity 55–59
    Nugent equation 55, 59–60
apparent matrix density ($RHOmaa$)
    125–128, 138, 140, 143, 150
apparent matrix photoelectric absorption
    ($Umaa$) 108, 125, 138–140, 150
Archie equation 1–4, 7, 19, 57
    formation factor, F 2–8, 10–12, 24–28,
        58–59
    porosity exponent, m 2–8, 10–11, 13,
        24–33
    "tortuosity factor", a 4
    resistivity index, I 34–36, 39
    saturation exponent, n 3, 17, 33–37
    in sandstones 2–10
    in carbonates 23–29
    sensitivity analysis 8–10
    alternative models 38–40
"Archie rock" 2, 11–13, 37–39

Bayesian probability 106, 149, 153, 162–163
buoyancy pressure 173, 217–218, 221–222
BVW height functions 225–230

capillary pressure
    measurements 172–175
    Washburn equation 173, 240
    buoyancy pressure 173, 217–218,
        221–222
    entry (or displacement) pressure
        218–219, 236–238
    Free-water level (FWL) 218–234
    oil-water contact (OWC) 36, 218–220
    pore-throat 172–175
    $r35$ 72, 176–178, 180 188, 221
    Winland equation 72–73, 176–178
    Thomeer hyperbolas 235–238

carbonate electrofacies 123–124,
    126–128, 132–133,135,138–145,
    150–151, 156–157, 159–160
carbonate porosity 55–64
cerebellum model articulation controller
    (CMAC) 153–154
clastic electrofacies 163–165
clay
    clay and shale 48
    wet clay 20–21,
    dry clay 20–21, 45
    cation-exchange capacity (CEC) 11, 13,
        15–16, 18, 20
    clay-mineral estimation 106–113
clustering, depth-constrained 128–131,
    141–142, 144
    Ward's method 74–77, 84–87, 129–130
Coates equation 80, 81, 84–86, 88
Compositional evaluation 94–119
    matrix algebra 94–95
    inverse solution 94–99
    underdetermined systems 99–102
    overdetermined systems 102–103
    optimization models 103–105
    multiple model solutions 105–106
    X-ray diffraction 105, 108, 110–112
    geochemical logs 113–117
    colinearity 115–116
    inversion mapping 117–119
crossplot
    formation factor – porosity 5–6, 8, 25,
        27, 28, 30
    m – porosity 7, 11, 28,
    m - pore parameters 33
    $m - n$ 35
    neutron-density porosity 21, 46, 133
    Thomas – Stieber 52–53
    $\Delta tmaa$ – pore parameters 57
    permeability – porosity 68–70, 73,
        76–77, 82, 83, 86, 181
    $RHOmaa$-$Umaa$ 125–127
    Capillary pressure – saturation 174,
        181, 185, 192, 219, 222, 223

Darcy's law 172
density
   grain density 105, 123–124
   apparent matrix density (*RHOmaa*)
      125–128, 138, 140, 143, 150
Discriminant analysis (DFA) 148–151
   non-parametric 151–157
dispersed shale 22, 46, 51–54
Dunham textural classes 81–84, 123–124, 186, 189, 191, 205–208, 237
   mudstone 83–84, 87, 123–124,143, 195, 205–212
   wackestone 81–83, 123–124, 145, 185–186, 188, 190–191, 195, 207–212, 235
   packstone 81–83, 123–124, 145, 185–186, 188, 190–191, 195, 207–212, 235
   grainstone 81–83, 123–124, 145, 185–186, 188, 190–191, 195, 207–212, 235
dynamic reservoir model 217

electrofacies 122–162
   clastic 163–165
      carbonate 123–124, 126–128, 132133,135,138–145, 150–151, 156–157, 159–160

facies 122–123
   electrofacies 123–124, 131–157
   lithofacies 124–128
   petrofacies 171–172, 179–182, 185–186, 225–233
   NMR sandstones 200–205
   NMR facies carbonates 205–213
   process facies 162–163
Faciolog 135
Flow unit 71–73, 176–180, 183
Flow zone indicator (*FZI*) 69–73
   and permeability prediction 73–77
formation factor, F 2–8, 10–12, 24–28, 58–59
forward modeling 94, 133, 233
fractal models 39–40
fractures 23–27, 31, 55, 131
Free-water level (*FWL*) 218–234
fuzzy logic 90–91

gamma-ray
   API units 47
   Gamma-ray index (GRI) 48–49
   spectral gamma-ray 48, 107–109, 113, 138, 150, 165
gas effect 46–47, 189–190
gas shales 116–117

geomodel 122, 157–161, 241–242
grain-supported carbonates 84–87, 123–124, 126, 188, 204, 206, 211
grainstone 81–83, 123–124, 145, 185–186, 188, 190–191, 195, 207–212, 235

Hannai-Bruggeman equation 2
Humble equation 4–7, 11

Indonesia equation 14–15

karstic breccias 61–62, 98, 118–119, 193, 210
Kozeny-Carman equation 69–72
Kozeny constant 69

laminar shale 22, 46, 50–54
Larionov equation 48–49
Leverett *J*-function 225, 227–228, 232
Lucia model 81–82, 235

Markov chain 162, 165–166
matrix photoelectric absorption (*Umaa*) 125
Maxwell's equation 1–2
moments, statistical 200–204
   center-of-gravity 201
   dispersion 201
   skewness 201
   kurtosis 201
mud-supported carbonates 84–87, 123–124, 126, 188, 204, 206, 211
mudstone 83–84, 87, 123–124,143, 195, 205–212

neural networks 89–91, 157–162
   CMAC 153–154
"non-Archie rock" 10–13
nuclear magnetic resonance (NMR)
   relaxivity 199, 204, 239–240
   bulk diffusion term 199, 204–205
   diffusive coupling 62
   T2 relaxation distribution 62–64, 80–81, 84–88, 198–213
   NMR facies in sandstones 200–205
   NMR facies in carbonates 205–213
   Coates equation 80–81, 84–86, 88
   SDR equation 80–81, 201
Nugent equation 55, 59–60

oil-water contact (OWC) 36, 218–220

packstone 81–83, 123–124, 145, 185–186, 188, 190–191, 195, 207–212, 235
permeability estimation

from porosity 67–69
from porosity and Swi 77–80
Timur equation 78–80
from *FZI* 73–76
from NMR 84–88
in dual and triple porosity systems 82–88
neural networks 89–91
fuzzy logic 90–91
Permeability-height functions 230–233
petrofacies 171–172, 179–182, 185–186, 225–233
"petrophysics," term 171
photoelectric factor (PeF) 96–97, 100–101, 108–109, 125–126, 139, 159–160
Umaa 108, 125, 138–140, 150
Pittman equation 197
pore-body
size from NMR 198–199, 239
pore-throat
Winland equation, $r35$ 72–73, 176–178
in sandstones 172–183
in carbonates 183–197
porosity
spherical packs 44
effective porosity 45, 46, 50–51
shale correction 50–51
gas correction 50–51
carbonate 55–64
dual-porosity 26–30
vugs 24, 26, 31, 32, 55–64, 84, 86–87, 96–98, 195, 205, 209
oomoldic 28–30, 57–60, 189–194
fractures 23–27, 31, 55
triple-porosity 31
porosity exponent, m 2–8, 10–11, 13, 24–33
grain effects 3
as a function of porosity 5, 6–7, 10–13, 26–31
variable-m 5, 25, 28, 32
dual-porosity system 26–30
triple-porosity system 31
dielectric logging measurement 31–32
Nugent equation 55, 59–60
petrographic evaluation 32–33
error analysis 8–10
principal component analysis (PCA) 134–136
PCA electrofacies 134–146
process facies 162–163

$r35$ 72, 176–178, 180–188, 221
regression analysis 51, 68, 73, 76–78. 84, 88–89, 115, 176, 189–191, 223, 225, 229, 231–232

weighted 88–89
relaxivity 199, 204, 239–240
resistivity
resistivity index, I 34–36, 39
tensor resistivity 53

saturation
sensitivity analysis 8–10
height function 217–241
saturation exponent, n 3, 17, 33–37
wettability effects 36–37
saturation-height
Leverett J-function 225, 227–228, 232
BVW-height functions 225–230
Permeability-height functions 230–233
sensitivity analysis 8–10
SDR equation 80–81, 201
shale
from gamma-ray 47–50
Gamma-ray Index (GRI) 48–49
Larionov equation 48–49
Stieber equation 48–49
from neutron-density 45–46
morphology 51–54
laminar 22, 46, 50–54
dispersed 52–54
structural 52
Thomas-Stieber plot 52–53
North American Shale composite (NASC) 47
shaly sandstone saturation
analysis 13–23
Simandoux equation 14
Indonesia equation 14–15
Double-layer models 15–19
Waxman-Smits equation 15–19
Dual-water 20–22
Shingled block lattice (SBL) 153–157
Simandoux equation 14
size and permeability 72, 172, 175, 183, 187, 189, 191, 193–194
sonic *see acoustic*
static reservoir model 241–243
Stieber equation 48–49
structural shale 52
supervised methods 147–167

T2 relaxation time distribution 62–64, 80–81, 84–88, 198–213
Thomas-Stieber plot 52–53
Thomeer hyperbolas 235–238
Timur equation 78–80
"Tixier" equation 6
tortuosity 1–2, 32, 69

unsupervised methods 131–147

vuggy porosity 24, 26, 31, 32, 55–64, 84, 86–87, 96–98, 195, 205, 209
   acoustic methods 56–62
   NMR methods 62–64

wackestone 81–83, 123–124, 145, 185–186, 188, 190–191, 195, 207–212, 235
Walther's law 162–163, 165
Ward's method 74–77, 84–87, 129–130
Washburn equation 173, 240
Water saturation, Sw 3–4, 8–10, 11–23, 33–40, 54

"irreducible" 77–79, 84,
   clay-bound 17, 20–21, 81, 210
Waxman-Smits equation 15–19
wettability
   water-wet 34, 37, 173, 199
   oil-wet 37, 39
   mixed (Dalmatian) 37
Winland equation 72–73, 176–178
   $r35$ 72, 176–178, 180–188, 221
Wyllie equation 55–56, 60

X-ray diffraction 105, 108, 110–112

zonation 128–131, 141–142, 144